The Politics and Everyday Practice of International Humanitarianism

The Politics and Everyday Practice of International Humanitarianism

Miriam Bradley

OXFORD
UNIVERSITY PRESS

OXFORD
UNIVERSITY PRESS

Great Clarendon Street, Oxford, OX2 6DP,
United Kingdom

Oxford University Press is a department of the University of Oxford. It furthers the University's objective of excellence in research, scholarship, and education by publishing worldwide. Oxford is a registered trade mark of Oxford University Press in the UK and in certain other countries

Published in the United States of America by Oxford University Press
198 Madison Avenue, New York, NY 10016, United States of America

British Library Cataloguing in Publication Data
Data available

Library of Congress Control Number: 2023930854

ISBN 978–0–19–884069–5
ISBN 978–0–19–884070–1 (pbk.)

DOI: 10.1093/oso/9780198840695.001.0001

Printed and bound by
CPI Group (UK) Ltd, Croydon, CR0 4YY

Contents

List of Abbreviations

AHDEN	Association Haïtienne de Droit de l'Environnement
ALNAP	Active Learning Network for Accountability and Performance in Humanitarian Action
AMISOM	African Union Mission in Somalia
ASEAN	Association of Southeast Asian Nations
AU	African Union
BBB	build back better
BBS	build back safer
BRR	Rehabilitation and Reconstruction Agency, Aceh-Nias
CALP	Cash Learning Partnership
CAP	Consolidated Appeals Process
CBPF	country-based pooled fund
CBRWG	Cash Based Response Working Group
CCC	community care centre
CDC	Centers for Disease Control and Prevention
CEAS	Common European Asylum System
CERF	Central Emergency Response Fund
CFW	cash for work
CHF	Swiss francs
CRASH	Centre de Réflexion sur l'Action et les Savoirs Humanitaires
CRC	1989 Convention on the Rights of the Child
CRRF	Comprehensive Refugee Response Framework
CVA	cash and voucher assistance
DAC	Development Assistance Committee, OECD
DANIDA	Danish International Development Agency
DARA	Development Assistance Research Associates
DFID	Department for International Development, United Kingdom
DRC	Democratic Republic of the Congo
DRR	disaster risk reduction
EC	European Community
EPLF	Eritrean People's Liberation Front
ERA	Eritrean Relief Association
ERD	Emergency Relief Desk
ETU	Ebola treatment unit
EU	European Union
EUNAVFOR MED	European Union Naval Force Operation Mediterranean
FAO	Food and Agriculture Organization
FAR	Forces Armées Rwandaises
FEWSNET	Famine Early Warning Systems Network

FMG	Federal Military Government, Nigeria
FRA	EU Agency for Fundamental Rights
FSNAU	Food Security and Nutrition Analysis Unit
GAJP	Global Health Justice Partnership
GAM	Gerakan Aceh Merdeka, Sumatra
GDP	gross domestic product
GHD	Good Humanitarian Donorship
HPG	Humanitarian Policy Group
HPN	Humanitarian Practice Network
HRP	Humanitarian Response Plan
ICC	International Criminal Court
ICG	International Crisis Group
ICISS	International Commission on Intervention and State Sovereignty
ICJ	International Court of Justice
ICRC	International Committee of the Red Cross
ICTR	International Criminal Tribunal for Rwanda
ICTY	International Criminal Tribunal for the Former Yugoslavia
ICU	Islamic Courts Union, Somalia
ICVA	International Council of Voluntary Agencies
IDMC	International Displacement Monitoring Centre
IDNDR	International Decade for Natural Disaster Reduction
IDP	internally displaced person
IDRC	International Dispute Resolution Centre
IDRL	international disaster response law
IED	improvised explosive device
IFAD	International Fund for Agricultural Development
IFOR	Implementation Force, Bosnia
IFRC	International Federation of Red Cross and Red Crescent Societies
IGO	intergovernmental organization
IHL	international humanitarian law
IHR	international health regulations
IHRL	international human rights law
ILC	International Law Commission
IMC	International Medical Corps
IMF	International Monetary Fund
INALWA	International Air Lift West Africa
INGO	international non-governmental organization
IOM	International Organization for Migration
IPC	Integrated Food Security Phase Classification
IRC	International Rescue Committee
IRIN News	Integrated Regional Information Networks News
IRL	international refugee law
IRRC	International Review of the Red Cross
ISAF	International Security Assistance Force, Afghanistan
JCA	Joint Church Aid/'Jesus Christ Airlines'
JNA	Yugoslav People's Army

LRRD	Linking Relief, Rehabilitation, and Development
LTTE	Liberation Tigers of Tamil Eelam
MDG	Millennium Development Goals
MDM	Médecins du Monde
MINUSTAH	United Nations Stabilization Mission in Haiti
MOAS	Migrant Offshore Aid Station
MONUSCO	United Nations Organization Stabilization Mission in the Democratic Republic of the Congo
MSF	Médecins sans Frontières
NACLA	North American Congress on Latin America
NATO	North Atlantic Treaty Alliance
NGO	non-governmental organization
NWoW	New Way of Working
OAS	Organization of American States
OAU	Organisation of African Unity
OCHA	Office for the Coordination of Humanitarian Affairs
ODI	Overseas Development Institute
OECD	Organisation for Economic Co-operation and Development
OHCHR	Office of the High Commissioner for Human Rights
OIC	Organisation of Islamic Cooperation
OSE	Office of the Special Envoy
PHEIC	public health emergency of international concern
PMSC	private military and security company
PPE	personal protective equipment
PR	public relations
PRT	provincial reconstruction team
RC	Red Cross and/or Red Crescent
REST	Relief Society of Tigray
RPF	Rwandan Patriotic Front
RRC	Relief and Rehabilitation Commission
RRN	Relief and Rehabilitation Network
RTE	real-time evaluation
SAR	search and rescue
SDG	Sustainable Development Goal
SIPRI	Stockholm International Peace Research Institute
TFG	Transitional Federal Government, Somalia
TDC	Transnational Development Clinic
TPLF	Tigrayan People's Liberation Front
UAE	United Arab Emirates
UCDP	Uppsala Conflict Data Program
UN	United Nations
UN OCHA	United Nations Office for the Coordination of Humanitarian Affairs
UNAMA	United Nations Assistance Mission in Afghanistan
UNAMIR	United Nations Assistance Mission in Rwanda
UNCT	United Nations Country Team
UNDP	United Nations Development Programme

UNDSS	United Nations Department of Safety and Security
UNEOE	United Nations Emergency Office for Ethiopia
UNGA	United Nations General Assembly
UNHCR	United Nations Refugee Agency (Office of the United Nations High Commissioner for Refugees)
UNICEF	United Nations Children's Fund
UNISDR	United Nations Office for Disaster Risk Reduction
UNITAF	Unified Task Force for Somalia
UNMEER	United Nations Mission for Ebola Emergency Response
UNMIL	United Nations Mission in Liberia
UNOSOM	United Nations Operation in Somalia
UNPROFOR	United Nations Protection Force in Yugoslavia
UNRRA	United Nations Relief and Rehabilitation Administration
UNRWA	United Nations Relief and Works Agency for Palestine Refugees in the Near East
UNSC	United Nations Security Council
USAID	United States Agency for International Development
USD	US dollar
VSO	Voluntary Service Overseas
WASH	water, sanitation, and hygiene
WCC	World Council of Churches
WFP	World Food Programme
WHO	World Health Organization
WHS	World Humanitarian Summit

1
Introduction
Humanitarian Emergencies

Central to both the politics and the practice of international humanitarianism is the idea of the 'humanitarian emergency'. In the most general terms, a humanitarian emergency can be said to occur when people are in a dire, urgent, and exceptional situation such that they need help from strangers. While suffering and charity are by no means new, it is the contemporary configuration of them that gives shape to the idea of the humanitarian emergency or humanitarian crisis (Redfield and Bornstein 2010). The term 'humanitarian' has come to have a particular meaning in international law and international relations, where it is mainly used to refer to a legitimate form of 'organized assistance and protection for people who are suffering, or who are likely to suffer, from armed conflicts or disasters' (Slim 2015: 45; see also Gordon and Donini 2015: 81).

This kind of organized assistance and protection is the focus of this book. Such humanitarian action is often understood as something different from, and separate from, politics, narrowly defined. Humanitarianism is seen as a sphere of moral action, while politics implies self-interest. The nature of the humanitarian project, and the interaction between humanitarians and political actors, is the subject of a great deal of normative debate. Rather than making an argument for what humanitarianism *should* be and do, however, this book focuses on explaining what humanitarianism *actually* looks like and why. The central message is that humanitarianism is deeply entwined with politics. Economic and political factors play a major role in determining the vulnerability of people to conflict and disaster, the ways international humanitarian agencies respond with protection and assistance, and the impacts of such responses.

This introductory chapter proceeds in three main parts. The first section discusses the main types of humanitarian emergency and problematizes the labels and definitions we use to classify them. The second section examines the process and consequences of constructing any given event or situation as a humanitarian emergency as opposed to, for example, a political crisis.

The Politics and Everyday Practice of International Humanitarianism. Miriam Bradley, Oxford University Press.
 DOI: 10.1093/oso/9780198840695.003.0001

It argues that the very representations that are most effective at attracting the attention of Western publics, and hence at building pressure for an international response, may also have the somewhat paradoxical effect of prescribing an international response that is insufficient or inappropriate. The third section turns to the question of how humanitarianism is defined and explains the focus and structure of the rest of the book.

Crises in the contemporary world

Humanitarian emergencies come in many forms. There are disasters provoked by natural phenomena or technological failure, epidemics, famines, armed conflicts, and even urban violence. But what makes an event a disaster, an epidemic, a famine, or a war? Defining each of these different types of events and conditions is less straightforward than it might at first seem. Furthermore, the human consequences and needs that result from such events are not determined simply by the nature and magnitude of the event or condition in question but also by the level of vulnerability of the people affected by it and the ability of their community to adapt and respond. Differences in the vulnerabilities and capacities of different communities and countries mean that an earthquake in Haiti may constitute a humanitarian emergency while an earthquake of similar magnitude in Italy may not.

This section critically analyses the ways in which different kinds of humanitarian emergency are defined and highlights three points. First, humanitarian emergencies are invariably at least partially human-made because even when we consider the most natural of events, their impacts are shaped by the political, economic, and social context. Second, assessments of the capacity of a community to adapt and respond are not necessarily objective, and other factors and preconceptions play a role in constructing a given event as a humanitarian emergency. Third, despite the importance of political, economic, and social factors, the Western public imagination mostly does not understand humanitarian emergencies as political in nature. Even when emergencies are seen to be caused by human factors, the complexity of their political, economic, and social causes is rarely well understood.

Disasters

Disasters encompass what were once called 'natural disasters' (earthquakes, droughts, floods, etc.) and technological disasters, such as nuclear accidents.

There is an emerging academic and policy consensus that disasters are never 'natural' and should not be labelled as such, even when triggered by environmental hazards. Heat waves, droughts, and even earthquakes, tsunamis, and volcanoes may occur as a result of human-made climate change (McGuire 2012; Walker et al. 2012). And what makes one of these events *disastrous*? In general, we consider something to be a disaster if its impact exceeds the community's capacity to adapt and respond, because it is only when the community lacks sufficient capacity that outside assistance is required. This has two important implications. First, the salience of capacity means that the factors that determine whether any given event is disastrous are not always, or even mostly, natural. They depend on the vulnerability of people to that event, and vulnerability is shaped by political, economic, and structural factors (Alexander 1997: 291–292). In other words, even so-called natural disasters are shaped by politics. Second, the threshold for 'sufficient capacity' is not easily measured and offers scope for a subjective judgement as to whether or not a given community or country needs external help.

Disasters invariably entail both an economic cost and a human cost, and while the economic cost of any given disaster is generally much higher in richer countries, the human cost is generally much higher in poorer countries (Alexander 1997: 285, 287; Strömberg 2007: 204). In 2010, for example, Haiti and Chile were both struck by major earthquakes. Out of 169 countries ranked in the 2010 Human Development Index, Chile was 45 (high human development) while Haiti was 145 (low human development) (UNDP 2010: 143, 145). The earthquake in Haiti, of 7.0 magnitude on the Richter scale, left between 100,000 and 316,000 people dead, while the earthquake in Chile, of significantly greater magnitude at 8.8 on the Richter scale, left 526 people dead (IFRC 2013: 3, 2016: 12). Furthermore, in any given country or in respect of any given disaster, different geographic areas and different demographic groups tend to experience differential impacts. Individual vulnerability to a disaster often depends on economic or social status.

Disasters can be categorized as rapid- or slow-onset. Rapid-onset disasters—such as earthquakes or tsunamis—are usually characterized by a peak in mortality during and immediately after the event.[1] Consequently, the initial lifesaving response can only be undertaken by those already close at hand when the disaster strikes, which tends to exclude international humanitarian agencies unless they were already operational nearby. Rapid-onset disasters are not generally followed by outbreaks of communicable diseases

[1] For examples of rapid-onset disasters, and international humanitarian responses to them, see Chapter 7 on the 2004 Indian Ocean tsunami and Chapter 9 on the 2010 earthquake in Haiti.

but, under certain conditions, they may be. For example, local health services may be affected by a disaster, with infrastructure damaged or personnel killed, and if normal health-care activities such as vaccination programmes are halted, infectious diseases may spread. Alternatively, the concentration of people displaced by a disaster in camps or other dwellings with insufficient water and sanitation facilities may lead to an increase in the transmission of infectious diseases.

Additional humanitarian needs may arise if homes and livelihoods are damaged or destroyed by the disaster. Rapid-onset disasters may destroy livelihoods in an instant, as when fishing boats are washed away by a tsunami or a shop is buried by an earthquake. Slow-onset disasters, such as drought, may erode livelihoods over time, meaning there is time to mitigate their impacts. However, this requires action on the part of those with the power to mitigate those impacts. Thus, drought may lead to famine, for example, but this is not by any means automatic and depends on economic, social, and political factors.[2]

Famine

In the Western public imagination, famine has been largely understood in terms of mass starvation, as a consequence of too little food. According to this perspective, famine implies food shortage, severe hunger, and mass mortality. However, closer examination of empirical cases reveals that famine can occur—and be recognized as famine by people both within and outside the relevant region—without food shortage, as in Bangladesh in 1974, without starvation, as in Darfur in 1984, or without excess mortality, as in the Sahel 1972–1974 (de Waal 1989; Sen 1981). This calls into question how we conceptualize famine, how we know when famine is occurring, and how we understand the causes of famine, all of which, in turn, has consequences for how we respond to famines in practice. Alex de Waal has argued that 'less severe famine is not theoretically distinct from acute poverty, and severe famine is distinct largely because of the severity of social collapse' (de Waal 1990: 469). However, once the famine label comes into play, the situation is understood as a humanitarian emergency requiring an international humanitarian response.[3] By contrast, a situation defined in terms of poverty is

[2] See 'Famine' below, Chapter 4 on Ethiopia 1983–1985, and Chapter 10 on Famine in Somalia, 2011–2012.

[3] This was the case with the 2011 declaration of famine in Somalia, for example, which triggered a massive increase in funding overnight after months of reliable warnings of impending famine had failed

more likely to attract a developmental response (or no international response at all).

The publication of Amartya Sen's landmark book, *Poverty and Famines*, in 1981, decisively shifted the focus away from supply-side explanations for famine. Based mainly on analysis of South Asian examples, Sen argued that famines are not caused by a lack of food and have often occurred where the food supply was plentiful but where people had differential access to it (Sen 1981). In other words, famine is about access to food—or livelihoods—rather than aggregate food supply and, to the extent that access to food and livelihoods is political, so is famine. With a focus on famines in Africa, de Waal has built on Sen's work and critiqued aspects of it. Whereas Sen has argued that famine does not occur in democracies, de Waal has emphasized that democracy is neither necessary nor sufficient for the prevention of famine, and that what matters is the existence of an anti-famine political contract, through which citizens hold authorities accountable for the prevention of famine (de Waal 1997; Sen 1999: 16). The clear message from both of them is that famine is preventable and where—due to some form of political contract—it is in the interests of the government (or other authority) to prevent it, it is prevented.

In some cases, authorities have an interest in preventing famine in some parts of the territory under their control but not in others. If a government depends on support from the urban population, for example, it may take great care to ensure the food supply to the cities with little regard for elsewhere. In addition, drought, conflict, and other factors contributing to famines often occur inconsistently across territory. Thus, famine can affect countries unevenly, with some parts of the population (e.g. the rural population, members of a particular ethnic group, opposition supporters, etc.) suffering disproportionately.[4] In conflict contexts in particular, it may actually serve the interests of authorities to have particular areas suffering from famine, and recent history is replete with examples of governments restricting the supply of food into opposition-held areas in an effort to starve rebels into submission.[5] Famines have both winners and losers, not only creating victims but also generating benefits for others (Keen 1994). With all this in mind, it is not surprising that, in many cases, famine is not merely a consequence but an actual goal of conflict (Macrae and Zwi 1992). Despite the fact

to prompt a significant response. See 'Early warning, late response' in Chapter 10 on Famine in Somalia, 2011–2012.

[4] For examples, see Chapter 3 on Drought and Famine in Ethiopia 1983–1985 and Chapter 10 on Famine in Somalia, 2011–2012.

[5] For examples, see Chapter 2 on the Nigerian Civil War, 1967–1970 and Chapter 3 on Drought and Famine in Ethiopia 1983–1985.

that famine is inherently political, and may even be a deliberate strategy, it is often constructed and understood as apolitical and 'natural'.[6]

Health crises

Epidemics and outbreaks of disease can also constitute humanitarian emergencies, often—but not always—in tandem with some other kind of crisis, such as famine or armed conflict. Once again, we are faced with threshold or definitional questions. According to the Oxford English Dictionary, an epidemic is 'a widespread occurrence of an infectious disease in a community at a particular time'. The World Health Organization (WHO) defines a disease outbreak as 'the occurrence of cases of disease in excess of what would normally be expected in a defined community, geographical area or season'. But at what point do we decide that any given disease has exceeded 'normal' levels to become an outbreak or an epidemic? Under what conditions is an outbreak or epidemic considered a humanitarian emergency?

A comparison with 'normal' diseases and their impacts is instructive. In 2015, it is estimated that there were 429,000 deaths from malaria and 134,200 deaths from measles worldwide (WHO 2016b: xvi, 2017: 206). By contrast, the Ebola outbreak in west Africa in 2013–2016 caused far fewer deaths, estimated at 11,310 (WHO 2016a: 1). Of course, in the absence of a concerted international response, the number of deaths may have been much higher. However, it was not expected to reach the annual malaria or measles fatality figures. A meaningful international response did not begin until mid-September 2014, at which point it was predicted that, in the absence of such a response, the outbreak would eventually exceed 140,000 cases—with a case fatality rate of over 50%, the outbreak could thus be expected to claim more than 70,000 lives (DuBois et al. 2015; Fisman et al. 2014). Measles and malaria claim more lives, but they are 'normal' diseases that occur every day, every year. It would appear, then, that a key factor in making a health crisis a humanitarian emergency is its 'exceptionality'. Furthermore, malaria and measles deaths are concentrated in poorer parts of the world, whereas avian flu and Ebola had the potential for further-reaching international impact and elicited major international responses. The sense of emergency created by a disease can also depend on who is affected by that disease.

While the scale of the international response to a health crisis is likely to be larger when richer countries are affected (or at risk of being affected), the

[6] For more on this, see Chapter 3 on Drought and Famine in Ethiopia 1983–1985.

scale of the outbreak itself is likely to be larger when less wealthy countries are affected. This is because the factors which determine the transmission rates of a communicable disease relate not only to the disease itself but also to the societies affected. Structural conditions and the state of health systems in affected countries are major determinants of how fast a disease spreads, to whom it spreads, and how deadly it is. The unprecedented scale of the Ebola epidemic in west Africa in 2014 can be attributed to 'the synergy of several factors that created a perfect storm', including the legacies of civil wars, dysfunctional health services, and a scarcity of health workers (Piot et al. 2014). Within any given country or society, there is also variation in vulnerability to disease. Some people are at higher risk than others. As with disasters and famine, then, vulnerability to epidemics and disease outbreaks is determined by political, economic, and social factors.

Conflict and violence

Wars, armed conflicts, genocides, and other forms of large-scale violence are more obviously complex and human-made than earthquakes or famines but, equally, remain open to different understandings and representations. Two discourses have been particularly influential in terms of how we understand post-Cold War conflicts.[7] According to one, conflict is the inevitable outcome of ethnic diversity, caused by ancestral or primordial hatred. On this view, ethnic conflicts are inevitable—and so international intervention to prevent or resolve them is of limited utility.[8] According to the other, the more influential of the two, underdevelopment is the main driver of conflicts, and the kind of response required is one that links relief and development.[9] A third discourse, one that emphasized 'the legacy of colonialism, declining terms of trade and an inequitable trading system' as drivers of conflict, has been almost completely sidelined since the end of the Cold War, with the consequence that the 'problem' is seen as coming entirely from within the global South (Duffield 2001: 115). According to the dominant perspectives, then, geopolitics and the global North bear no responsibility for conflicts in the global South.

Increasingly, large-scale violence that does not meet the threshold of armed conflict is also understood to constitute a humanitarian emergency and

[7] For a fuller discussion of these discourses, see Duffield (2001: 108–117).

[8] For examples of how this kind of discourse has affected international responses, see Chapter 4 on the Bosnian War, 1992–1995 and Chapter 5 on Genocide in Rwanda and Its Aftermath, 1994–1996.

[9] On efforts to link relief and development, see Chapter 16 on the Nexus Concept.

to demand an international humanitarian response. Several humanitarian agencies have recently sought to address urban violence, for example (Bradley 2020; Reid-Henry and Sending 2014). Here, the dominant discourse implies crime- and greed-based, rather than political or grievance-based, explanations for violence. The work of international humanitarian agencies in this regard is limited to the global South. In 2015, four US cities numbered among the fifty with the highest murder rates in the world (*The Economist* 2016), but it is hard to imagine urban violence within the United States being characterized as a humanitarian emergency. We are more likely to see violence in an African or Latin American city than a US or Western European city labelled as urban violence requiring an international humanitarian response, even if the murder rates were similar and the authorities equally ineffective in preventing or responding to that violence.

The motivations and dynamics of conflict and violence have important implications for their humanitarian consequences and for the impacts of different kinds of international response. Like famine, conflict and violence have winners and losers. Some people and organizations benefit from armed conflict itself and seek to prolong the conflicts in which they are involved (Keen 1998b). Yet, the kind of detailed contextual analysis that unpacks the motivations and dynamics of conflict and violence in different settings is largely absent from representations by international humanitarian agencies.

Constructing humanitarian emergencies

Disasters, epidemics, famines, conflicts, and other situations of violence: all these kinds of crisis include a political dimension. While this may be obvious when it comes to conflict, it is also the case that the causes of famine are overwhelmingly political, and this has implications for the kind of response that is required (and the effects different kinds of response will have). Likewise, disasters are not natural, because even those caused by natural phenomena have consequences that are shaped by political, economic, and social structures. When these crises are presented as humanitarian emergencies, however, their political dimensions are largely obscured from view.

Labelling events as humanitarian emergencies shapes how we understand those events and has significant material consequences for the way in which international actors respond. In any given context, both the humanitarian element and the emergency element are socially constructed (Calhoun 2009: 1). This is not to deny that earthquakes, technological failures, wars, and other conflicts really do happen and that they have very real consequences in terms of casualties, disease, displacement, loss of livelihoods, and shelter, etc. Rather, it is to suggest that these events and their consequences can be viewed through many different lenses and that viewing them through a humanitarian emergency lens implies a particular interpretation of the problem and the corresponding solution.

Insights from securitization theory can help us to understand the process, which I term 'humanitarianization', through which humanitarian emergencies are constructed or, in other words, through which a humanitarian emergency lens is selected over alternative ways of viewing an event or context. Securitization theory posits that any given issue is not objectively a security threat but rather may be constructed as such through a process involving three main steps: discursive practices through which certain privileged actors claim that a particular object faces an existential threat which can and must be defended against, acceptance of that claim by a relevant audience, and the consequent use (or at least legitimation) of exceptional measures to so defend against the threat (Balzacq et al. 2016; Buzan et al. 1998). Humanitarianization can be understood as a similar process involving discursive (or, very often, visual) practices through which certain privileged actors claim that the well-being or lives of particular people are at risk and that their suffering can, and should, be alleviated by distant others; acceptance of that claim by a wider public; and the consequent legitimation of international humanitarian assistance.

The success of such a humanitarianization process depends on a concern for the suffering of others and the mobilization of empathy, on the one hand, and a belief that effective action can be taken to alleviate that suffering on the other (Calhoun 2008: 74). Even then, though, there may be compassion fatigue, with the consequence that, while an international humanitarian response is seen as legitimate, the necessary resources for such a response are not mobilized, and so the response is not provided or is provided on a smaller scale than that legitimized by the humanitarianization process. Arguably, the public and the media only have the appetite for one emergency at a time (Moeller 2006; Olsen et al. 2003).

Privileged actors in humanitarianization

Just as political elites have a privileged position in the articulation and construction of security threats (Balzacq et al. 2016: 8; Buzan et al.1998: 31, 40), so certain actors have a privileged position in the construction of humanitarian emergencies. The big international humanitarian agencies—the International Committee of the Red Cross (ICRC), the International Federation of Red Cross and Red Crescent Societies (IFRC), the International Rescue Committee (IRC), Médecins Sans Frontières (MSF), Oxfam, Save the Children, the UN Refugee Agency (UNHCR), the UN Children's Agency (UNICEF), and World Vision, among others—that operate in contemporary crises, largely in the global South, yield expert authority (Barnett and Finnemore 2004) and associated productive power (Barnett and Duvall 2005) to 'call' a humanitarian emergency. Since 'aid agencies rarely define problems in a way that automatically disqualifies their involvement' (Duffield 2001: 109), it is unsurprising that they represent disasters, famines, health crises, and conflicts in such a way as to elicit support for a humanitarian response.

International humanitarian agencies have a special place and power in the humanitarianization process, but they are not the only actors with the power to label and call attention to humanitarian emergencies. The media and celebrities also play a role here.[10] We know relatively little about the extent to which the media *directly* influences policy decisions, and the evidence we do have suggests that such influence exists only under certain conditions (Olsen et al. 2003; Robinson 2000). However, the media are still important as gatekeepers, determining which events and crises are publicized to public audiences in those distant, predominantly Western, countries whose governments provide the lion's share of funding to international humanitarian agencies (Moeller 2006). Increasingly, celebrities publicly support and advocate on behalf of particular causes, and this can also have an important impact. Both the media and celebrities can influence *which* crises we have awareness of and *how* those crises are (re)presented to us. In this way, they help to determine which crises capture wider public attention and how that public understands those crises and the kind of response they require.

Notably absent from the process of constructing a 'humanitarian emergency' are the voices of those most affected by it, those whose suffering is to be alleviated and whose lives are to be saved (Malkki 1996). Instead, humanitarian emergencies are constructed by international humanitarian agencies, media, and celebrities. They seek to mobilize empathy and compassion and, through the use of particular narratives and visual representations, to convert those sentiments into action.

[10] See Chapter 17 on Media and Celebrities.

Discursive practices and visual imagery

The representations favoured by these humanitarianizing actors share a number of characteristics. Again, a comparison with securitization is instructive. In its original Copenhagen School incarnation, securitization theory emphasized the speech act, the utterance of 'security' as a decisive step in the process. This emphasis subsequently gave way to the notion of discursive practices, and more recent work has focused on how visual images can 'speak security' (Hansen 2011). When it comes to the construction of a humanitarian emergency, visual imagery is the main form of representation (Kurasawa 2013; Malkki 1996). These representations seek to evoke emotions in their audiences; in the case of securitization, fear is the primary emotion being evoked, in the case of humanitarianization, it is pity.

In many cases, the visual imagery employed in processes of humanitarianization takes the form of a 'mass' or 'sea' of humanity depicting many, many bodies, all more or less alike and lacking individual personalities and biographies (Malkki 1996: 388). In other cases, imagery takes the form of an individual woman, an individual child, or a woman and child. These images are designed to depict 'exemplary victims', figures of 'innocence', in order to provoke pity and mobilize empathy. Whereas reality tends to be complex and messy, these kinds of images 'eliminate the nuances, inconsistencies, and complexities that are essential components of political society. Such pictures choose visceral emotion over perception, signifiers over sense' (Moeller 2002: 43).

Humanitarian compassion is 'reserved for those who only suffer but do not act' (Feldman 2009: 31). This has two important implications: an emphasis on physical evidence of suffering; and the silencing of victims, or at least a devaluation of their own accounts of their lived experiences (and even more so when it comes to their analysis of the contexts that contributed to those experiences). In this way, 'all that is retained of people's words is what contributes to a telling image in the public space: both the Chechen fighter and the Palestinian stone thrower become suffering beings who can only be described in terms of their physical injuries and psychological trauma' (Fassin 2007: 517). In short, 'wounds speak louder than words' (Malkki 1996: 384). Successful humanitarianization requires the mobilization of compassion internationally, and this, in turn, may entail 'some degree of exploitation of people's suffering' (Feldman 2009: 24). The individuals concerned may consent and conform to this one-dimensional representation which depicts them as suffering beings, and as nothing more than suffering beings, because they know that such a representation can be effective in generating public support (Fassin 2007: 517).

Photographs and narratives from 'expert' as opposed to personal testimony are well suited to depicting those represented as 'bare life' (Agamben 1998). Theirs are lives to be saved, without biography or history, and the focus is on their suffering rather than on their political lives or on the geopolitical context in which they live. Thus, the process of humanitarianization 'essentializes the victims: against the thickness of biographies and the complexity of history' (Fassin 2007: 512). Likewise, the humanitarian emergency is abstracted from its historical and political context.[11]

Consequences of humanitarianization

The idea of the humanitarian emergency, and the associated ways in which contemporary crises are constructed in the richer countries that are primarily (though not always exclusively) donors rather than recipients of humanitarian aid, have consequences. The discourse of humanitarianism, and the labelling of an event or context as a humanitarian emergency, legitimizes action—and a particular kind of action. The construction of a 'humanitarian emergency' depoliticizes, de-historicizes, and decontextualizes the event. Such a construction calls principally for a response that focuses on symptoms rather than causes. The primary and central objective of a humanitarian response is alleviating distant human suffering, and the extent to which action can additionally seek to address the causes of that suffering and still be labelled 'humanitarian' is disputed.[12]

The successful securitization of an issue legitimizes extraordinary measures, often including the suspension of democratic practices or of other 'normal' political processes, checks and balances, and, in many cases, also involving military action (Buzan et al. 1998: 21). Humanitarianization follows a similar logic, but the actors and practices it legitimizes are different. Humanitarianism is understood as charitable, altruistic, and principled, and this understanding can serve to delegitimize actors, like governments, who we tend to think of as self-interested. By contrast, international non-governmental organizations (INGOs) and United Nations (UN) agencies are

[11] For a more general discussion of one-dimensional representations of Africa and Africans in the global north and the reasons one-dimensional representations of anyone or any place are problematic, see Chimamanda Ngozi Adichie's excellent Ted Talk, The Danger of a Single Story: https://www.ted.com/talks/chimamanda_ngozi_adichie_the_danger_of_a_single_story?language=en.

[12] This dispute is discussed in detail in Chapter 13 on Politics, Principles, and Humanitarian Action and Chapter 16 on the Nexus Concept.

thought of as principled actors with altruistic motivations and expertise in humanitarian response. They are called on to act in response to humanitarian emergencies. Where a securitization process is successful, it empowers certain elites, often the very same (usually political) elites who designated the issue as a security threat to deal with that issue (Balzacq et al. 2016: 8). Similarly, there is often significant overlap between those whose voice carries most weight in the humanitarianization process and those whose action is legitimized by that process—that is, operational humanitarian agencies.

The construction of a humanitarian emergency, through the process I have labelled humanitarianization, can be highly effective in provoking an emotional response among Western audiences, in converting emotions into action (particularly in the form of donations), and in legitimizing a practical humanitarian response. Images and narrative techniques that 'sentimentalize' crises and the experiences of victims, and serve to make Western audiences invest emotionally, are deemed important in generating public support (Brauman 2009: 114; Kurasawa 2013: 202, 209). By emphasizing the immediacy of the event or situation, the idea of the emergency is able to command attention and mobilize resources (Calhoun 2009). The 'suddenness of an event' that 'requires immediate action' (Fassin and Vasquez 2005: 394) gives humanitarian response, with its focus on saving lives, a high level of legitimacy (Müller 2013a: 64).

A problem arises because those same characteristics that work to provoke pity and mobilize empathy and financial donations for a humanitarian response may also contribute to limiting the terms of that response. So-called *humanitarian* crises have political causes and, ultimately, require political solutions. The simplification and depoliticization of these crises that help to legitimize and generate support for a humanitarian response can obscure this reality, and the 'humanitarianization' of events or contexts may, through various mechanisms, make a political response less likely. A focus on suffering rather than the causes of suffering in part reflects the humanitarian concern to respond to suffering regardless of its causes, and in an ideal world, we might hope for a humanitarian response to address the symptoms of a crisis at the same time as a political response to address its causes. Liisa Malkki claimed that it is 'neither logically nor practically necessary that humanitarian intervention in and of itself dehistoricize or depoliticize' (Malkki 1996: 398). Yet, in practice, it seems that those constructions that generate the most support for a humanitarian response serve also to depoliticize the event or context.

Depoliticization is, itself, an inherently political process, in which the political character of politics is deliberately minimized or obscured, with policymaking presented as a neutral and necessary process in which technical choices are made and the scope for disagreement, contestation, and political alternatives is limited or denied (Cuttitta 2018: 634; Louis and Maertens 2021). The language of humanitarianism casts the primary problem as humanitarian rather than political or military (Feldman 2009). Use of the word 'emergency' focuses attention on the immediate event and not its causes. It implies sudden, unpredictable, urgent, exceptions to some sort of order, even if many so-called humanitarian emergencies actually develop over long periods of time and are predictable (Calhoun 2008: 83). The focus on 'crisis' or 'emergency' can also have the effect of making ongoing, more systemic issues appear of lesser importance and implicitly revising the definition of 'normal' to make acceptable the kind of conditions that would elsewhere be deemed unacceptable simply because they are less bad than those of the acute, crisis moment (Feldman 2009).

However common such 'emergencies' are, we characterize them as exceptions to the rule, as aberrations rather than constituent elements of the global order (Calhoun 2008: 87). Such a characterization calls for a humanitarian response rather than political analysis and struggle or economic analysis and development (Calhoun 2009: 2). Treating emergencies as the exception rather than the rule additionally implies that the causes of each emergency are local, and obscures the role of external actors (Calhoun 2008: 88). Thus, humanitarian representational practices serve to reproduce rather than challenge the existing *international* order. They 'tend to hide the political, or political-economic, connections that link television viewers' own history with that of "those poor people over there"' (Malkki 1996: 389). The ways that humanitarian emergencies are represented by international humanitarian agencies (as well as the media and celebrities) tends to reproduce the 'symbolic and material asymmetries between Western (and most often, white) rescuers and non-Western (and most often, racialized) victims of emergencies and crises in which they intervene' (Kurasawa 2013: 202).

The humanitarian response itself may further undermine pressure for a political response. For example, Alex de Waal has argued that famine prevention requires a political contract that allows citizens to hold governments accountable for famine (de Waal 1997). Treating famine as a technical rather than a political problem, something to be solved by foreign technical experts, international humanitarian agencies have unintentionally undermined accountability mechanisms in African countries. The famine response provided by these agencies has not only often failed to address (or even acknowledge) the fundamental political causes of famine, but has also, in many cases, involved compromises that served to strengthen authoritarian regimes (de Waal 1997). A similar logic can be at work in non-famine contexts. In the West Bank and Gaza, for example, the provision of humanitarian assistance to meet the most basic economic and social needs has sometimes had the effect of diffusing or undermining political demands (Feldman 2009). Elsewhere, providing humanitarian assistance while ignoring the politics and dynamics of conflict has served to exacerbate conflict or to facilitate the abuse of the civilian population by governments or rebel groups (Anderson 1999).[13]

As with humanitarian representational practices, humanitarian action may contribute to political inaction, both in terms of international responses to specific crises and in terms of addressing global inequality as a systemic cause of those crises. The notion of the 'humanitarian alibi' describes those situations in which governments, in particular the small number of Western governments who provide the lion's share of funding for international humanitarian action,[14] use their support for a humanitarian response to conceal or detract attention from their failure to take any more concerted political action to address the causes of the emergency.[15]

The humanitarian ethic and ethos is primarily grounded in notions of pity, compassion, and empathy rather than, for example, anger, solidarity, and justice.[16] Correspondingly, the kinds of action it demands—charitable giving to relieve suffering—leave intact, and may even reinforce, global hierarchy and inequality. Contrasting participation in public protests against global injustice with attendance at Live8 music concerts, Costas Douzinas claims that humanitarianism 'has turned into the ultimate political ideology bringing together the well-being of the West with the hardships of the global South' (Douzinas 2007: 11). In a similar vein, B. S. Chimni has described humanitarianism as the ideology through which hegemonic states 'establish and sustain global relations of domination' (Chimni 2000: 244).

Humanitarianization is not always successful, and crises are not always fully humanitarianized. On the one hand, the affected populations rendered by the humanitarianization process as victims rather than agents or actors may resist this construction and show themselves to be anything but passive. For example, efforts to render refugees in camps as bare life have often been contested by those refugees.[17] Camp residents are cared for in terms of their security and their biological needs and, in return, are expected not to

[13] For examples of this kind of dynamic, see Chapter 2 on the Nigerian Civil War, 1967–1970, Chapter 3 on Drought and famine in Ethiopia, 1983–85, and Chapter 8 on Sri Lanka, 2008–2009.

[14] See Chapter 18 on Donors and the Funding of Humanitarian Action for an extended discussion of the politics of humanitarian finance.

[15] For an example, see the 'Humanitarian alibi' in Chapter 4 on the Bosnian War, 1992–1995.

[16] See 'The Boundaries of Humanitarianism', the introduction to the second part of the book.

[17] See 'Politics of camps' in Chapter 22 on Material Assistance and Direct Service Provision.

make political demands, with camp authorities denying their political will and agency (Turner 2016: 143). However, the depoliticization inherent in camp management can create its own opposite—hyper-politicization—as the denial of residents' history and politics 'creates a gap in the social and symbolic order of life in the camp, which in turn creates room for the creation of new competing orders and identities' (Turner 2016: 145). Camps offer opportunities for the forging of new identities and can provide fertile ground for radicalization and violence. On the other hand, where powerful actors—including dominant states and the political organs of the UN—have a stake in the outcome of a crisis, they may resist humanitarianization or seek to subordinate the humanitarian response to their political goals. In this way, then, humanitarianization efforts can be seen to provoke a counter-response—whether by the supposedly powerless or the more powerful—which serves to re-politicize or to hyper-politicize.

The meaning of humanitarianism and the focus of this book

This discussion of the construction of humanitarian emergencies—of the process of humanitarianization—brings us to the question of what it means to be humanitarian. Humanitarianism is a contested concept,[18] and this question is at the heart of the more theoretical debates in the field of humanitarian policy. Pretty much everyone can agree that humanitarianism is about 'saving strangers' or alleviating the suffering of distant others. In other words, there are two crucial elements: first, that of helping people or saving their lives—or at least intending to do so; and second, that of distance, in the sense that these are not people known or closely connected to those doing the helping (Calhoun 2008: 79). In addition to these two elements, the focus and analysis of this chapter suggest a third: contemporary understandings of humanitarianism largely imply that humanitarian action is not about alleviating the suffering of distant others *wherever* suffering may be found, but rather is limited to alleviating the suffering of distant others *in the context of a humanitarian emergency*. Thus, it is about responding to famine but not to generalized poverty, to Ebola but not to measles.

Yet, the precise boundaries between what does and does not constitute a humanitarian emergency are both historically contingent and contested.

[18] The main lines of contention are discussed in Chapter 13 on Politics, Principles, and Humanitarian Action. See also Barnett and Weiss (2011: 9–17).

There is no single answer to the question of 'what state of affairs qualifies as legitimate emergency' and hence legitimates or demands an international humanitarian response (Barnett 2018: 325–328; Slim 2015: 8).[19] Furthermore, there is fierce debate on two linked questions. First, what is the scope of humanitarian ambition? Is it simply saving lives, or should humanitarians also seek to address the causes of suffering, promoting structural change and a more transformative agenda? Second, what is the proper role of principles and politics in humanitarian action? The so-called principles of humanitarian action are discussed in detail in Chapter 13, and suffice to say here that there is contestation over whether it is possible, or desirable, to adhere to these principles in practice or whether being more political is either unavoidable or preferable.

Like the concept of security, the concept of humanitarianism is malleable, and the adjective 'humanitarian' is appropriated by many in an effort to legitimize their actions (see, e.g., Chimni 2000). Many object to the way the adjective 'humanitarian' is used in order to justify or legitimize military interventions in particular. If we reject the idea that anyone can define what it means to be humanitarian, then who gets to define the concept and police its use? The question of what to include and exclude within the category 'humanitarian' is a recurring theme among humanitarian scholars and practitioners.[20]

In this book, I do not take an explicit position on the question of what it means to be humanitarian, but something of a definition of humanitarianism is implicit in my choices to include or exclude particular issues and actors. This book focuses on organizations like the ICRC, the IRC, MSF, Oxfam, Save the Children, UNHCR, UNICEF, the World Food Programme (WFP), and World Vision. All of these are widely understood to be international humanitarian agencies, some of them are what are referred to as mixed- or multi-mandate agencies in that their work encompasses humanitarian programming as well as other kinds of work (primarily development and human rights programming), and across these agencies we also see different understandings of humanitarian programming itself.[21] I write as a critical scholar of international institutions, examining what these different organizations

[19] See Chapter 15 on the Shifting Parameters of Humanitarian Emergencies.

[20] For accessible overviews of these debates and the reasons for the 'fuzziness' of the field, see Barnett (2018) and Slim (2015: 7–10).

[21] Excluded from the remit of this book are organizations which focus exclusively or primarily on development work (e.g. the United Nations Development Programme or the World Bank) or human rights work (e.g. Amnesty International or Human Rights Watch). Moreover, I do not discuss military interventions, even when their purported purpose is 'humanitarian', except in as far as they interact with the work of civilian humanitarian agencies (discussed in Chapter 20 on Armed Actors).

do on a day-to-day basis (the everyday practice of the title) and how their interests, and those of other actors, shape this practice (the politics of the title). My goal is to show how 'actually existing humanitarianism' functions, a goal which presupposes a disjuncture between theory and practice. This disjuncture, I argue throughout the book, cannot be explained only through reference to external forces but rather requires attention to the identities, interests, and idiosyncrasies of humanitarian organizations themselves.

The rest of this book is structured in four main parts. The first contains case studies of eleven humanitarian emergencies from 1967 to the present day. Each of these provides a narrative account of the humanitarian response to the relevant emergency, highlights the main challenges to, and criticisms of, that response, and outlines the effects of these experiences on humanitarianism and humanitarian action going forward. The second part of the book comprises four thematic chapters, which collectively outline the ethical and legal foundations of contemporary humanitarian action, discussing how ethics, law, and politics function and interact in theory and in practice. The third part turns to the actors that are most relevant to humanitarian action and humanitarian outcomes. While the main unit of analysis in this book is the international humanitarian agency, this third section considers the roles of governments, local and national NGOs in affected states, national and international military actors, and the relationships between these different actors. Finally, in the fourth part of the book, the nuts and bolts of humanitarian response are examined, with chapters on needs assessment and evaluation, and different dimensions of humanitarian operations and activities. Throughout, I show how politics shapes humanitarian needs, response, and outcomes—and, conversely, how humanitarian action affects political outcomes.

I

CASE STUDIES

Introduction to Part I

A Brief History of Modern Humanitarian Action

Through in-depth case studies of eleven humanitarian emergencies, this part of the book provides rich empirical material which illustrates the main debates and critiques regarding international humanitarianism, many of which are then developed further in the thematic chapters in the rest of the book.

The first humanitarian emergency presented here is the 1967–1970 Nigerian Civil War, but modern humanitarianism predates that war by more than 100 years, and the main concerns and debates surrounding contemporary humanitarian action are not new. There is a popular narrative that the challenges and dilemmas that have characterized humanitarianism since the 1990s are unprecedented, but a more historical perspective shows that this is not the case (Barnett and Weiss 2008). For as long as modern humanitarianism has existed, it has been affected by politics, and there has been debate over the proper role of politics and principle in humanitarian action. Furthermore, humanitarian action has always had political effects such that there have been concerns about its unintended consequences and the potential for it do more harm than good.

Many accounts of humanitarianism start with the establishment of the International Committee of the Red Cross (ICRC). Horrified by the suffering of soldiers dying on the battlefield without medical care or assistance in the Battle of Solferino in 1859, visiting businessman Henri Dunant responded by sending for medical supplies and recruiting local women to tend the wounded. This notion of providing humanitarian assistance to the victims of war was institutionalized by Dunant and four fellow Genevans with the establishment of the ICRC in 1863 and the negotiation of the Geneva Convention for the Amelioration of the Condition of the Wounded in Armies in the Field in 1864. If the principles of humanitarian action—humanity, impartiality, neutrality, and independence—are part of the definition of humanitarianism, and if humanitarianism is necessarily limited to treating symptoms rather than addressing causes, then the ICRC is rightly regarded as the father of modern humanitarianism.[1] However, an alternative perspective identifies William Wilberforce and the abolitionist movement as the 'original humanitarians', working fifty years before Henri Dunant witnessed the Battle of Solferino and pursuing much more transformative goals—seeking to address the causes of suffering and not just the symptoms (Barnett 2011: 57–60).

As early as the nineteenth century, then, there were both limited and more transformative approaches to 'saving strangers', and there was also some recognition of the risk that humanitarian action could inadvertently do more harm than good. When Dunant invited Florence Nightingale to promote the Red Cross cause in England, she refused, on the basis that, in taking on what was properly the responsibility of the military, the efforts of voluntary societies would make it easier for militaries to wage war (Hutchinson 1996: 40).

In the First World War, advances in technology led to mass warfare with humanitarian consequences on a scale previously unthinkable, at the same time as modern communications meant that the progress of war and destruction was well known in the home countries of the soldiers. These two factors helped to generate a great deal of international support for humanizing war. This translated into support for immediate relief and the first major expansion of the Red Cross, and also provided the impetus to develop the two 1929 Geneva Conventions—international law that seeks to protect certain categories of person during war. The first was an updated version of the 1864 Convention. The second, on the Treatment of Prisoners of War, was drafted by the ICRC, and built on its work during the First World War, which had developed to encompass inspecting prisons and negotiating with authorities to obtain improvements in conditions for detainees, operating a tracing agency to exchange information about prisoners, and providing relief assistance to prisoners (Baudendistel 2006: 246–247).

In the immediate aftermath of the First World War, new organizations were formed, primarily to deal with the fallout of the war, and the normative framework further developed. Save the Children was established in the United Kingdom in 1919 and—in an early articulation of what would come to be called the humanitarian principle of impartiality—insisted that all children, including the children of former enemies, were eligible for relief (Barnett 2011: 85). More Save the Children sections were subsequently founded in different countries, and in 1920, the International Save the Children Union was formed in Geneva. The ICRC pushed the League of Nations to coordinate intergovernmental cooperation for refugees, and the

[1] See 'Principles of humanity, impartiality, neutrality, and independence' in Chapter 13 on Politics, Principles, and Humanitarian Action.

latter established the office of its High Commissioner for Refugees in 1921, appointing the Norwegian Fridtjof Nansen to the post (Krill 2001: 607). In 1924, the League adopted the Declaration of the Rights of the Child, which had been drafted by the founder of Save the Children (Save the Children n.d.).

Several major events of the inter-war years also conditioned developments in international humanitarian action in this period, notably in expanding its remit to encompass civil war and civilians. In the aftermath of the Russian Revolution, for example, the ICRC took a central role in providing material assistance to the one million or more Russian refugees (Skran 2018). When famine broke out in Russia in 1921–1922, Nansen organized a relief programme for millions of its victims. In the Spanish Civil War, acceptance of the roles of international humanitarian law (IHL) and international humanitarian action (specifically, the work of the ICRC) in internal conflict made significant advances, even if the impact of the ICRC on the ground was negligible (Forsythe 2005: 39). While not always successful, practical efforts were made to protect and assist civilians in these settings.

During the Second World War, a number of Western states became more willing to get involved in the organization of relief. There may have been an element of compassion inspiring their involvement, and there was certainly recognition that their political, economic, and strategic interests were at stake. A co-dependent relationship between humanitarian agencies and states began to emerge, although the humanitarian agencies were arguably more dependent on states than the other way around (Barnett 2011: 31). The humanitarian effort during the war itself was largely dominated by the United States, and from 1942 to the end of the war, US organizations were tightly regulated by the government's War Relief Control Board (Barnett 2011: 109). Organizations such as the American Red Cross and Catholic Relief Services were favourites and thrived under this system, while others were effectively closed down by refusal of an export licence (Barnett 2011: 109). The US government saw how relief could be used to further its own war aims, and some of the private US agencies became para-statal (Barnett 2011: 109). This heavy-handed regulation demonstrates that challenges to the independence and autonomy of international humanitarian agencies are nothing new.

The 1940s also saw the emergence of the basic structure of the humanitarian architecture that persists today. For example, the Oxford Famine Relief Committee (now Oxfam) was set up in 1942 to respond to famine in occupied Greece. At the end of the war, Oxfam expanded from its initial focus on Greece to elsewhere in Europe and subsequently on to the 'Third World' (Black 1992: 24, 48). In 1943, forty-four countries, led by the United States and the United Kingdom, set up the United Nations Relief and Rehabilitation

Administration (Barnett 2011: 110). The Food and Agricultural Organization was also established in 1945 and the United Nations Children's Fund and the World Health Organization the following year. In 1947, the International Relief Organization was established and was succeeded in 1950 by the United Nations Refugee Agency (UNHCR).

In the first half of the Cold War, the principles of humanitarian action were increasingly professed, debated, and eventually institutionalized but not always adhered to. In the years after the Second World War, the ICRC was fiercely criticized for remaining silent about the Holocaust—silence that was justified in the name of neutrality. In the period, the language of impartiality was increasingly used, and the principle of aid based on need was often espoused. However, in practice, many of the relief agencies set up at this time were about helping very specific groups of people, with Lutheran World Relief focused on providing aid to Lutherans, and UNRRA set up to help the victims of *German* aggression, for example. In the early post-war years, Oxfam distinguished itself from many other agencies by its insistence that Germans were as deserving of aid as everyone else. Continuing the pattern set in the Second World War, states wanting to use humanitarian relief to further their interests continued to compromise the independence of humanitarian agencies. Through regulation, funding, or, indeed, setting up intergovernmental organizations, states succeeded in reducing the autonomy of humanitarian relief providers.

Two major trends characterized the evolution of international humanitarianism in the Cold War era. First, the aid architecture became increasingly permanent and professionalized. Prior to the Second World War, organizations tended to be established on an ad hoc basis to respond to a specific event and to disband after that event, but many of the organizations that were established during or immediately after the Second World War still exist today. Several new organizations formed, including World Vision and Médecins Sans Frontières (MSF), which are now two of the largest humanitarian NGOs. In 1950, inspired by missionary zeal and also the hope that its work would help to counter the appeal of Communism, World Vision started with a child sponsorship programme in Korea and then elsewhere in Asia. MSF was established subsequent to the 1967–1970 Nigerian Civil War, the subject of the first case study chapter in this part of the book. The second trend was an expansion from relief into development. In part, this was perhaps the logical move for organizations established in emergencies but remaining in existence after the emergency phase was over. However, it also stemmed from a desire to try to address the causes of suffering rather than just its symptoms. Again, then, we see shifts and questioning of the scope of humanitarian ambition.

There were also some significant developments at the legal and normative level, with several treaties signed in the Cold War era. Beginning in February 1945, the ICRC quietly orchestrated a systematic review of IHL, leading to the 1949 Geneva Conventions (Best 1997: ch. 4). The four Geneva Conventions, all of which are still in force today, pertain to (i) wounded and sick in armed forces in the field; (ii) wounded, sick, and shipwrecked members of armed forces at sea; (iii) prisoners of war; and (iv) civilian persons. In 1977, they were supplemented by two Additional Protocols, also still in force, relating to: the protection of (i) victims of international armed conflicts; and (ii) victims of non-international armed conflicts.

The case studies in this part of the book have been selected because each provides a clear example of some of the key issues and dilemmas of contemporary humanitarianism and because collectively they illustrate most of the main issues and dilemmas. This is not to suggest that any of them depicts the first example of those issues or dilemmas, nor that they are representative of contemporary international humanitarian response as a whole. Many other case studies would have made for interesting additions to the book—Operation Lifeline Sudan, Cambodia under the Khmer Rouge, and Kosovo in 1999, to name just a few—but the limits of space forced me to be selective.

2

Nigerian Civil War, 1967–1970

The declaration of independence by the military governor of Nigeria's Eastern Region in mid-1967 led to a brutal secessionist conflict that lasted three-and-a-half years. The Federal Military Government (FMG) imposed a blockade with the aim of starving the breakaway region into submission, and severe famine ensued. By the time Biafra surrendered on 12 January 1970, up to two million people—mostly civilians on the Biafran side—had died, some as a direct result of military action but most due to starvation and disease as a result of the famine.

For the first time, there was major international humanitarian mobilization on behalf of black Africans (Pérouse de Montclos 2009: 71). Private relief agencies played a central role in this mobilization, and while many of them publicly sided with the Biafrans, the conflict was understood first and foremost as a humanitarian crisis. However, the support generated in response to the humanitarian crisis served to strengthen the Biafran secessionists and hence to contribute to political outcomes, almost certainly lengthening the war. In this way, private actors—church groups and other aid agencies—had a significant impact on the course of a major civil war and, hence, on international relations (Stremlau 1977: 238).

Relief agencies working in this context faced enormous challenges in terms of getting access to provide relief within Biafra and dilemmas over whether to provide relief in the absence of permission from the Nigerian government, to take sides in the conflict, and to speak out about what they witnessed. As well as being a salutary lesson for humanitarian actors in the potential for aid to do more harm than good, the crisis and the international response contributed to longer-term outcomes in terms of representations of postcolonial Africa in the global North, the rise of non-governmental organizations (NGOs) as central actors in humanitarian response, and the birth of some important international humanitarian NGOs, notably Concern Worldwide and Médecins Sans Frontières (MSF).

The Politics and Everyday Practice of International Humanitarianism. Miriam Bradley, Oxford University Press.
 DOI: 10.1093/oso/9780198840695.003.0002

Background

When Nigeria was granted independence from Britain in 1960, there were high hopes for democracy, prosperity, and stability in Africa's most populous country. Within ten years, however, the country had suffered two military coups, the massacre of perhaps tens of thousands of people, and then a brutal secessionist conflict that cost up to two million lives and caused enormous suffering and hardship for many more.[1]

Causes of the war

The seeds of discontent were sown by colonial-era politics and policies. In 1914, the Northern and Southern Nigeria protectorates were amalgamated to create the single colony of Nigeria, in order to use the resource-rich Southern Protectorate's budget surpluses to offset the Northern Protectorate's deficit (Omaka 2016: 10; Uche 2008: 114–115). The Richards Convention of 1943 then divided Nigeria into three regions, each of which was home to a majority ethnic group (Hausa in the Northern Region, Ibo in the Eastern Region, and Yoruba in the Western Region) and a number of ethnic minorities. The British policy of indirect rule ensured that each region had its own system of local government and law (and, hence, a distinct experience of colonial rule), and allowed the three majority ethnic groups—which collectively accounted for only around 60% of the country's total population—to dominate the minorities in their respective regions (Omaka 2016: 10; Stremlau 1977: 30–32). As part of the same policy, respect for Islamic traditions in the North led the colonial power to shield that region from the Christian missionaries and their Western-style schools, whose presence in the south had a significant impact on education levels there (Omaka 2016: 10).

In 1958, commercial quantities of oil were discovered in the Eastern Region, prompting a shift in the national fiscal structure such that mining rents and royalties would no longer be returned to the regions but distributed between the region of origin, the federal government, and the other regions (Uche 2008: 115–116). The oil discovery accelerated economic development in the Eastern Region, and the new fiscal policy marked the beginning of a centralization of power as well as a struggle for control

[1] For a more detailed discussion of the factors that led to the attempted secession of Biafra and the ensuing war, see Heerten (2017: ch. 1); Omaka (2016: chs 1 and 2); Stremlau (1977: ch. 2)). On conflict dynamics and the international politics of the war, see Stremlau (1977), which includes a helpful chronology of key events from January 1966 to January 1970 on pp. xv–xix.

of national revenue (Uche 2008: 116). By independence in 1960, then, there was significant inequality across the regions, with higher levels of education and economic development in the southern regions (of which there were three from 1963, when a new Mid-Western Region was separated from the Western Region), and fierce competition for control of the national government. Regional inequality was a source of tension in itself and additionally meant that most of those considered best qualified for positions in the administration of the country were of southern origin, generating some resentment among northerners (Omaka 2016: 13–14; Stremlau 1977: 32).

The more immediate triggers for Biafran secession were two coups and three waves of associated violence in 1966. On 15 January, a military revolt overthrew the First Republic and brought to power an Ibo-dominated group. After a cabinet crisis over succession, Major General Aguiyi-Ironsi, an Ibo and head of the army, who had remained loyal to the federal government, assumed power (Stremlau 1977: 33). Yakubu Gowon, a Christian from the Northern region, became army chief of staff (Stremlau 1977: xv). The coup was initially popular, but support for Ironsi's government eroded as he implemented policies interpreted as favouring Ibos (Stremlau 1977: 34; Uche 2008: 118–119). Rightly or wrongly, many Nigerians ascribed ethnic motivations to the coup, which served to fuel inter-ethnic tensions (Anthony 2014: 205; Omaka 2016: 18–22). On 24 May, a military decree eliminated Nigeria's regions, creating a unitary state, fuelling fears that the Ibos planned to dominate the military and the bureaucracy and kicking off the first of three waves of anti-Ibo violence, with riots in the North (Stremlau 1977: xv, 34). The second wave was a counter-coup, beginning on 29 July, in which more than 200 mostly Ibo soldiers were systematically killed by their comrades and which brought General Gowon to power on 1 August (Anthony 2014: 210; Stremlau 1977: 35).

In September 1966, a third wave of rioting and massacres began, mostly directed at Ibos living in the North. An estimated 5,000–50,000 people were killed, and up to two million fled to the Ibo-dominated East of the country, swelling the population there (Anthony 2014: 216; Stremlau 1977: 38). Attempts to reconcile the Eastern Region and the Lagos-based FMG were unsuccessful, with Lieutenant Colonel Ojukwu, military governor of Nigeria's Eastern Region since the first coup in January 1966, refusing to accept anything short of full autonomy for the region (Omaka 2016: 36–40; Stremlau 1977: 45–55). On 30 May 1967, Ojukwu unilaterally declared the former Eastern Region's independence from Nigeria, naming the seceding enclave the 'Republic of Biafra.'

Conflict dynamics

The FMG's first response was to ban foreign currency transactions, to stop incoming post and telecommunications, and, crucially, to impose a blockade, which other countries largely respected (Stremlau 1977: 74). The blockade had the explicit aim of starving the enclave into submission, a military strategy which was, at the time, not prohibited in international law.[2] Gowon needed time to mobilize federal forces, who then attacked on 6 July. Fighting over the next five weeks produced gains and losses on both sides, and on 10 August, Gowon declared that what had been 'police action' against secession was now 'total war' (Stremlau 1977: 76–78). Biafra was quickly surrounded by the Nigerian army, which then advanced to take control of land claimed by the secessionists. Between one-quarter and one-third of the 12 to 14 million inhabitants of the former Eastern Region were ethnic minorities, many of whom had resisted Ibo domination and chose to remain in their homes as the FMG took control of the territory on which they lived (Heerten 2017: 58–59; Stremlau 1977: 57, 59). Ibos, on the other hand, generally fled into the ever-shrinking space of Biafra. Since the Eastern Region had not been self-sufficient in terms of food and relied heavily on supplies from Northern Nigeria and Iceland for protein sources in particular, the blockade and the subsequent loss of major food-producing areas led to famine with high mortality rates associated with protein deficiency and starvation (Omaka 2016: 62).

On 19 May 1968, Port Harcourt—the principal economic and commercial centre of the secessionist state, including an airport, docks, and an oil refinery—fell to the Nigerian military, leaving the secessionists completely surrounded and without access to the sea (Heerten 2017: 95). From that moment on, there was probably no real chance of the Biafrans winning the war (Heerten 2017: 95; Pérouse de Montclos 2009: 70). By August 1968, around two-thirds of the original Biafran territory was in Nigerian hands, with up to 8 million people in the remaining 9,000 square miles, in which fertilizer and food were scarce (Anthony 2014: 216).

The Biafrans were not only at a military impasse but also blocked from making any real progress in terms of political recognition by other governments (de Waal 1997: 73; Heerten 2017: 55; Stremlau 1977: ch. 9). Many

[2] Starvation as a method of warfare was prohibited in the 1977 Additional Protocols to the 1949 Geneva Conventions and defined in the International Criminal Court (ICC) Statute as a war crime in international armed conflict. For a general discussion of international humanitarian law (IHL) and the ICC, see 'International humanitarian law' and 'International justice' in Chapter 14 on Law and International Humanitarianism.

countries, but especially newly independent African states, worried that the Biafran example might inspire secessionist movements within their own borders and serve to destabilize the region as a whole (Stremlau 1977: 275). Four African countries did eventually recognize Biafra (Tanzania in April 1968 and Gabon, Ivory Coast, and Zambia the following month), though their recognition appears to have been more an act of protest against the way the FMG was conducting the war than support for Biafran sovereignty in the long run (Stremlau 1977: 136–139, 273). The Organisation of African Unity (OAU) issued two resolutions on the war and led some efforts to mediate between the two parties, expressing concern for the war's human toll and a commitment to the territorial integrity of Nigeria as well as to the OAU principle of non-interference in the internal affairs of member states (Stremlau 1977: 93, 272, and ch. 7).

Both the United States and the United Kingdom declared the war an internal Nigerian matter, a stance which echoed that of the OAU, and implicitly lent support to the FMG (Stremlau 1977: 62–65). Furthermore, the UK government allowed British arms manufacturers to sell weapons to Lagos, even as other countries enacted arms embargoes, and argued for the importance of maintaining the territorial integrity of Nigeria (Heerten 2017: 115; Stremlau 1977: 76; Uche 2008: 113). The British position was not disinterested, with Cold War politics and oil interests influencing its decisions. Gowon was buying Soviet aircraft and the British wanted to limit Russian involvement and influence in Nigeria, a concern which helped to ensure the continued flow of arms from the United Kingdom (Stremlau 1977: 80, 267; Uche 2008: 127–128). The British also had significant oil interests at stake, and the UK government was concerned to protect Shell–BP investments in Nigerian oil as well as the supply of Nigerian oil to the United Kingdom, particularly in the context of the six-day Arab–Israeli war in June 1967, which disrupted the supply of oil from the Middle East (Uche 2008: 113).

As France competed with the United Kingdom for influence in their former West African colonies, it had an interest in breaking up a major Anglophone country and, hence, in supporting the secessionists (Griffin 2015: 119; Pérouse de Montclos 2009: 72; Uche 2008: 128). Furthermore, unlike other oil companies operating in Nigeria, the French state-based oil company held permits for reserves located in the Ibo heartland (Uche 2008: 128–129). At the same time, division among Francophone African leaders on the issue, concern over Soviet influence, and a desire to maintain cordial relations with Nigeria to protect French investments in federal territory all combined to prevent France from fully embracing the Biafran cause (Stremlau 1977: 225; Uche 2008: 128–129). Thus, while never officially recognizing Biafra, France announced an embargo on the supply of arms to both sides in June 1968, and the following month, a French cabinet statement supported claims to Biafran self-determination (Griffin 2015: 124; Stremlau 1977: 228). Unofficially, the French provided the secessionists with mercenaries, weapons, and access to foreign exchange with which to purchase the weapons. However, the effectiveness of the mercenaries was limited, and while facilitating arms shipments likely had a bigger impact, the monetary value of this indirect military assistance paled in significance next to French contributions to the Nigerian treasury through trade (Griffin 2015: 121–123; Stremlau 1977: 230–231, 233).

In summary, the international reaction was mixed. Some states supported Lagos, while others (officially or unofficially) sided with Biafra. Most either ignored Biafra or hedged their bets (Stremlau 1977: 64). The United States and several European countries maintained an explicit policy of separating the political from the humanitarian (Stremlau 1977: 292–297). Such a separation was, ultimately, wishful thinking, but it allowed them to support Lagos through trade, investment, maintenance of diplomatic relations, and, in the case of the United Kingdom, the sale of arms, at the same time as funding the humanitarian response. Biafra was at a major disadvantage in terms of trained military manpower, arms supply, and money with which to buy arms. However, it was at an advantage in terms of global public sympathy, which helped to sustain a major international humanitarian response, which, in turn, helped to sustain the Biafran fighters.

Global public response

In 1968, the civil war in Nigeria became a major news story and a global cause célèbre. It became 'Biafra'—a humanitarian emergency; that is to say, it was seen largely through a humanitarian lens, focused on the suffering of the Biafran population, rather than through a political or military lens.[3] A public opinion poll in the summer of 1968 revealed that Biafra was the number one international issue of concern to the French people, and French public sentiment was overwhelmingly pro-Biafran (Stremlau 1977: 227). In Ireland, there was huge public (though not official) support for Biafra, generating a response 'unequalled by the reaction to any foreign conflict between that of Spain in the 1930s and that of Yugoslavia in the 1990s' (Staunton 1999: 513; see also, O'Sullivan 2014: 302–303). There was an outpouring of compassion and sympathy in many other countries too, notably Germany,

[3] See 'Constructing humanitarian emergencies' in Chapter 1 on Humanitarian Emergencies.

the United Kingdom and the United States, with citizens donating money for relief and criticizing the inaction of their governments (Black 1992: 121; Heerten 2016: 250).

A number of factors helped to generate a response of this scale and nature. First, there was a proactive effort on the part of the secessionists to develop this kind of transnational support. When it became clear that efforts to secure significant recognition and support from states and intergovernmental organizations were not bearing fruit, Biafran leaders turned to non-state actors and Western publics as alternative sources of international support (Heerten 2017: 90). Their strategy included contracting the services of two public relations (PR) firms—the Californian Robert S. Goldstein Enterprises, which had expertise in TV and film-based campaigns, in December 1967, and Geneva-based Markpress, which had the capacity for mass international circulation of press releases, in January 1968 (Heerten 2017: 90). The Biafran leadership also sent envoys abroad on pseudo-ambassadorial missions, although their reception was mixed. Government officials in many countries were hesitant to meet with them for fear of damaging their relationship with Lagos, but they had some impact on Western publics, eventually generating a transnational Biafra lobby (Heerten 2017: 91–4).

Second, first-hand accounts by foreigners had a significant impact on world opinion (Stremlau 1977: 117). The presence and activism of Christian missionaries in the Biafran enclave was key both to making 'Biafra' a major international news item and in shaping the way that the crisis was understood internationally (Omaka 2016: 64–68; Stremlau 1977: 119–120; Waters 2004). The Irish Holy Ghost order was the largest bloc of missionaries in Nigeria and had a long-standing presence, primarily in Ibo areas. By 1965, the Catholic Church as a whole had 424 foreign priests in the Eastern Region, and Norwegian and Dutch missionary groups were also well-established (Omaka 2016: 53, 64). Most missionaries remained after the outbreak of war, and they played a crucial role in defining and disseminating the message, with a symbiotic relationship developing between aid agencies and the media (Heerten 2017: 98, 117; Stremlau 1977: 119; Waters 2004). Many of the missionaries had a close relationship with the Biafran leaders and, in several cases, adopted an overtly political stance. Journalists, in turn, were dependent on aid agencies and/or Biafran officials for transport into and within Biafra, and this gave those working for the Biafran cause significant influence over what the journalists saw and how they understood the conflict.[4]

[4] On the role of the media in setting and framing the humanitarian agenda and on the symbiotic relationship between the media and aid agencies, see Chapter 17 on Media and Celebrities.

Third, from mid-1968 onwards, the message disseminated by the secessionists, their PR firms, and the missionaries and other aid workers focused on the humanitarian consequences of the conflict in Biafra and not on the political or military dimensions of the conflict itself (Griffin 2015: 124–125; Stremlau 1977: 119). By the end of 1967, the Biafran leadership could no longer maintain an image of itself as a military power capable of swift victory against the FMG and changed its discourse from one of fighting for political self-determination to one of defence against an existential threat (Stremlau 1977: 109). Markpress PR agency considered the Biafra account to be non-political and circulated press releases without substantial editing, allowing the Biafran leaders to provide the media with reports and material from their own perspective (Doron 2014: 241; Heerten 2017: 90; Omaka 2016: 69; Stremlau 1977: 116–117). In presenting the human-interest angle, the war was portrayed as one between a clear oppressor (the Nigerian government) and its innocent victims (the Biafran people). This oversimplified depiction of the conflict was contested, with others seeking to portray the complexity of the situation and to promote alternative views. However, the coverage that resonated most with the public was that which tugged at heartstrings and not that which presented a more reasoned political or military analysis (Heerten 2016: 253).

Closely linked to the human-interest angle was a discourse of genocide, with parallels drawn between the Biafrans and the Jews, on the one hand, and between the Nigerian government and Nazi Germany on the other (Anthony 2014: 217–218; Heerten 2016: 257–263; Waters 2004: 704–705). Charges of genocide were an important part of Biafran propaganda—and the contestation of those charges an important part of the counter-narrative (Desgrandchamps 2014; Doron 2014; Heerten 2016). The secessionists went to great lengths to ensure that their propaganda was powerful, consistent, and credible (Doron 2014). The messages they disseminated, including the claims of genocide, were not just aimed at securing international support (Heerten 2016: 254–255; Stremlau 1977: 112). If the Biafran civilian population had not believed the war was one of survival, most would have been unlikely to accept such extreme hardship—and such high mortality rates—and would instead have pushed the secessionist leaders to surrender much sooner than they did.

Fourth, visual imagery played an important role in supporting this simplified narrative and in generating global public support for the Biafrans (Heerten 2016). The 'Biafran babies' were exemplary victims,[5] whose

[5] See 'Discursive practices and visual imagery' in Chapter 1 on Humanitarian Emergencies.

malnourishment, illness, and physical suffering was visible in their skeletal bodies and bulging stomachs (Black 1992: 124; Heerten 2016: 256; O'Sullivan 2014: 306–307). This was the first time that the public around the world had been confronted with media images of a major famine on colour television and in a new kind of photojournalism, creating both the 'event' and the iconography of Biafra (de Waal 1997: 74; Heerten 2016: 255; Omaka 2016: 63; Pérouse de Montclos 2009: 72; Waters 2004: 697, 709). This kind of coverage was important in driving public concern and the humanitarian response, yet it also emphasized the distance between Biafra and the world of the television viewer (O'Sullivan 2014: 302, 309).

Fifth, the war between Nigeria and Biafra took place in a particular moment in time when world sentiment was uniquely tilted to favour the oppressed (even if such sentiment was often not converted into action). The student and labour protests in France and elsewhere in May 1968 were emblematic of that moment, and the rise of the Biafran cause must be understood in the context of that moment (Heerten 2017: 206). Domestic pressure on de Gaulle may have been a significant factor behind the French statement of support for Biafran secession in July 1968—although it also seems that the French government sought to shape public opinion on Biafra and that de Gaulle himself was involved in the decision to send a French Red Cross mission to Biafra (Davey 2015: 30; Griffin 2015: 125, 127; Stremlau 1977: 227). The moment was also characterized by the civil rights movement in the United States, and Martin Luther King had actually planned to go to Nigeria/Biafra, together with a small group of fellow civil rights activists, on a 'peacekeeping mission' in April 1968 but was assassinated less than two weeks before their scheduled departure (Heerten 2017: 94). It is probably no coincidence that public pressure over Biafra in Western democracies peaked in summer 1968 (Stremlau 1977: 255).

Finally, Biafra was largely Christian and Northern Nigeria was largely Muslim, with the consequence that the conflict could be, and frequently was, portrayed as a religious war (Heerten 2017: 85–89). Many within Biafra saw the war as the realization of longstanding fears of Muslim domination, and the Biafran propaganda directorate promoted this view (Stremlau 1977: 113). Such a perspective failed to capture the complexity of the conflict (never mind the fact that the Nigerian head of state, Gowon, was a Christian), but it nonetheless contributed to foreign sympathy and support (de Waal 1997: 74).

Global sympathy for the Biafrans translated into financial contributions from individuals as well as pressure on several governments to donate food and funds themselves. This, in turn, led to a significant humanitarian response. The basic imagery and simple messaging led to an equally simplistic response that focused on the humanitarian imperative to get relief to Biafra in any way possible, and flattened the complexity of the political crisis (O'Sullivan 2014: 306). Much of the relief was provided and delivered not by established, professional aid agencies but by missionaries and church groups or newly formed organizations set up specifically to respond to the crisis in Biafra. Even the professional agencies did not all have experience of operations on the ground and certainly not of this scale and difficulty—and the relative inexperience of foreign aid workers may have contributed to the unintended consequences of the aid they provided (Black 1992: 126; Davis 1975: 505).

International humanitarian response

Apart from some minimal input from the United Nations Children's Fund (UNICEF), United Nations (UN) agencies were absent, but the International Committee of the Red Cross (ICRC) and a number of NGOs launched a relief operation that was unprecedented in many respects. Through their combined efforts, around 7,800 flights airlifted up to 100,000 metric tons of food and medicine into Biafra (Stremlau 1977: 242). Never before had the ICRC undertaken an operation of that scale or nature, and never before had international NGOs played the dominant role in humanitarian response (Davis 1975: 505; de Waal 1997: 73; Stremlau 1977: 207). The ICRC budget jumped from USD 500,000 per year worldwide to USD 1.4 million per month for the Biafran war alone, totalling CHF 158 million over the course of the conflict (Davis 1975: 503; de Waal 1997: 76). Oxfam initially provided funding and supplies to missions and operational agencies working in Biafran territory and in areas that had been re-taken by the Nigerian army, and then in the second half of 1968, dispatched a team of seven medical and relief workers, becoming operational for only the second time in its history (Black 1992: 119, 126; de Waal 1997: 75).

In the context of the blockade, getting aid into the Biafran enclave was extremely difficult, and there was much political wrangling over relief deliveries. On the one hand, the Nigerian government refused consent for air delivery of food unless transported on its own planes or—initially—those of the ICRC, ostensibly due to concerns that other planes also carried weapons (Anthony 2014: 216; Waters 2004: 710). On the other hand, the Biafrans would not accept food arriving from Nigeria, ostensibly due to concerns that it would be poisoned (Anthony 2014: 216; Omaka 2016: 133–134;

Waters 2004: 705). ICRC flights into Biafra from neighbouring countries with inspections by FMG officials before departure offered a feasible compromise but for a further sticking point: Lagos refused to authorize night flights into Biafra, and the Biafran leadership refused to accept daytime flights. Planes carrying arms into Biafra flew by night, and in darkness, the Nigerians could not distinguish the planes carrying relief from those carrying arms. Since international public opinion would not tolerate the FMG cutting off relief altogether, it could not seriously target the arms supplies as long as the relief supplies were also coming in at night (Stremlau 1977: 246).

Relief agencies were thus caught between wanting to provide aid and needing approval from Lagos, and different organizations adopted different stances and strategies. The first international relief was sent by church groups, who bought space on military supply flights (de Waal 1997: 75). By contrast, with a mandate that required it to operate with the consent of the states in which it operated, the ICRC initially negotiated with the Nigerian government for permission to provide aid to Biafra (Davis 1975: 509). After around four months of negotiations with the FMG and the Biafran leaders, in which it proved impossible to get both sides to agree to the same conditions, on 9 April 1968, the FMG confirmed it would tolerate ICRC night flights, at the ICRC's own risk (ICRC 1969: 13). The ICRC then began operating night flights from Fernando Po, an island of Equatorial Guinea (a Spanish colony at the outset of the Biafran war, independent from October 1968), but suspended the flights on 10 August, citing FMG antiaircraft fire, a move interpreted by the Biafrans as an effort to pressurize them to accept a federally approved plan for overland relief (Stremlau 1977: 208–210).

As the ICRC suspended its night flights into Biafra in August 1968, Swedish Count Carl Gustav von Rosen was defying all warnings from the FMG to break the blockade, carrying relief into Biafra on behalf of the German churches, launching what would become the biggest privately run airlift in history (Stremlau 1977: 210). The German churches were soon joined by around thirty more mainly Catholic and Protestant groups from Europe and North America, together forming Joint Church Aid (JCA), nicknamed 'Jesus Christ Airlines' (Heerten 2017: 99; Omaka 2016: 95; Stremlau 1977: 244). Many of the groups and agencies within JCA were set up specifically to respond to the humanitarian crisis in Biafra, but what they lacked in experience of relief work they compensated for in 'a chaotic combination of zeal and daredevilry' (Black 1992: 124), undertaking risky night flights from Sao Tomé from November 1968 through to January 1970 (Omaka 2016: 95–97, 111–113). The church-based organizations were generally seen as quicker and more efficient than the ICRC, and by the end of the war, the JCA operation had delivered a total of 66,000 tons of relief in 5,310 flights at a cost of USD 116 million (Davis 1975; Desgrandchamps 2012; Stremlau 1977: 244).

With the church groups complying with the demands of the Biafran leaders to fly relief in only at night, the ICRC lost any leverage to pressurize the secessionists to accept daytime flights or overland relief deliveries and, in September 1968, mounted the International Air Lift West Africa (INALWA), a much-expanded all-night air shuttle from Fernando Po and Dahomey (now Benin) including planes chartered by the ICRC itself and by other organizations (Stremlau 1977: 211–212). While the ICRC lacked official permission from Lagos, the Nigerian consul in Fernando Po was privately instructed to inspect ICRC planes before take-off, and although the FMG declined any responsibility for the safety of the flights, it privately agreed to refrain from doing anything to jeopardize them, a move that was taken by the ICRC as tacit permission (Stremlau 1977: 211–212). The ICRC additionally provided relief to those in FMG-held territory. In 1968, with ever-increasing numbers of people fleeing Biafran territory and the interruption of economic activity in regions devastated by the war, the situation of the civilian population in territory under federal control was deteriorating rapidly (ICRC 1969: 11). Providing relief there was important in light of the ICRC's recent public commitment to the principle of impartiality[6] and also served to improve its credibility and ease relations with the FMG (Desgrandchamps 2012: 1418). Over time, however, tensions mounted between the ICRC and the FMG, with the government blaming humanitarians for inhibiting their ability to bring the war to an end (Black 1992: 128). A Swedish Red Cross plane, flying as part of INALWA, was shot down on 5 June 1969, and the ICRC suspended the airlift (Davis 1975: 508; Desgrandchamps 2012: 1414; Stremlau 1977: 246; Waters 2004: 711).

Other airlifts continued to fly relief into Biafra until the end of the war. Some of the secular agencies, including Oxfam, that had been assisting the ICRC airlift quietly switched their support to JCA (Black 1992: 129). In addition, two smaller airlifts operated from Libreville in Gabon. Africa Concern (now Concern Worldwide) was founded in Ireland in March 1968 and delivered 3,203 tons of relief in 306 flights, and the French Red Cross delivered 3,075 tonnes of food and medicines in 439 flights between the summer of 1968 and the end of the war (Davey 2015: 31; O'Sullivan 2014: 309; Stremlau 1977: 245).

[6] Impartiality was one of the seven 'Fundamental Principles of the Red Cross', declared in 1965 (Pictet 1979). For more on the principle of impartiality, see 'Principles of humanity, impartiality, neutrality, and independence' in Chapter 13 on Politics, Principles, and Humanitarian Action.

Both the ICRC and Oxfam—for different reasons—made some foolish decisions, which compromised their commitment to providing relief to both sides in the conflict, and to avoid taking sides. In May 1968, the ICRC launched a campaign somewhat tactlessly called 'SOS Biafra', in which it called on some thirty national societies to mobilize material support from their respective governments and publics, as well as advocating for the FMG to lift the blockade (Desgrandchamps 2012: 1415–1416; ICRC 1969: 9). This not only hindered negotiations about relief deliveries to Biafra but also affected the management of the operation on the FMG-controlled side (Desgrandchamps 2012: 1416). At the end of 1968, when the authorities in Equatorial Guinea began to obstruct the ICRC airlift, August Lindt, a Swiss diplomat and former ICRC delegate who had been brought in to run the ICRC operation in Nigeria, tried to transfer the airlift to Libreville (Gabon), a move which was totally unacceptable to Lagos, given that Gabon had recognized Biafra and not only relief but also arms were being sent from its capital (Desgrandchamps 2012: 1419). For Oxfam's part, although it sought to avoid taking sides and to provide relief where it was needed in territory controlled by the FMG as well as in territory controlled by the secessionists, it went ahead with its work in Biafra without permission from Lagos and made statements that were indisputably partisan (Black 1992: 123). The FMG declared Lindt *persona non grata* in mid-1969 and, at the end of the war, expelled all overseas relief workers and all international aid agencies bar UNICEF from the territory that had been claimed by the secessionists (Black 1992: 128, 130).

Political impacts: Supporting the secessionists and prolonging the war

After Port Harcourt fell to the Nigerian military in mid-1968, the secessionists would have been unable to continue fighting without the international support—support that was a response to the humanitarian crisis. With the support of international aid, however, the secessionists were able to keep fighting until they were finally defeated by the Nigerian army in January 1970. As such, the famine benefited the secessionist leaders in Biafra. It was the humanitarian crisis that kept the Biafrans in the news internationally and made them seem a worthy political cause. Images of skeletal, famine-stricken children—the 'Biafran babies'—were effective for internationalizing the Biafran cause. The representation of the conflict in terms of powerful oppressors, on the one hand, and weak and starved victims, on the other, served to legitimize the Biafran struggle for independence, and the discourse of genocide undoubtedly lent additional power to such legitimization (Pérouse de Montclos 2009: 72).

The aid provided in response to the famine also brought material benefits to the Biafran leadership. Denied income from exports by the blockade, they desperately needed cash in order to finance arms purchases, especially after the FMG introduced new paper money in January 1968, rendering that held in Biafra worthless (Stremlau 1977: 220–221). It is impossible to know how much covert funding the secessionists received from governments who were sympathetic to the Biafran cause and/or stood to gain from the break-up of Nigeria, but the foreign exchange component of private relief was probably the most decisive and reliable source of hard currency (Pérouse de Montclos 2009: 74; Stremlau 1977: 223). Aid agencies exchanged hard currency for local currency in order to buy local food for their feeding programmes (at least until the airlift began), to hire transport for the distribution of relief, and to make cash grants for local projects (Stremlau 1977: 239–241).

The provision of relief and logistical support also brought additional tangible benefits to the secessionist cause. The Biafran leaders had significant control over the distribution of aid within the enclave, and the army and bureaucrats were prioritized over the rest of the population (Stremlau 1977: 247–248, 280). In the last months of the conflict, Biafran soldiers were trying to divert food aid by force and aid agencies hired armed guards to protect their stores (Pérouse de Montclos 2009: 73). The provision of relief to the civilian population also made life within the Biafran enclave more bearable than it would otherwise have been and, hence, can be expected to have reduced popular demands for surrender. In addition to carrying relief items, the JCA cargo included spare parts for trucks in Biafra—trucks that could equally well transport troops as the food they were supposed to deliver (Pérouse de Montclos 2009: 73). In September–October 1968, JCA enlarged and improved the only airstrip within Biafra, something the ICRC had refused to do (Davis 1975: 508; Pérouse de Montclos 2009: 73; Stremlau 1977: 243–244). Accusations that the aid agencies smuggled arms in their flights appear unproven (although that does not mean they were untrue), but both the spare parts and the expanded runway were dual-use goods that could be—and, indeed, were—used for military purposes as well as for the delivery of food and other relief items.

In light of all these benefits, secessionist leaders manipulated and exaggerated the effects of famine. Knowing that an insistence on night relief flights ensured cover for the flights carrying weapons, the Ojukwu government refused to compromise on relief shipments, either by allowing daytime flights or a land corridor from Nigeria, even though this endangered those flying the relief in and reduced the quantity that could be delivered (Anthony 2014: 216; Black 1992: 128; Terry 2002: 42–43). Indeed, Ojukwu instrumentalized both hunger and relief with the explicit aim of prolonging and internationalizing the war (de Waal 1997: 75–76; Forsythe 2005: 67).

In the absence of international relief and support for the Biafrans, the war would have likely been much shorter and with far fewer casualties. When the secessionists were finally defeated, fears of government retaliation against the Biafran people did not materialize, undermining the arguments that the secessionists were defending the enclave against a genocidal central government. Of course, it is easy to see all this with hindsight, but the risk of aid prolonging the conflict, the manipulation of aid, and the mis-representation of the situation in the media were already apparent during the conflict. For example, the British government—while obviously also having other motives for supporting Lagos—argued that more lives would be saved by letting Nigeria win and hence shortening the war (Pérouse de Montclos 2009: 74). Robert Goldstein, of the Californian PR agency hired by the Biafran government in December 1967, quit the account in August 1968, accusing Okukwu of 'using the starving hordes as hostages to negotiate a victory' (Waters 2004: 710). In the OAU summit of September 1968, the Presidents of Mali and Cameroon blamed Ojukwu for starving his people, through both his continued insistence on secession and the conditions he imposed on relief deliveries (Stremlau 1977: 274). By late 1969, some members of the World Council of Churches (WCC) were also arguing that aid was prolonging the war (Omaka 2016: 146). Moreover, by the time of the massive expansion of the relief operation in September 1968, there was significant evidence to suggest that there would be no genocide: federal troops had taken control of large tracts of Biafran territory without massacring the inhabitants (de Waal 1997: 76–77; Smillie 2012: 30). Indeed, by late 1968, half of the Ibo population was living in relative safety in FMG-controlled territory, and the property of those who had fled to Biafra was protected (Black 1992: 129).

Longer-term significance of the humanitarian response to the war in Biafra

Perhaps less foreseeable were the ways in which the international humanitarian response to the war in Biafra contributed to broader shifts, specifically in terms of constructing 'Africa' in the humanitarian imaginary, elevating NGOs in world politics, and the founding of MSF.

The Biafran crisis shaped understandings of Africa and humanitarian aid in the global North. The humanitarian response in Biafra, and the visual imagery that accompanied and drove that response, served to transform the secessionist region from 'a potential country into little more than an object of pity' (Smillie 2012: 31). The iconography and the relief effort contributed not only to a one-dimensional view of Biafra but also to global North perceptions of Africa more broadly. The Biafran child, with a grotesquely swollen stomach, became the foremost icon of suffering in a postcolonial world and shorthand for African misery (Black 1992: 124; Heerten 2017: 141). In this way, the war and the humanitarian response in Biafra were significant in constructing postcolonial Africa as a continent of suffering. Even if the imagery was new to many, the message conveyed—that Africa was an object of pity, in need of perpetual help from the global North—was a continuation of the rationales for the colonialism and missionary work of the not-too-distant past, and contributed to the primacy of emergency and relief in Western understandings of the Third World (O'Sullivan 2014: 306, 310). The representations of Biafra and the humanitarian response thus served to reinforce preconceptions about Africa (and, indeed, about the Third World more broadly), preconceptions that were further entrenched with the famine in Ethiopia in the 1980s and which remain widespread today.[7]

In addition, the humanitarian response in Biafra had long-term significance for being the first to be dominated by NGOs (de Waal 1997: 73). This is not to say that it *caused* the rise in importance of NGOs in international humanitarian response, but it may have accelerated it and certainly demonstrated the potential for such non-state actors to play a central role in the provision of relief and to mediate between governments and publics in the global North and life and authorities on the ground in humanitarian crises. The crisis in Biafra prompted the creation of new NGOs, including Africa Concern, and thrust existing agencies, such as Oxfam, into a spotlight they have rarely left since (O'Sullivan 2014: 299). Biafra gave these and other NGOs—including Caritas, Christian Aid, Save the Children, and War on Want—the opportunity to take centre stage, and in subsequent years, they consolidated their role (O'Sullivan 2014: 303).

Indirectly, the Nigerian Civil War also gave birth to MSF in 1971, and to the origin myth subsequently developed by and around MSF. A group of French doctors, veterans of the 1968 student rebellion, worked for the Red Cross in Biafra, where they witnessed Nigerian soldiers massacre unarmed

[7] See 'Wider impact of Western media and celebrity portrayals of the famine' in Chapter 3 on Drought and Famine in Ethiopia, 1983–1985.

men, women, and children (Barnett 2011: 143). They had signed a commitment to discretion which prohibited them from speaking publicly about what they saw (Desgrandchamps 2014: 291) According to the myth, the doctors found this policy of silence abhorrent and opted to violate it and make public statements about what they had witnessed, leading to a rupture with the ICRC and the formation of MSF in 1971, with témoignage—or bearing witness—as a central institutional principle (Brauman 2012: 2–3; de Waal 1997: 76). However, this version of events involves some post hoc reconstruction of the narrative. In fact, the public statements in question did not present any allegations that were not already public (Heerten 2017: 323). Furthermore, the ICRC actually considered some of the French doctors' statements to be good publicity, and one of their articles was even reproduced in the International Review of the Red Cross, the ICRC's in-house journal (Desgrandchamps 2014: 291–292). For their part, the doctors had not renounced the ICRC and continued to work through the Red Cross in other emergencies after the end of the Nigerian Civil War before MSF was established (Barnett 2011: 144). Moreover, MSF's founding charter explicitly stipulated that members must 'refrain from passing judgement or publicly expressing an opinion—favourable or unfavourable—with regard to events and to the forces and leaders that accept their aid' (cited in Brauman 2012: 4). While témoignage subsequently became an important part of MSF's institutional identity, the role of public statements has been contested among the senior management, and more often than not, the organization undertakes its medical work without making public statements about abuses in its zones of operations (Weissman 2011a).

Conclusions

The conflict and resulting famine in Biafra were inherently political events, with famine the direct outcome of the policies of the secessionist leaders in Biafra and the FMG in Lagos. The politics of decolonization and of Cold War coalition-building shaped patterns of support for warring parties and additionally impacted the stances and strategies of aid agencies. While some agencies took an overtly political stance, many sought to depoliticize their work and emphasize their neutrality with respect to political and military matters (O'Sullivan 2014: 309). The overarching narrative in the global north was one of a humanitarian rather than a political crisis, but aid that was justified and driven by humanitarian motives had political consequences. Aid was manipulated by the warring parties, with the Biafran secessionists using the famine to maintain support internationally, and the aid to keep going in a war of attrition. Central to the founding myth of MSF, the Nigerian Civil War—or more accurately 'Biafra', the humanitarian emergency and response rather than the war itself—is often seen as a turning point in modern humanitarianism, prompting reflection and debates about the unintended consequences of aid, the rights and wrongs of working without the consent of the state, and the appropriateness of speaking out in the face of major violations of human rights and IHL.[8]

[8] These debates reappear in several of the other case studies in this book and are discussed in more general terms in Chapter 13 on Politics, Principles, and Humanitarian Action.

3
Drought and Famine in Ethiopia, 1983–1985

Between 1983 and 1985, Ethiopia suffered drought and a famine that cost more than a million lives (Keller 1992: 616; Kumar 1987: 36, 42–43). Over 300,000 people died in 1984 alone, and by 1985, around 2.5 million—of a total population of 40 million—were starving (Keller 1992: 616). Ethiopia, and indeed the whole Horn of Africa, has experienced periodic drought and famine for centuries, and while drought is often understood as a 'natural' disaster, the frequency and severity of droughts in the region are not determined purely by natural phenomena—deforestation and human-made climate change have also contributed (Keller 1992: 609; Kumar 1987: 6). Moreover, while drought often contributes to famine, the relationship between them is not straightforward: drought does not always result in famine, and famine is not always the result of drought.[1]

A comparison of Ethiopia and neighbouring countries in the early 1980s reveals a number of interesting points. First, similar weather conditions in different places did not lead to the same outcomes in terms of famine and mortality. Kenya, for example, suffered a brief but severe drought and managed to avoid serious famine. Second, in Ethiopia, indicators of famine (specifically, increasing grain prices, starvation, and excess mortality) appeared *before* drought and crop failure (de Waal 1997: 114–115). Third, not all countries that experienced severe famine attracted the enormous international response that was seen in Ethiopia. Following the failure of rains in parts of north Kordofan, north Darfur, and the Red Sea Hills in 1983 and 1984, and the failure of the Sudanese government to respond appropriately, famine in Sudan was estimated to have cost 250,000 lives by January 1985 (de Waal 1997: 91). Yet, Sudan did not attract anything like the level of international attention and assistance that was directed at Ethiopia.

This chapter advances a number of arguments to explain these puzzles. Famine in Ethiopia was not the direct outcome of weather conditions. Rather, it was the consequence of weather conditions combined with government

[1] See 'Famine' in the introductory chapter on Humanitarian Emergencies.

The Politics and Everyday Practice of International Humanitarianism. Miriam Bradley, Oxford University Press.
 DOI: 10.1093/oso/9780198840695.003.0003

policies and armed conflict. While political factors were undeniably central causes of the famine, the way in which the drought and famine in Ethiopia were portrayed by Western media and celebrities served to obscure these political causes from view. This decontextualized and depoliticized portrayal helped to generate widespread support for the cause and, hence, an enormous international response, but it also had consequences for the nature of that response. Humanitarian aid was diverted and manipulated by the government, with negative consequences for many of those it was intended to help. Arguably, the simplified and depoliticized understanding of the context made the aid particularly susceptible to such manipulation.

Causes of the famine

In Ethiopia in the early 1980s, it was not the drought alone but the drought in conjunction with war, and with particular government policies, that led to famine. The revolutionary government of Mengitsu Haile Mariam had come to power in 1974 partly as a consequence of Emperor Haile Selassie's mismanagement of famine in the north-eastern province of Wollo in 1973 and with the promise of eradicating famine in Ethiopia (de Waal 1997: 106–107). The establishment of the Relief and Rehabilitation Commission (RRC) in 1974 ensured that the Ethiopian government had a capable early warning system for drought and famine by the time drought spread across the Horn of Africa in the early 1980s. However, by the early 1980s, several government policies combined to cause, rather than prevent, a famine which lasted from 1983 to 1985.[2]

Mengitsu's government imposed Stalinist-model agricultural policies involving forced collectivization and villagization (Keller 1992: 618). Radical reforms, including the nationalization of industry and land, and the massive redistribution of land usage rights, had provided many benefits for the peasantry (Kumar 1987: 7). At the same time, however, grain quotas served to extract food from rural Ethiopia at very low prices to supply the cities, forcing even drought-affected regions with insufficient food to meet their own needs to provide grain to the central Agricultural Marketing Board (de Waal 1997: 110). Compounding the shortage of food, restrictions on non-farm activities, including the private hire of seasonal agricultural labour, served to reduce rural household income (Clapham 1991: 251–252; de Waal 1997: 111). In addition, in late 1984, the government initiated a (forced) resettlement

[2] Some analysts date the start of the famine to 1982 (Kumar 1987) or 1984 (Hendrie 1989), and some date the end to 1986 (Keller 1992), but most characterize 1983–1985 as famine years.

programme that moved people from the rebel-controlled northern regions to the government-controlled central and southern regions of the country (Hendrie 1989: 353).

In the northern regions, popular resentment against government policies exacerbated the long-standing armed struggle of the Eritrean secessionist movement (Eritrean People's Liberation Front, EPLF) and led to further armed rebellion with the formation of the Tigrayan People's Liberation Front (TPLF) in Tigray in 1975 and its expansion into the neighbouring province of Wollo in 1982. In response to these rebellions, the Ethiopian army and air force employed colonial-style counter-insurgency policies in Tigray, Wollo, and Eritrea.

These counter-insurgency strategies led to famine both directly and indirectly. Crops, livestock, and food stores were systematically destroyed by military offensives as part of a scorched earth policy aimed at denying cover and a regular food supply to the armed opposition (de Waal 1997: 117; Keller 1992: 614, 620). Restrictions on migration and on the transportation of food were implemented as part of a determined effort to stop food reaching the TPLF and the civilians in areas under its control, and prevented the movement of grain from surplus-producing areas to deficit areas (de Waal 1997: 118). Thus, the famine was not only an incidental consequence of agricultural and counter-insurgency policies but also a deliberate strategy employed by the government, even if the architects of this strategy may not have predicted the severity of its impacts (de Waal 1997: 117). The counter-insurgency strategies also contributed indirectly to famine as they implied enormous military expenditure, requiring resources that could otherwise have been used to prevent or alleviate famine (Keller 1992: 614–615).

Undoubtedly, the drought played a role, but it was not the only—or even the primary—cause of famine in Ethiopia in the early 1980s. The causes were overwhelmingly political, and the government used famine as part of its strategy to combat the secessionist movements in the north of the country. However, the way the famine was covered by mainstream media in the west served to obscure the political nature of the famine, presenting it instead as a largely natural occurrence.

Media coverage

The 'quality' press in the United Kingdom and other European countries had been running the story of the famine since the beginning of 1984 or earlier (Philo 1993: 105). In the United Kingdom, television appeals by household names Esther Rantzen (on the BBC) and Jonathan Dimbleby (on ITV) had been broadcast in early 1983 but without stirring much public attention (Franks 2006a: 293). On 17 July 1984, a television documentary, *Seeds of Despair*, went out on ITV, at the same time as a report by Michael Buerk on the BBC, and in coordination with a funding appeal from the Disasters Emergency Committee (Philo 1993: 105–106). The story was not given lead status but nonetheless generated significant public interest, with donations from the British public totalling over £9 million by October, despite quickly fading from the news (Franks 2006a: 299; Philo 1993: 106). For most of 1984, however, aid agencies reported struggling to get much attention from governments, journalists, or the public in several Western countries (Franks 2006a: 294–295).

Everything changed on 23 October 1984, when the BBC broadcast a second television news report by Michael Buerk, which was subsequently seen by an estimated 470 million people on news media around the world (Franks 2006a: 291).[3] This report presented a depoliticized version of events, very different from the account given above. Filmed at the Korem relief camp, Buerk referred to 'a biblical famine', 'the closest thing to hell on earth', 'wasted people [. . .], dulled by hunger, driven beyond the point of desperation'. This dramatic narrative was accompanied by powerful imagery from skilled cameraman Mo Amin (Franks 2006a: 304). The report omitted to show the rocket launchers or the Ethiopian fighter jets on their frequent flights to bomb rebel positions (Vaux 2001: 52). Instead, the report mainly depicted women and children, 'exemplary victims', abstracted from the political context of the famine (Müller 2013b: 473).

To the extent that the mainstream media took an overtly political angle, it was not with respect to the causes of the famine in Ethiopia but rather focused on the contrast between the grain surpluses rotting in warehouses in Europe, on the one hand, and the shortages in Ethiopia on the other. Suzanne Franks suggests that this angle, which was the focus of Peter Gill's film, *Bitter Harvest*, broadcast on 25 October 1984, was an important factor in generating an unprecedented public response from Western publics (Franks 2006a: 303–304). However, it was Buerk's report that appears to have catalysed public opinion and that won a number of prestigious awards. Buerk did not explain the famine as a consequence of a particular distribution of power or of particular power relationships, something that would require a political response. Rather, it was explained as something non-political, 'biblical', requiring a charitable response from international non-governmental organizations (INGOs) rather than political or economic analyses and responses.

[3] To view the report, see https://www.youtube.com/watch?v=GkLPx8mQ-t0.

Celebrity humanitarianism

The issue of the Ethiopian famine was subsequently taken up by celebrities. This was not the first time that celebrities had taken up a humanitarian cause—in 1971, Indian sitar player, Ravi Shankar, and former Beatles guitarist, George Harrison, organized the Concert for Bangladesh at Madison Square Garden, raising USD 240,000 for United Nations Children's Fund (UNICEF) operations in the war in East Pakistan (de Waal 2008: 51). However, the famine in Ethiopia arguably marked the beginning of continuous high-profile celebrity engagement with humanitarian affairs. Moved by Buerk's report for the BBC, Bob Geldof, then lead singer of the Boomtown Rats, enlisted a number of other celebrity musicians to join him in producing a charity record to raise funds for Ethiopia. Band Aid was formed, and on 29 November 1984, barely a month after the BBC report first aired, the single 'Do they know it's Christmas?' was released and went straight to number one in the United Kingdom. Clearly, the cause and the single caught the public's imagination, even if the condescending title and lyrics of the track have subsequently been much criticized (Müller 2013a: 66–67). Similar initiatives followed in countries including Canada, France, Spain, and the United States, and on 13 July 1985, the Live Aid rock concert was held simultaneously in stadiums in London and Philadelphia and broadcast live in around sixty countries (Müller 2013a: 67).

The characterization of the situation in Ethiopia by the Band Aid initiative was in keeping with the biblical imagery of Buerk's report, with Geldof describing the famine as 'the worst natural disaster mankind has ever seen' (Müller 2013a: 66). Their representations were not intended to explain realities but to evoke emotions. Corresponding to the simple, depoliticized, and decontextualized 'natural disaster' narrative, the response called for was one of 'compassion and apolitical commitment to alleviate suffering' (Müller 2013a: 62). Rather than engaging with the complex causes of the famine, those who participated in Band Aid/Live Aid instead took the easier option of blaming Western governments and relief agencies for their inaction.

In addition, the response sought from the public—charitable donations and the consumption of Band Aid products—made addressing the famine a lifestyle choice rather than an obligation and did nothing to challenge the global economic and political order, serving instead to reinforce that order and its contradictions through the notion of 'compassionate consumption' (Müller 2013b: 474). It is important to note that some critical media coverage drew attention to the man-made causes of the famine, and some celebrities—notably the British band Chumbawamba in 1986—opted both to critique the

Band Aid approach and to focus on global capitalism as a major determinant of recurring crises (Müller 2013a: 67). However, these perspectives and initiatives existed largely on the sidelines. They failed to capture the public imagination in the way that the Buerk report and the efforts of Bob Geldof and his celebrity collaborators did.

In terms of raising funds, the Band Aid initiative was extraordinarily successful. By September 1986, Band Aid had raised USD 110 million and, in addition, Western aid agencies reported massive increases in donations from the public (Philo 1993: 122). This mobilization of empathy not only resulted in unprecedented financial donations from Western publics but also put effective pressure on Western leaders to provide a more generous official response. For example, the Reagan government in the United States responded by adding USD 10 million to its existing commitments, and the Thatcher government in the United Kingdom added £5 million and 6,000 tonnes of grain (Philo 1993: 121). Thus, food aid and money poured into Ethiopia.

International humanitarian response

Several international humanitarian agencies were operating in Ethiopia prior to the media coverage and celebrity involvement described above. However, the timing, scale, and nature of the international response were undoubtedly affected by this coverage and involvement. As outlined in the previous section, Western media had covered the famine, first in newspapers, and subsequently in the July 1984 television reports in the United Kingdom. Thus, there was knowledge and media coverage of the famine well before Buerk's second report for the BBC in October 1984. What, then, explains the failure to provide any significant response prior to that report and the sudden change after it?

Underestimating the need for external assistance

The messages coming out of Ethiopia were mixed. The Ethiopian government wanted additional food aid but did not want to raise international awareness of the worsening famine. Indeed, in 1984, when the situation was already severe, and having prohibited foreign journalists from travelling outside Addis Ababa during the months of August and September, the Ethiopian government delayed acknowledgement of the existence of famine until 3 October, after the celebration in September of the tenth anniversary of the revolution (Franks 2006a: 298; Kumar 1987: 26, 28–29; Müller 2013a: 70). On the other hand, the Ethiopian RRC, some of the international agencies operating in Ethiopia, and research organizations such as the International Disasters Institute and the British-based Food Emergencies Research Unit were warning of the forthcoming famine as early as 1982 and continued to do so with increasing frequency and intensity throughout 1983 and into 1984 (Franks 2006a: 302; Kumar 1987: 28; Müller 2013a: 69; Philo 1993: 104–105).

Believing that Ethiopia tended to overstate the need for food aid, and reluctant to assist a Soviet client state, Western donor governments ignored the warnings (Franks 2006a: 295). Given that there was evidence that food aid was being diverted and abused to support counter-insurgency strategies, such reluctance was not completely unreasonable (de Waal 1997: 121). However, by mid-1984, the level of need was both real and apparent, yet Western governments continued to prevaricate. Undoubtedly, some in the Reagan administration saw food aid to Ethiopia primarily as a foreign policy tool rather than as a means to alleviate hunger and poverty, and US food aid went from 8,172 metric tons in 1982 to zero in the first ten months of 1984 (Keller 1992: 615; Poster 2012).

UN agencies also underestimated the need for food aid until at least mid-1984, although for different reasons. In a particularly extreme example of presenting as an assessment of needs what would more accurately be described as an assessment of an organization's capacity to respond,[4] both the Food and Agriculture Organization (FAO) and the World Food Programme (WFP) estimated needs in accordance with what they believed was the maximum the Ethiopian ports and in-country distribution channels could handle. Thus, the FAO 'repeatedly slashed the Government's requests for help between 1981 and 1984, based upon projected needs' (Keller 1992: 615; see also Kumar 1987: 29–30). Likewise, in 1984, 'WFP estimated (wrongly) that the ports could handle only 125,000 mt of relief and then took the shameful step of estimating the actual needs in Ethiopia at 125,000 mt—perhaps one-tenth of the real needs' (Keen 1998a: 322). It would have been more honest and accurate to emphasize that needs far exceeded distributional capacity and to argue for an increase in that capacity, but neither the WFP nor the FAO did this, and as the foremost global authorities on food aid, this was especially problematic—their underestimations unwittingly lent support to those who sought to play down the crisis (Kumar 1987: 30).

[4] See 'Assessing needs, contexts, and capacities' in Chapter 21 on Needs Assessment, Evaluation, and Response Decisions.

All this may explain the delay in responding and the insufficiency of the initial response, but it does not explain why an enormous response began at the end of 1984, even as parts of Sudan were experiencing record numbers of famine deaths and failing to make the news headlines or attract anything like the same magnitude of international response.

Importance and impact of depoliticizing the famine

Part of the explanation surely lies in the fact that in Sudan there were no relief camps comparable to Korem, with the dramatic imagery it provided for the Buerk report, and the famine there could not be so easily presented as a 'natural disaster' (Müller 2013a: 66–67). Undoubtedly, the dramatic imagery from Korem played an important part. Greg Philo quotes Russell Johnston, a British Member of Parliament, in the House of Commons on 22 November 1984, as saying, 'The entire aid world has been screaming from the rooftops for the last eighteen months that what has happened in Ethiopia was about to occur, yet it was only when we saw it in colour on the screens in our living rooms that the Government acted' (Philo 1993: 104–105).

Not only the timing of the response but also its nature was shaped by the media coverage and celebrity engagement. As Alex de Waal has argued, 'for Western governments, the political priority became to avoid embarrassment at the hands of figures like Bob Geldof' (de Waal 1997: 122–123). This implied providing a large-scale and visible, media-friendly response, even if that meant focusing on areas where there was lesser need. When the international response was finally scaled up at the end of 1984, it centred on the provision of in-kind food aid.[5] While this may be the most obvious response to a food-related crisis, it was not necessarily the best response, given that there were severe limits on distribution capacity (Kumar 1987: 47–48). Providing cash to those in need could have enabled them to buy food that was already in the country but which they could not afford, thus stimulating the economy and circumventing logistical difficulties involved in importing large quantities of food aid (Kumar 1987: 48–49). Given the scale of the famine, importing food was also necessary, but a combination of food aid and cash transfers could have maximized food access of those in need.

The response comprised two main parts, by far the largest of which was undertaken via official channels and consisted of the provision of (mainly food) aid to people living in government-held areas and the operation

[5] For more on the pros and cons of food aid versus alternatives in different contexts, see 'Politics of food aid' in Chapter 22 on Material Assistance and Direct Service Provision.

of feeding centres in government-held towns, to which people living in contested or rebel-held areas were supposed to go to collect dry rations each month (Hendrie 1989: 352–353). The second component involved a number of European church-based NGOs, under the auspices of the Emergency Relief Desk (ERD) and in close collaboration with the relief arms of the EPLF and the TPLF, engaging in cross-border operations from Sudan into rebel-held areas (Hendrie 1989: 352). However, this component was much smaller. Despite the fact that people living in the rebel-held areas of Eritrea and Tigray accounted for between a third and a half of the famine-affected population, only 90,000 of the 1.25 million tonnes of food aid received by Ethiopia in 1985 was distributed across the border into these areas (de Waal 1997: 125).

Ethiopia in 1983–1985 has become known as a classic example of the manipulation of aid by political actors. Its notoriety may be attributed in part to the scale of the response: all other things being equal, if aid is to be diverted or manipulated, the more aid that is provided, the more that will be diverted or manipulated. However, it was arguably also the case that the depoliticized portrayal of the context and causes of the famine contributed to a response that was particularly susceptible to misuse and prone to doing harm. Ignoring the fact that the famine was at least in part a crime perpetrated by the Ethiopian government, some relief agencies became complicit in that crime.

Non-cooperation and access restrictions imposed by the government

Given that denial of access to food was a major element in the government counterinsurgency strategy in the northern regions, it is not surprising that the government also restricted the access of international agencies to rebel-held areas in Eritrea and Tigray (Jansson et al. 1987: 49–50). Since both Eritrea and Tigray bordered Sudan, and parts of each province were held by the EPLF and TPLF, respectively, it was feasible to provide aid across the Sudanese border without the cooperation of the Ethiopian government. However, donors were initially unwilling to violate national sovereignty by providing aid on Ethiopian territory without the government's consent, and no UN agency consigned food to the cross-border operation at any point in the 1980s (de Waal 1997: 125; Macrae and Zwi 1992: 302).

The EPLF and the TPLF had established their own relief arms in the 1970s, the Eritrean Relief Association (ERA) and the Relief Society of Tigray (REST). Working from a logistics base in eastern Sudan, they sought to provide assistance to civilians living in the areas under their control. Both the TPLF and the EPLF 'on occasion used the inhabitants of the drought-affected areas as pawns in their respective military strategies' (Keller 1992: 618), but for the most part, their military strategies and objectives were linked to the survival and well-being of the civilian populations in the territories they controlled (de Waal 1997: 127–132; Müller 2013b: 475).

During the famine of 1983–1985, the Sudan-based ERD provided support to the Eritrean and Tigrayan populations via ERA and REST. Since the Ethiopian government refused consent to the provision of aid in rebel-held areas, the identities of the individual INGOs that participated in the ERD were concealed so that they could continue working in government-held territory (Hendrie 1989: 354). Given that the operational agencies involved, as well as the Western donors, sought to keep the cross-border operation low profile and confidential, 'the rural-based delivery systems of ERA and REST were never publicly acknowledged as alternative channels of aid that could complement the access of the official operation', with the consequence that, until late 1985, the international funding for the cross-border operation was insufficient relative to needs and distribution capacity (Hendrie 1989: 353).

Use of food aid for non-humanitarian ends

During the famine in Ethiopia, the government taxed the import of relief supplies and used food aid to force the resettlement of much of the population from the northern provinces as well as to feed its own armed forces. The Ethiopian government levied charges on all food and other relief supplies coming into the country. According to Edmond Keller, as 'much as $30 million was raised in 1985 by imposing an import fee that was initially set as high as $50.50 per metric ton for all donors (except the United Nations, which had to pay $49). However, this levy was reduced to $20 following an outcry from all the affected agencies' (Keller 1992: 621). These revenues helped to finance the government's military strategies.

Food aid was instrumental in forcing the resettlement of populations from the rebellious northern provinces to the southern lowlands, where they could more easily be controlled by the government. This was ostensibly to move famine-affected people from the most drought-affected regions for humanitarian reasons, but resettlement was often accompanied by harassment and force, and the air force continued its bombardment of rural villages during the peak months of the famine, leading many to conclude that its motivations were, in fact, political and strategic (Hendrie 1989: 353). A Médecins

Sans Frontières (MSF) report explains that when people arrived at food distribution centres, 'they were loaded onto trucks, often requisitioned from aid organisations, and transported like livestock [. . .] Conditions *en route* were appalling and no preparations were made to resettle the families when they arrived in the malaria-infested regions. At least 100,000 people were estimated to have died in 1985 during resettlement operations' (Binet 2013: 8). Some families were separated in the process of resettlement (Binet 2013: 81; Forsythe 1997: 242). Fear of the government resettlement programme led to massive migrations in the opposite direction, to assistance centres run by ERA and REST, and to refugee camps in eastern Sudan, with some peasants who lived within three or four hours' walk from government feeding centres opting instead to walk for four to six weeks to get to Sudan (Hendrie 1989: 353).

In addition, the government diverted food aid to feed the soldiers it conscripted to fight the rebellions in the northern regions. Government auxiliary forces were paid with UN food rations to the extent that they were dubbed 'wheat militias' (Keller 1992: 621). In Tigray, for example, the majority of food aid went to militiamen and their families (de Waal 1997: 125).

International responses to restrictions on, and abuse of, aid

Despite the innovative cross-border operation, the overall international humanitarian response to the famine in Ethiopia in 1983–1985 was characterized by deference to the Ethiopian government and an emphasis on discretion, especially by the UN agencies.

The UN Emergency Office for Ethiopia (UNEOE), created to coordinate international relief efforts with the Ethiopian government, ended up serving as the government's mouthpiece, remaining silent on the government's role in creating the famine and on the access restrictions imposed by the government, and suppressing information on how the government was manipulating international humanitarian actors, using relief supplies to feed its own army and militia and to force the resettlement of civilians (de Waal 1997: 123–124). According to Edmond Keller, 'the FAO went so far as to support the resettlement and villagisation programmes, while those voluntary organizations that agreed to operate in certain areas also indirectly aided the Derg in its political objectives' (Keller 1992: 621).

There was a trade-off between advocacy and access—the government granted access subject to a number of conditions and would not permit public criticism of its policies—and most agencies prioritized access.[6] In October 1985, MSF France publicly criticized government policies and then withdrew from Ethiopia just before the government issued an expulsion order (Binet 2013: 8; de Waal 1997: 124). Other international aid agencies kept quiet, some criticizing the MSF France position for being 'political' and others agreeing with the analysis but choosing to avoid public criticism of the government themselves (Binet 2013: 8). The International Committee of the Red Cross (ICRC) eventually withdrew in 1988 on the grounds that the use of relief in the resettlement programmes was harsh and unacceptably political (Forsythe 1997: 242). However, the International Federation of Red Cross and Red Crescent Societies (IFRC) stepped in to take its place, providing assistance on the same governmental terms that the ICRC had rejected—underscoring the limited negotiating power of individual humanitarian agencies in the absence of a common position.

Wider impact of Western media and celebrity portrayals of the famine

Depicting the famine as a 'natural' disaster and Ethiopian famine victims as passive and lacking in agency, the media and Band Aid representation contributed not only to a response to the famine that largely ignored the political context in which it was being implemented but also to longer-term constructions of 'Africa' and other parts of the world perceived as destitute (Müller 2013b: 470). The Band/Live Aid celebrities represented 'Africa' in a particular way, with a particular place in the world system—as a continent of suffering, in need of perpetual charity from the global North. In doing so, they continued and amplified the message conveyed by the 'Biafran babies' in the Nigerian Civil War, which served to portray Africa as a place of poverty and misery and an object of pity. Such representations of Africa are still central to Western public understandings of Africa decades later, and while dominant contemporary understandings are inevitably the product of multiple influences, media and celebrity portrayals of the famine in Ethiopia in the mid-1980s are likely to have been central among them (Müller 2013b: 475).

The lyrics of the Band Aid single refer to Africa:

> Where nothing ever grows,
> No rain or rivers flow,
> Do they know it's Christmas time at all?

[6] For a detailed discussion of the relationship between humanitarian access and advocacy, see 'Neutrality, access, and humanitarian space' in Chapter 23 on Dialogue, Negotiation, and Advocacy.

The song treats Africa as a homogeneous entity and focuses on Christmas celebrations in the West, apparently ignorant of the fact that Ethiopian Orthodox Christians celebrate Christmas on 7 January following forty days of fasting (Müller 2013a: 67). At the same time, in asking whether 'they know it's Christmas time at all', it condescendingly assumes the ignorance of Ethiopians—or, indeed, of all Africans. These representations are reinforced by the response demanded by the Band Aid celebrity humanitarians—one of charity rather than justice, which imagines the recipients of the revenues of 'compassionate consumption' by Western publics as African victims without voice or agency (Müller 2013b: 474). Such representations serve to reinforce, rather than redress, global inequality.

A one-dimensional view of the global South based on stereotypes of drought and famine in Africa continues to dominate perceptions in the global North. Voluntary Service Overseas (VSO) carried out research in the United Kingdom in 2001 which demonstrated the persistence of stereotypes of Africa and the global South more broadly, with a large majority of respondents associating war, famine, debt, starving people, natural disaster, poverty, and corruption with the 'developing world' and believing that 'developing countries' are dependent on resources and knowledge from the global North for development and progress (VSO 2002: 5). The idea of Africa as the archetypical continent of suffering that requires perpetual charitable intervention from the global North has been reinforced in recent decades (Müller 2013b: 475). Western media representations of the Rwandan genocide in 1994 and the Ebola outbreak in West Africa in 2013–2016 have much in common with those of Ethiopia in the 1980s.[7]

Conclusions

The actions of the Ethiopian government and armed forces were central causes of famine, but Western media and celebrity advocacy portrayed the famine as a natural disaster for which blame could not be apportioned. This apolitical framing increased public support for the cause, but by obscuring the role played by the government in causing and aggravating the famine, it made for a response focused on symptoms and lacking any grounding in political analysis of the causes of famine. This, in turn, facilitated the manipulation of international food aid by the government, raising the question (not for the first or last time in the history of humanitarian action) of whether

[7] See, respectively, 'International inaction' in Chapter 5 on Genocide in Rwanda and its Aftermath, 1994–1996, and 'Stereotyping affected populations' in Chapter 11 on Ebola in West Africa, 2014–15.

aid was doing more harm than good for those it was intended to assist. Almost all the international humanitarian agencies operating in Ethiopia chose not to speak out about the access restrictions, the diversion of food aid to government militias, or the instrumentalization of food aid as part of the government's forced resettlement programme. They prioritized access over advocacy, but as time went on, more and more questioned that prioritization, and the experience stimulated a wider debate in the sector over the conditions under which aid agencies should withdraw (Barnett and Weiss 2011: 57).

4

Bosnian War, 1992–1995

When war broke out in Bosnia in 1992, the international community mounted a huge and high-profile humanitarian relief effort, led by the United Nations Refugee Agency (UNHCR) and supported by a United Nations (UN) peacekeeping force. UNHCR was the lead agency for the humanitarian response in the former Yugoslavia as a whole, but it was in Bosnia that it faced most challenges. For the first time in its history, UNHCR engaged in a large-scale response in the midst of armed conflict and ethnic cleansing, seeking to protect people in situ instead of waiting for them to flee across an international border.

The scale of the relief operation represented an impressive accomplishment, but it was also the principal accomplishment of the UN in Bosnia, when the name of the peacekeeping force—the UN Protection Force in Yugoslavia (UNPROFOR)—promised so much more. UNPROFOR was not a 'protection' force: it lacked the capacity to offer serious military protection, and its response was limited to the provision and protection of humanitarian relief without any concomitant effort to protect the civilian population, much less to end the conflict and build peace (Roberts 1995: 18). It did not even always do a good job of protecting aid.

The inadequacy of the relief effort was obvious to all involved, but it was nonetheless used by powerful states as a substitute for more concerted political or military action. Moreover, the relief operation was sometimes in tension with a stronger military response, with the North Atlantic Treaty Organization (NATO) refraining from air strikes on several occasions because they would endanger international aid workers and expose them to retaliatory attacks. Knowing that the relief effort was the centrepiece of the international response, Serb forces were able to exercise significant leverage, denying access to aid convoys, threatening to attack the humanitarian operation in retaliation for NATO air strikes, and several times taking UNPROFOR troops hostage.

The Politics and Everyday Practice of International Humanitarianism. Miriam Bradley, Oxford University Press.
 DOI: 10.1093/oso/9780198840695.003.0004

Background

For much of the twentieth century, Yugoslavia was a unified, multi-ethnic state populated by Serbs, Croats, Slovenes, Macedonians, Montenegrins, Albanians, Bosniaks,[1] Kosovars, and others. Under the Communist leadership of Josip Tito, from the Second World War until his death in 1980, the Socialist Federal Republic of Yugoslavia was a stable state with negligible levels of ethnic violence. However, the 1980s saw economic problems and a rise in ethnic nationalism linked to the crumbling of the Soviet Union and other Eastern European Communist States (Gagnon 1990: 19–21). In 1988, the Serbian politician Slobodan Milošević introduced constitutional reform that tipped the balance of power within Yugoslavia towards Serbia, intensifying nationalist sentiments, spurring separatist movements, and ultimately leading to the disintegration of Yugoslavia in the early 1990s.

On 25 June 1991, declarations of independence by Croatia and Slovenia effectively broke Yugoslavia into Serbia, Croatia, and Slovenia but were rejected by the Yugoslav federal government. After a ten-day war in Slovenia, the Yugoslav People's Army (JNA) withdrew, but in Croatia (a more ethnically heterogeneous country, with a significant Serb minority), the war was longer and bloodier (UN Secretary-General 1999: 8). War between Croatia and Yugoslavia caused sectarian violence and massive displacement of the Serb minority in Croatia. By the end of 1991, 10–15,000 people had been killed and around 350,000 displaced, prompting the involvement of UNHCR to provide relief for the refugee populations (Ogata 2005: 54). In February 1992, the UN Security Council authorized the establishment of UNPROFOR, initially mandated to monitor a ceasefire between Serb and Croat parties in those parts of Croatia with significant Serbian minorities (UN Security Council 1992a).

With Slovenia and Croatia on track for full independence, Bosnia-Herzegovina (hereafter, Bosnia) and Macedonia (now North Macedonia) faced a choice between remaining in a Yugoslav federation dominated by Serbia or pursuing independence themselves (Hansen 2006: 117). On 8 September 1991, Macedonia broke off peacefully, and in March 1992, following a referendum, Bosnia also declared its independence. On 6 April, the European Community recognized Bosnia as an independent state, and the United States followed suit the next day (UN Secretary-General 1999: 9).

[1] In English, the terms 'Bosniak' (or 'Bosniac') and 'Bosnian Muslim' are often used interchangeably despite the fact that not all Bosniaks are Muslim by religion, and neither are they all nationals of Bosnia. In this chapter, I use both terms, following the work I am citing for each given instance.

Bosnian Serbs, who accounted for around 31% of the population,[2] had not voted for independence, having mostly either boycotted the referendum or been prevented from voting, and the government of Serbia declared it would fight to protect their interests (Hansen 2006: 117).

Beginning on 6 April, Serb paramilitaries, together with the former Yugoslav army, went about ethnically cleansing Bosnian Muslims and Croats from Serb-majority areas of eastern Bosnia. That same month, Serb forces blockaded the predominantly Muslim Bosnian capital, Sarajevo, in a siege that would continue until early 1996, shortly after the end of the war. Across Bosnia, tens of thousands of people, mainly Bosnian Muslims, were killed in the first two months of the war (UN Secretary-General 1999: 7). Huge levels of displacement resulted from ethnic cleansing and general targeting of the civilian population such that ethnic enclaves were created and sometimes forcibly contained and blockaded (Duffield 1994: 2). With the support of the former Yugoslav Army, Bosnian Serb forces were militarily much stronger and better equipped than the Bosnian government side, and by mid-June 1992, Serb forces controlled two-thirds of Bosnia and around one million people had been displaced (Ogata 2005: 51).

The next three years were characterized by varying levels of violence, mainly between Serb paramilitaries and Bosnian government forces, with several incidents of mass atrocities perpetrated by the Serb forces and a series of peace plans brokered by different configurations of third countries but never accepted by all warring parties (Hansen 2006: 118–121). In early 1993, an alliance between the Bosnian Muslims and Croats fell apart, and bloody clashes between all three main communities ensued, with Croat forces also pursuing a policy of ethnic cleansing, leading to a new wave of displacement, this time of Muslims from Croat territory in central Bosnia (Hammerstad 2014: 193; Ogata 2005: 79). This lasted until early 1994 when, under pressure from the United States, Muslim and Croat forces stopped fighting each other and created a Muslim–Croat Federation (Cutts 1999: 1; Ogata 2005: 101).

By the time the war ended at the end of 1995, tens of thousands of Bosnians had been killed and nearly 2.6 million had been displaced, around half of them internally within Bosnia, 500,000 in neighbouring countries, and 700,000 in the European Union (EU) (Barnett 2011: 175; Hammerstad 2014: 193).

[2] Before the war, Bosnia's population was roughly 44% Muslim, 31% Serb, and 17% Croat (Ogata 2005: 51).

International discourses

Until decisive action by NATO in the second half of 1995 brought the war to an end, the international response was largely limited to the provision of humanitarian relief. On the one hand, Western states wanted to (be seen to) do something. On the other hand, for more than three years they were not prepared to make concerted political or military efforts to bring the conflict to an end. These conflicting desires generated an impressive-looking but half-hearted response (Barnett 2011: 174). In its first major operation outside its refugee protection mandate, UNHCR led what was then the world's largest relief operation, involving over 250 aid agencies providing assistance to some 2.7 million people in Croatia and Bosnia (Barnett 2011: 178; Ogata 2005: 52). In addition, UNPROFOR became what was then the largest ever UN peacekeeping operation, with around 40,000 troops—mostly from West European NATO members, with France and the United Kingdom providing the largest numbers—at its peak (Hammerstad 2014: 118). However, UNPROFOR's mandate was largely restricted to delivering and protecting humanitarian relief. It was also aimed at containing would-be refugees in situ—ostensibly to counteract ethnic cleansing but additionally to prevent them arriving on the territory of EU states, whose proximity made refugee arrivals a real likelihood (Hammerstad 2014: 196).

Such a policy response both relied upon, and served to reinforce, a particular set of ideas about Bosnia that were reflected in Western policy discourses. While these discourses changed over time and were not identical in every country, key Western policymakers promoted a 'humanitarian responsibility discourse' (Hansen 2006: 115–147). This dominant construction of the Bosnian war in the West was underpinned by two basic discourses: the 'Balkan discourse', according to which it was a war driven by barbarism and ancient hatreds and for which all parties to conflict were equally to blame, implying that Western states should not intervene unless it was in their self-interest (Hansen 2006: 106–111); and the 'genocide discourse', which characterized Serbian and Bosnian Serb conduct towards the 'Bosnian' or 'Bosnian Muslim' population as genocide and hence implied a legal and ethical obligation on the rest of the world to act (Hansen 2006: 112–114). In Western Europe in particular, the 'humanitarian responsibility discourse' dominated, modifying the basic 'Balkan discourse' by distinguishing civilians as 'innocent victims' from the 'responsible leaders' of the parties, a discourse which maintained the equality of the parties to conflict but separated the 'innocent victims' from the crimes of their leaders in a move which both excluded the possibility of civilian agency and implied an ethical responsibility to assist the innocent (Hansen 2006: 123–128).

There was some contestation of this discourse by the media, particularly following major atrocities by Serb forces that challenged the characterization of all parties as equal, but the 'humanitarian responsibility discourse' proved resilient to criticism for most of the war (Hansen 2006: 128–131). This discourse and a Western policy of minimal engagement were mutually constitutive.

International military response

In response to the siege of Sarajevo, the mandate of UNPROFOR—originally created to respond to the war in Croatia—was extended to Bosnia. On 8 June 1992, the UN Security Council authorized the force to secure and take control of Sarajevo airport (UN Secretary-General 1992a; UN Security Council 1992b). On 13 July, the Security Council authorized additional UNPROFOR troops 'to ensure the security and functioning of Sarajevo airport and the delivery of humanitarian assistance' (UN Security Council 1992c). A month later, on 13 August, acting under Chapter VII of the UN Charter, the Security Council called on states to take 'all measures necessary' to facilitate the delivery of humanitarian assistance—the first time it had authorized the use of force for the delivery of humanitarian aid (UN Security Council 1992d; Wheeler 2000: 252). In September 1992, the Security Council additionally authorized the use of force to protect aid convoys (UN Security Council 1992e; UN Secretary-General 1992b).

In April and May 1993, the UN Security Council designated first Srebrenica and then another five predominantly Muslim towns—Sarajevo, Goražde, Žepa, Tuzla, and Bihać—as 'safe areas', declaring that they 'should be free from armed attacks and from any other hostile act', that all Bosnian Serb military or paramilitary units should withdraw from them, and that UNPROFOR and international humanitarian agencies should have free and unimpeded access to them (UN Security Council 1993a, b). A month later, the Security Council expanded UNPROFOR's mandate to 'deter attacks against' these six 'safe areas' and authorized states to use air power to support UNPROFOR in this regard (UN Security Council 1993c). The Secretary-General requested an additional 34,000 troops to implement this expanded mandate, but the Security Council only accepted a so-called light option of 7,600 troops, which presupposed the compliance of the parties to conflict, and many UN member states anyway resisted providing additional forces[3] or redeploying their existing troops to the safe areas such that it took a year for the full 7,600 to arrive (Landgren 1995: 445; Ogata 2005: 92).

Many Bosnians sought assistance and protection in the UN-designated safe areas, but these areas did not prove very safe. Serb bombardments persisted at roughly the same rate as before the zones were designated as 'safe', with shells continuing to land in Sarajevo at an average rate of 1,000 per day, for example (UN Secretary-General 1999: 25). Intense attacks in April 1994 on Goražde prompted NATO air strikes against a Serb artillery command facility and military vehicles on 10 and 11 April (UN Secretary-General 1999: 34). In response, the Bosnian Serb forces took around 150 UN personnel, mostly UNPROFOR troops, hostage. They continued their advance on Goražde, and NATO responded with further strikes, which pushed the Serbs to agree to a ceasefire and the release of the hostages in exchange for a halt in the combat air patrols over Goražde (UN Secretary-General 1999: 35). In October and November 1994, after Bosnian government forces briefly took control of several hundred square kilometres between Bihać and the Croatian border, a counterattack by Serb forces reversed the Bosnian government's gains and additionally violated the safe area, to which NATO responded with limited air strikes (UN Secretary-General 1999: 38). On 25 May 1995, after Serb forces failed to comply with an order to remove their heavy weapons from the exclusion zone around Sarajevo, NATO aircraft attacked several ammunitions bunkers near Pale (UN Secretary-General 1999: 46). Serb fighters took around 300 UN military observers and UNPROFOR troops hostage in response, using some of them as human shields and chaining them to potential targets to deter further NATO attacks, only gradually releasing them between 2 and 18 June (UN Secretary-General 1999: 46, 49).

On 6 July 1995, the Bosnian Serb Army launched an attack on Srebrenica. The Dutch peacekeepers guarding the enclave refused requests from the Bosniak forces to give them back the weapons they had surrendered under a 1993 agreement to demilitarize the 'safe area' (UN Secretary-General 1999: 57, 61). However, the Dutch battalion itself lacked the capacity to prevent Serb advances, and the UNPROFOR headquarters in Sarajevo initially refused its requests for NATO close air support (UN Secretary-General 1999: 57–58, 61). The attack on the Srebrenica safe area appears to have begun with relatively limited objectives, but the Serb forces advanced with unexpected ease and, when they assessed that UNPROFOR was either unwilling or

[3] Several members of the Organization of the Islamic Conference did offer large contingents, but the Secretariat anticipated that the Bosnian Serbs would not agree to their deployment and that without Serb consent their capacity to fulfil their mandate would be significantly limited (UN Secretary-General 1999: 29).

unable to stop them, they decided to continue all the way to Srebrenica town, attacking UNPROFOR observation posts and capturing or driving out Dutch troops as they went (UN Secretary-General 1999: 61–62). Facing little resistance, Bosnian Serbs then carried out ethnic cleansing of Bosniaks, including the execution of approximately 8,000 men and boys, between 12 and 22 July 1995 (Toom 2020; UN Secretary-General 1999: 72–86).

The Srebrenica massacre and the subsequent fall of Žepa, another supposedly safe area, shamed NATO into a serious intervention. The final straw was a Bosnian Serb attack on a Sarajevo marketplace on 28 August. All UNPROFOR troops left Bosnian Serb-controlled territory, and Operation Deliberate Force was launched. In 3,515 sorties between 30 August and 21 September 1995, NATO allies dropped over 1,000 bombs on 338 Serb targets, at the same time as diplomatic efforts were intensified (Hendrickson 2005). In November, the Dayton Agreement—formally signed by representatives of Bosnia, Croatia, and the Federal Republic of Yugoslavia (Serbia and Montenegro) in December 1995—brought an end to the war by dividing control of Bosnian territory among Bosniaks, Bosnian Croats, and Bosnian Serbs, freezing borders, and establishing a new federal constitution for the country. The NATO-led Implementation Force (IFOR), a larger and better equipped force than UNPROFOR, with 60,000 NATO peacekeepers plus additional peacekeepers from other countries, including Russia, was deployed to protect and guarantee the implementation of the Agreement (Hendrickson 2005).

It is estimated that up to 20,000 people, the vast majority of whom were Bosnian Muslims, were killed in and around the 'safe areas' (UN Secretary-General 1999: 6)

International humanitarian response

The humanitarian airlift to Sarajevo, coordinated by UNHCR with support from air force officers seconded from the United States, the United Kingdom, France, Canada, and Germany, ran from 3 July 1992 to 9 January 1996, making it the longest-running airlift in history (Ogata 2005: 57–58, 63). Despite being suspended several times due to security incidents and concerns, over three-and-a-half years, more than 150,000 metric tons of food and 14,000 of medical supplies as well as shelter materials were delivered in more than 12,100 sorties (Ogata 2005: 61, 78). In addition, UNHCR coordinated a massive logistical operation on the ground in Bosnia, with around 950,000 metric tons of humanitarian assistance delivered by land convoys (Ogata 2005: 64).

Convoys and relief staff were deliberately targeted, and UNPROFOR's effectiveness in protecting them was highly variable (Ogata 2005: 78, 80, 82). UNPROFOR could only travel through territories under Serb control with authorization from the Bosnian Serb authorities, severely restricting its ability to improve access for humanitarian agencies, who often had better access than UNPROFOR itself (Cutts 1999: 9; Ogata 2005, 78). In parts of the country, UNPROFOR provided continuous escorts, while elsewhere they stationed armoured vehicles at intervals along convoy routes and at combatant check points, counting off the trucks within each convoy as they passed, and searching if numbers at any point did not tally (Duffield 1994: 11). Even where UNPROFOR troops escorted convoys, they provided only 'passive protection', accompanying the convoys in armoured personnel carriers so that convoy personnel could shelter in the armoured vehicles should they come under attack, but not actively defending them (Cutts 1999: 9). The mandate to 'take all measures necessary' to facilitate the delivery of aid was interpreted by UNPROFOR as excluding the use of force, and objections by militia leaders frequently delayed delivery and sometimes implied virtual blockades (Duffield 1994: 11).

During the war, hundreds of thousands of Bosnians, especially those living under siege in Sarajevo, Srebrenica, Zepa, Goražde, and Bihać, were dependent on international assistance to meet their subsistence needs (Cutts 1999: 1). At the same time, in a context of ethnic cleansing, they desperately needed protection from violence. The international community did rather better at the former than the latter, such that the notion of the 'well-fed dead' became common currency (see, e.g. *New York Times* 1992).

In-country protection

It was with the aim of keeping Bosnians safe from violence (and of preventing them from leaving the country to seek asylum elsewhere) that UNHCR, for the first time in its history, engaged in protecting people in the midst of armed conflict on a massive scale. UNHCR's core mandate is for the protection of refugees who seek safety by crossing an international border, and until the 1990s, most of its limited experience with the internally displaced had involved providing material assistance rather than protection (Bradley 2019b: 627).

Working in the midst of armed conflict presented UNHCR with new dilemmas, and the fact of being present at the scene of ethnic cleansing that it was powerless to prevent haunted UNHCR throughout the war in

Bosnia. Should it assist with evacuations, and thus become complicit in ethnic cleansing, or leave people to their fate? In July 1992, for example, the mayor of Bosanski Novi organized transport to expel 4,000 Muslim inhabitants of the town, forcing them to give up their properties (Ogata 1992). UNHCR sought to negotiate an arrangement for them to be able to remain, but the local authorities instead killed and terrorized the population in a successful effort to blackmail UNHCR into evacuating 7,000 people from the town to Croatia (Ogata 1992). In March 1993, UNHCR was again pushed into evacuating Bosnians, this time from Srebrenica, when more than 5,500 people, desperate to flee the besieged area, forced themselves onto trucks from three convoys which had delivered food and medicine and were about to depart (Ogata 2005: 87). In this case, UNHCR was heavily criticized for its role in assisting ethnic cleansing, and the International Committee of the Red Cross (ICRC), UNHCR and UNPROFOR all then decided that they would only assist mass evacuations in 'emergency' cases but without specifying the criteria for an emergency and, hence, for mass evacuation (Cutts 1999: 17). Ethnic cleansing and the deliberate displacement of populations were not new, but UNHCR's presence at the scene was novel and exposed the agency to blackmail by the Serb authorities, ultimately making it complicit in the ethnic cleansing of Bosnian Muslims (Frelick 1992: 447–448; Ogata 1992).

UNHCR saw so-called preventive protection as a way to overcome the dilemma inherent in facilitating displacement when ethnic cleansing is a war aim (Ogata 2005: 55, 67). At the same time, however, the policy of preventive protection arguably undermined UNHCR's ability to press states to admit Bosnian asylum seekers in line with their obligations in refugee law. UNHCR was well aware that Western governments were motivated to support UNHCR's efforts to protect civilians within Bosnia in order to prevent them from fleeing and claiming asylum in the EU (Ogata 2005: 67). UNHCR's official line was that, by engaging in in-country protection, it was giving people the choice to stay or flee, but with European countries refusing entry to thousands, it was not much of a choice (Barnett 2011: 179). Moreover, some states saw in-country protection as a substitute for asylum, and in offering this 'alternative', UNHCR arguably tied its own hands when it came to insisting on the right to seek asylum (Barutciski 1996; Frelick 1992; Krever 2011; Turton 2011: 10). It is difficult to argue that individuals need international protection elsewhere when you are also claiming to be providing protection in-country. After being criticized for working to prevent Bosnians exercising their right to seek asylum, and especially after the Srebrenica massacre, UNHCR stopped using the language of *preventive* protection in favour of *in-country* protection (Hammerstad 2014: 204–205). Even so, to call it protection was problematic.

In Bosnia, UNHCR's efforts at protection aimed to bring safety to people, who would then not need to seek safety elsewhere, but in practice UNHCR and UNPROFOR brought material assistance, not protection from violence (Loescher 2001: 298–301). UNHCR understood protection in terms of 'monitoring the treatment of minority groups, mediating at the local level, and exposing practices of forced location' (Ogata 2005: 67). Such monitoring was limited because, despite the extensive relief operation, the warring parties found ways of preventing UNHCR's presence where ethnic cleansing was taking place (Cutts 1999: 16–17; Loescher 2001: 298). Where UNHCR was able to monitor and expose violations of human rights and humanitarian law, UNHCR staff compiled and published reports on ethnic cleansing and other abuses of human rights and international humanitarian law (IHL) (Cutts 1999: 17). The High Commissioner also maintained discussions with political leaders across the former Yugoslavia, urging them to reduce violence and stop targeting civilians (Ogata 2005: 77). However, it is not clear that such exposure and interventions led to any reduction in violence or the targeting of civilians.

The disaster of the so-called safe areas underscores the limits of in-country protection in Bosnia. While UNHCR had not intended for the idea of preventive protection to lead to the creation of safe areas or safety zones, elsewhere in the UN the safety zones idea was explicitly linked with preventing refugee flows (Ogata 2005: 67; UN Secretary-General 1999: 17). In Bosnia, UNHCR was working closely with the UN Security Council and secretariat for the first time and was tasked with providing humanitarian assistance within the safe areas. Even if the High Commissioner was not enthusiastic about the safe areas policy, UNHCR was ultimately a part of that policy (Ogata 2005: 93–95). The safety zones were not demilitarized, they lacked the consent of all parties to conflict, and they were not defended by force, since UNPROFOR lacked the strength to defend them. The poorly specified policy on evacuations exacerbated these fundamental weaknesses in the safe areas policy in that, although UNPROFOR was present in Srebrenica when Serb forces overran the enclave in May 1995, it failed to organize a mass evacuation, and the Bosnian Serb authorities did so themselves, separating out the men in the process and subsequently executing most of them (Cutts 1999: 17).

'Humanitarian alibi'

Despite the severe limitations of humanitarian action in this context, it served the interests of powerful states as a substitute for serious foreign policy engagement—a 'humanitarian alibi'. Emphasizing the centrality of the humanitarian mission enabled a moral failure to be presented as an organizational triumph: 'if UNPROFOR was judged according to how well it protected civilians, then its activities were a failure; if, however, it was judged by its delivery of humanitarian relief, then it could be judged a qualified success' (Barnett 2011: 177).

Bosnian leaders wanted military assistance over humanitarian assistance, but for most of the war, the UN Security Council and Western states opted for the opposite. In February 1993, Bosnian leaders banned international aid to Sarajevo in protest at the terrible conditions in the Muslim enclaves and the lack of a political solution or protection, with the vain hope of pressurizing the UN to act more decisively (Barnett 2011: 178; Ogata 2005: 82). Sadako Ogata, then UN High Commissioner for Refugees, decided to suspend the airlift and all UNHCR operations in Sarajevo and Serb-controlled areas of Bosnia until all parties to conflict honoured their commitments to allow safe passage for humanitarian aid (Cutts 1999: 5; Ogata 2005: 82–83). However, the UN Secretary-General quickly overturned this decision and ordered UNPROFOR to protect UNHCR (Loescher 2001: 300).

The humanitarian agencies providing this 'alibi' could be said to be complicit, even though they were calling out the failure of Western powers to take more concerted action to stop the violence. As early as August 1992, the ICRC president was bemoaning the fact that humanitarian interest among states and intergovernmental organizations was used to conceal their political inadequacy (Mercier 1995: 60–61). UNHCR, for its part, criticized the lack of political action but never gave up espousing the idea that its own humanitarian effort was contributing towards lasting peace and security in Bosnia (Hammerstad 2014: 207). It would have been difficult for UNHCR to contradict its main donors and the Security Council, who sought to use the humanitarian relief effort as a substitute for political action, particularly after Ogata's decision to suspend the airlift in February 1993 was overturned by the Secretary-General (Hammerstad 2014: 207). UNHCR had become heavily dependent on the Bosnia operation. Through most of the war, around one-quarter of UNHCR staff and one-third of its resources worldwide were committed to the relief effort there (Loescher 2001: 296). Furthermore, Ogata wanted to ensure that UNHCR maintained its relevance in a world in which she believed that there were reduced opportunities for asylum and that realistic refugee responses hence required attention to prevention as well as solutions (Loescher 2001: 296–297).

The humanitarian relief effort not only masked political and military inaction but was also, in many respects, in tension with political and military action such that the prioritization of relief curtailed efforts to address the causes of the humanitarian crisis and to bring the war to an end. UNPROFOR was mandated to protect aid but not people in a context in which protecting aid was arguably incompatible with protecting people. Such a mandate required UNPROFOR to maintain good relations with all sides in the conflict, which, in turn, prevented the UN from taking action against the side responsible for the worst violations (African Rights 1994: 22). Furthermore, the presence of thousands of peacekeepers and humanitarian workers spread thinly across Bosnian territory ensured that NATO would not engage in substantial air strikes and the Bosnian Serb forces could restrict access and violate IHL without fear of serious reprisals (Hammerstad 2014: 197; Hansen 2006: 118).

Manipulation of humanitarian assistance

As the displacement of civilians was a deliberate aim of warfare, the international humanitarian effort—whether it served to enable civilians to remain in situ, as it did in some places, or facilitated their displacement, as it did in others—inevitably contributed to war dynamics. For example, the besieged Muslim enclaves would not have survived without international aid, and their survival, in turn, ensured that Serb forces were engaged in sieges, which may have slowed Serb territorial gains (Hammerstad 2014: 197). On the other hand, there were occasions—including the incident at Bosanski Novi outlined above—where Serb forces blackmailed international humanitarian agencies into evacuating large numbers of people (Ogata 1992).

According to Ogata, 'relief assistance had to be given in accordance with needs', but all parties demanded more and ultimately UNHCR allocated aid to different communities in proportion to pre-war estimates of the population sizes rather than the level of need in each community (Cutts 1999: 14–15, 20; Ogata 2005: 69). Since the Bosnian Muslims had greater need relative to their population size, such a policy benefited the Bosnian Serbs (Hammerstad 2014: 197). Even so, the Serb forces frequently demanded an even greater share for the Serb population and, when UNHCR refused, they often blocked access to areas populated by Bosnian Muslims, such as besieged Sarajevo (Cutts 1999: 15–16). Indeed, the Sarajevo airlift was only able to operate on the condition that incoming aid was distributed both to the people in the besieged city and to those in surrounding areas controlled by Bosnian Serbs and not under siege (Cutts 1999: 19–20).

The Bosnian Serb forces were able to set most of the terms of UNHCR's operations in and around Sarajevo airport, inspecting planes on arrival and sometimes refusing to allow particular items to be offloaded or brought into the city with the consequence that hundreds of tonnes of food rotted on the airport tarmac, and sometimes preventing seriously ill or wounded civilians from being evacuated on UNHCR planes, with a number dying as a result (Cutts 1999: 20–21). The UN and the wider international community decided it was better to have the airlift operating under these conditions than not operating at all, but this decision further empowered the Bosnian Serbs and gave them enormous leverage (Cutts 1999: 21).

Across Bosnia, Serb forces were able to exploit the international community's commitment to, and emphasis on, the humanitarian response—not only denying access but also threatening to attack the humanitarian operation in retaliation for NATO air strikes and several times taking UNPROFOR troops hostage (Hammerstad 2014: 197). They attacked some convoys and imposed taxes on others (Cutts 1999: 11; Ogata 2005: 80). Moreover, knowing that the international community was obsessed with the airlift—a tangible and visible symbol of international compassion and engagement—even once land convoys offered a viable and massively cheaper alternative, Serb forces carried out some of the worst incidents of ethnic cleansing elsewhere in Bosnia while the international focus was on Sarajevo (Cutts 1999: 22).

It is never possible to calculate the precise impact of humanitarian action on the course, duration, and human costs of war. In some cases, at least with the benefit of hindsight, it is clear that aid lengthened conflict and resulted in many more casualties than would have been likely if that aid had not been provided.[4] In the case of Bosnia, the overall impact of aid is more difficult to ascertain. Certainly, aid was manipulated and instrumentalized by the warring parties and by external powers, and perhaps it both lengthened the war and kept some people alive during the war. At the same time, the involvement of international aid workers undoubtedly helped to ensure that the war and the plight of Bosnian Muslims was not forgotten by Western media, publics and policymakers, potentially contributing to the more active Western policy that emerged in late 1995 and ultimately ended the war (Hammerstad 2014: 197; Roberts 1996: 59).

Conclusions

The case of Bosnia highlights the link between Western disengagement and the hegemony of humanitarianism, with powerful states retreating into the language of humanitarianism and funding an enormous relief effort which served to conceal a policy vacuum on their part and to push aid workers to fill

[4] See, e.g. 'Political impacts: Supporting the secessionists and prolonging the war' in Chapter 2 on the Nigerian Civil War, 1967–1970.

that vacuum and to advocate policy (African Rights 1994: 22; Roberts 1996: 54). For example, the ICRC president, Cornelio Sommaruga, proposed the establishment of internationally protected safe areas in August 1992 (Mercier 1995: 61; Ogata 2005: 89). Médecins Sans Frontières (MSF) called on a more robust response from the UN in Goražde in April 1994 and demanded the resignation of the UN Special Envoy to the former Yugoslavia (Mooney 1995: 416). UNHCR was transformed from a specialized refugee protection actor into a major generalist humanitarian agency, running a huge relief operation in the midst of ongoing conflict and effectively becoming a global security actor (Hammerstad 2014: 199). Powerful states saw UNHCR as indispensable to their efforts to keep humanitarian assistance flowing into Bosnia and as a bridge between the humanitarian community and political and military actors (Cutts 1999: 4–5; Loescher 2001: 296). The case also highlights the limits of humanitarianism—and the scope for the manipulation and instrumentalization of aid, particularly when aid is the centrepiece of the international response. Emphasizing the centrality of the relief effort enabled the efforts of UNHCR and UNPROFOR to be judged a success, even if they failed to protect civilians or prevent ethnic cleansing. Yet, this same emphasis also facilitated the instrumentalization of the relief effort by Bosnian Serb forces, who knew that they could restrict access and violate IHL without fear of serious reprisals since they could always respond to any such reprisals by threatening the all-important relief effort.

5
Genocide in Rwanda and Its Aftermath, 1994–1996

Two years into the war in Bosnia, the world's attention shifted to Africa. From April to July 1994, up to a million Rwandans were killed in a state-led effort to destroy the Tutsi minority as well as many Hutus who refused to cooperate with that aim.

The international community did almost nothing to prevent or halt this genocide but played a significant—and much criticized—role in the aftermath. During and immediately after the genocide, around 2 million Rwandans fled to neighbouring countries, and international humanitarian agencies quickly arrived, setting up and running camps in Burundi, Tanzania, and Zaire (now the Democratic Republic of the Congo) until 1996. Tens of thousands died from disease in the refugee camps, but to a significant degree, this was due to factors beyond the control of aid agencies. Most assessments of their work suggest that the aid provided was effective in meeting basic needs, avoiding starvation, and limiting disease. The criticisms are not about the quality of the response in that sense but rather the disproportionate allocation of resources and, above all, the unintended consequences of providing long-term support to the camps, which sustained not only civilian refugees but also those who had perpetrated the genocide.

While in many ways exceptional, the Rwandan genocide and the humanitarian response in its aftermath also share some characteristics with other complex emergencies, and the case raises questions which point to wider issues. Why did the international community fail to prevent the genocide or move to stop it once it was underway? Why, in the aftermath of the genocide, was aid not always focused where it was needed most? At what point should aid agencies stop their operations when the assistance they are providing is being manipulated and diverted?

The Politics and Everyday Practice of International Humanitarianism. Miriam Bradley, Oxford University Press.
 DOI: 10.1093/oso/9780198840695.003.0005

Background

The genocide in Rwanda was the consequence of a combination of many factors, both historical and contemporaneous.[1] The politicization of Hutu and Tutsi identity began in the pre-colonial period with historical social positions used to define 'ethnic' categories, and the distinction between these two groups was then reinforced under German colonial rule (1899–1916) and Belgian trusteeship (1916–1961). Through the policy of indirect rule, the minority Tutsi held administrative and political power, and opportunities and fortunes were determined by one's identity as Hutu or Tutsi (or Twa)[2]—formalized with the introduction of compulsory identity cards in 1933—such that the Tutsi became the haves and the Hutu the have-nots (Eriksson et al. 1996: 11; Newbury 1998: 11).

In the context of decolonization in neighbouring countries, and with the aim of redressing longstanding injustices, both the colonial administration and the Catholic church shifted their support from the minority Tutsi to the majority Hutu in the late 1950s (Eriksson et al. 1996: 12). The abrupt reversal in the distribution of favour and fortune facilitated the Hutu or 'peasant' revolution of 1959–1961 (which, in turn, led to the exile of tens of thousands of Tutsi to Uganda and other neighbouring countries) and the establishment of an independent Hutu-led republic in 1962 (Eriksson et al. 1996: 12, 13). From 1963, Tutsi exiles in Burundi and Uganda launched raids into Rwanda, provoking retaliations against the Tutsi population within the country, which then prompted more Rwandan Tutsis to flee across the border (Guichaoua 2017: 7–8; Hammerstad 2014: 214).

Within Rwanda, intra-Hutu tensions increased under President Habyarimana, who came to power after engineering a coup in 1973 (Newbury 1995: 13). The main fault line was between groups from the northern prefectures of Gisenyi and Ruhengeri, which had been independent until the early twentieth century, when they were militarily defeated by German and Tutsi-led southern Rwandese troops, and those from the rest of the country. The northern Hutu sub-culture had a strong awareness of a pre-Tutsi past and was 'deeply suspicious of any reconciliatory gestures towards the exiled Tutsi community' (Eriksson et al. 1996: 12–13). Habyarimana's informal council (known as the *akazu*), which centred on his wife and brothers-in-law,

[1] For more on the causes and course of the genocide, see Guichaoua (2017), Prunier (1998), and Rever (2018).

[2] The Twa, who accounted for around 1% of the total population of Rwanda, were discriminated against even more intensely than the Hutu in the colonial period (Newbury 1995: 12, 1998: 11).

represented this northern tradition and gradually grew in influence, promoting a hard line against the Hutu moderates, who favoured dialogue with Tutsi exiles (Eriksson et al. 1996: 13; Guichaoua 2017: 21, 49–52). The hardliners depicted the Tutsi as an 'alien race' that sought to reinstate Tutsi power and subjugate all Hutus (Hammerstad 2014: 214).

Inter-ethnic tensions also increased in the 1980s, and an invasion by Tutsi refugees from Uganda in 1990 ultimately triggered a civil war. The 1980s and early 1990s saw economic deterioration, with fluctuations in the world price of coffee compounded by externally imposed structural adjustment programmes in 1990 and 1992, which hit the peasantry particularly hard and fuelled tensions between in-groups and out-groups (Eriksson et al. 1996: 13; Newbury 1995: 14–15; Uvin 2003: 79–80). On 1 October 1990, the Rwandan Patriotic Front (RPF), a Tutsi refugee army based in Uganda and led by Paul Kagame, invaded Rwanda. Only with military support from Belgium, France, and Zaire was the government able to contain the attacks (Guichaoua 2017: 23). The invasion strengthened the position of the hardliners in the government, police, and army and led to a civil war between the Hutu-dominated government and army, on the one hand, and the RPF on the other (Eriksson et al. 1996: 13; Newbury 1995: 14). Opération Noroît, an intervention by French forces, who remained in Rwanda until December 1993, reduced, but did not end, the fighting (Barnett 2011: 181).

In United Nations (UN)-backed peace accords signed in Arusha, Tanzania, in 1993, the government and the RPF pledged to end the civil war and cooperate, but it is not clear that either side believed that peace was sustainable or that power sharing was feasible. Notwithstanding some misgivings about the possibility of implementing the Arusha Accords, the UN Security Council authorized a peacekeeping operation, the UN Assistance Mission in Rwanda (UNAMIR) in October 1993 to oversee the agreement. However, neither side was implementing the basic elements of the Accords, and UNAMIR lacked the muscle to put sufficient pressure on them to comply, with the consequence that violence increased (Barnett 2011: 181; Eriksson et al. 1996: 17). Also in 1993, neighbouring Burundi's first Hutu president was assassinated, just months after his election, by elements of the Tutsi-dominated Burundian army, sparking widespread violence (Newbury 1995: 16). Tens of thousands of people were killed in what a subsequent UN-mandated commission of inquiry found amounted to genocide against the Tutsi minority, and thousands more fled the country. Given the close relationship between the two countries, these violent events in Burundi exacerbated anxiety and polarization in Rwanda.

Genocide

On 6 April 1994, President Habyarimana's plane was shot down, and within hours, Hutu extremists embarked on a spree of ethnically motivated killings. A ground-to-air missile caused the plane to explode in mid-air, killing the president together with some key army officers, government ministers, and foreign politicians, including the president of Burundi (Guichaoua 2017: 143). The following day, the Presidential Guard, together with army units linked to the presidency, and the *Interahamwe*, a Hutu paramilitary organization originally formed as the youth wing of the ruling party at the beginning of the 1990s, went on a rampage, attacking Tutsi and assassinating the head of government, political opposition, and the Belgian UNAMIR troops tasked with providing security for the prime minister (Guichaoua 2017: 148; Prunier 1998: 230). An interim government, which purported to speak on behalf of the 'Hutu people', was installed on 9 April and set about confronting the RPF and massacring the Tutsi population (Guichaoua 2017: xlv, 148).

No one has publicly admitted responsibility for shooting down the plane, and we may never know the truth. For some years, the dominant narrative—both within Rwanda and internationally—suggested that members of the Presidential Guard shot it down, unhappy with the concessions that Habyarimana had made in the peace talks and seeking an excuse to launch the genocide (see, e.g. Prunier 1998: 213–229). More recent research, by contrast, points to the RFP being behind the shooting down of the plane, with the aim of provoking a vicious response, which would justify RPF military operations, delegitimize those in power, and ultimately enable the RPF to replace them (Guichaoua 2017: 144–147; Rever 2018: 61–63). According to this second narrative, the leaders of the RPF were prepared to sacrifice lives among the Tutsi population they purported to represent in pursuit of the goal of gaining power in Rwanda, although they may not have realized how many lives would be sacrificed before they achieved this goal (Guichaoua 2017: 227–230).

Whether or not Hutu extremists were involved in shooting down the President's plane, the subsequent killing of Tutsi and moderate Hutu was state-led and deliberate and constituted genocide.[3] Massacres were carefully planned

[3] According to the 1948 Convention on the Prevention and Punishment of the Crime of Genocide:

> genocide means any of the following acts committed with intent to destroy, in whole or in part, a national, ethnical, racial or religious group, as such: (a) Killing members of the group; (b) Causing serious bodily or mental harm to members of the group; (c) Deliberately inflicting on the group conditions of life calculated to bring about its physical destruction in whole or in part; (d) Imposing measures intended to prevent births within the group; (e) Forcibly transferring children of the group to another group.

and well organized (Newbury 1995: 12; Prunier 1998: 224, 230–231, 244, 246). Those organizing them—predominantly members of the regime's political, economic and military elite—mobilized the wider Hutu population to participate in the killing (Newbury 1995: 12; Prunier 1998: 239–248). Over centuries, Rwandan society had developed with a high level of authoritarian social control by the state, stifling the emergence of non-governmental organizations (NGOs) or independent interest groups and facilitating the manipulation or mobilization of the subjugated rural population, who lacked the means and organization to disobey (Eriksson et al. 1996: 12; Guichaoua 2017: 15–17; Prunier 1998: 245). Radio Télévision Libre des Mille Collines, created by Hutu hardliners in April 1993, played an important role in mobilizing ordinary Hutu citizens to join the killing (Kellow and Steeves 1998). Those who resisted joining in were often killed themselves. Seeking to involve the entire Hutu population in the genocide not only enabled it to be carried out quickly, but also, in sharing guilt so widely, the instigators minimized the risk of accusations later on (Terry 2002: 170). An average of 5.5 people were killed each minute, more in the first four weeks when most of the deaths occurred (Barnett 2002: 1). The killings were brutal, often carried out by machete, resulting in slow and painful deaths, bodies were mutilated, and many of the women were raped before they were killed (Prunier 1998: 255–256).

The genocide was brought to an end by the military defeat of the *génocidaires* by the RPF, which had begun its offensive as soon as the genocide began, rapidly gaining territory from the Rwandan Armed Forces (Forces Armées Rwandaises—FAR) and taking control of Kigali on 4 July (Terry 2002: 156). As they gained territory, the RPF forces also committed war crimes including large-scale massacres of hundreds or even thousands of people in single days (Guichaoua 2017: 313; Prunier 1998: 266, 306; Rever 2018: 81–105). In contrast to the interim government, the FAR and the *Interahamwe*, all of whom committed their crimes openly, in broad daylight, and with little concern for repercussions, the RPF was more discreet, killing Hutus in secret and destroying evidence by incinerating the bodies (Rever 2018: 3–4, 74–76). It is estimated that the RPF killed between 25,000 and 40,000 people, perhaps more, between April and August 1994 (Terry 2002: 173–174).

In under 100 days, up to a million (best estimates suggest at least 800,000) Rwandans, of a prior population of around 7 million, were killed (Prunier 1998: 261–265). More than 2 million people fled the country, and over a million were temporarily displaced within Rwanda (Eriksson et al. 1996: 23).

International inaction

The international community failed to provide a political, diplomatic, or military response to prevent or halt the genocide.[4] At the outset, UNAMIR had only 2,500 lightly armed peacekeepers spread throughout Rwanda and had neither the mandate nor the capacity to counter the extremist forces (Barnett 2011: 181; Dallaire 2003). France and Belgium were close allies of the Hutu-led government of Juvenal Habyarimana and reluctant to take action to halt the killing. Other Western countries had less interest in Rwanda, and the United States in particular was unwilling to deploy ground troops in African conflicts after the killing of 18 elite troops (and at least 500 Somalis) in Mogadishu the year before (Hammerstad 2014: 216; Prunier 1998: 274). The Somalia experience contributed to a widely shared perspective among Western policymakers that outsiders could do little to resolve 'barbaric' African conflicts—a perspective that both depended on and reinforced particular representations of the violence in Rwanda (Power 2002).

Two aspects of the way the violence in Rwanda was represented in Western countries were crucial to facilitating inaction on the part of the UN Security Council and the major powers. First, the characterization of the violence as spontaneous and uncontrollable tribal bloodletting served to construct a situation that appeared impossible to solve and, hence, made a humanitarian, as opposed to a political, response more likely (African Rights 1994: 31). In the first weeks after the shooting down of President Habyarimana's plane, mainstream Western media mostly represented the violence in Rwanda as the inevitable consequence of ancient tribal hatreds (on coverage in the *New York Times*, for example, see Chari 2010: 340–342). Such media representations reinforced—and were reinforced by—similar representations by international policymakers (Barnett 2002: 51). Second, both Western media and Western policymakers largely avoided classifying the violence as genocide (Melvern 2001). Under the 1948 Convention on the Prevention and Punishment of the Crime of Genocide, states parties—which, in 1994, included all five permanent members of the UN Security Council and most countries worldwide—have an obligation to 'prevent and punish' the crime of genocide. Because of this obligation, the Security Council and Western powers avoided the language of genocide, talking instead in terms of ethnic cleansing and civil war, even once it was clear that genocide was happening in Rwanda (Barnett 2002: 3; Hammerstad 2014: 217). For the most

[4] For more comprehensive accounts of the role of the United Nations, see Barnett (2002) and Dallaire (2003). Roméo Dallaire was the Canadian General in command of UNAMIR.

part, the mainstream media remained silent on the question of genocide, reinforcing and facilitating the political discourse (Melvern 2001: 103–105).

In the first days and weeks of the genocide, the Security Council knew that violence was spiralling, but instead of increasing the size or strengthening the mandate of UNAMIR, it did the opposite. Ten Belgian peacekeepers were killed on 7 April, and the Security Council ordered UN troops not to take risks to save or protect Rwandans (Barnett 2011: 184). On 10 April, by which time Belgium, France, and the United States had sent a combined total of close to 2,000 elite military forces to Rwanda to rescue foreign nationals, the UNAMIR Force Commander told an assistant to the Secretary-General that with 4,000 effective troops he could stop the killing (Dallaire 2003: 284, 289). However, the option of using the foreign troops engaged in evacuation missions to reinforce UNAMIR was not being considered. Indeed, Belgium reassigned troops from UNAMIR to its national evacuation force and then announced its unilateral withdrawal from Rwanda (Dallaire 2003: 287, 294). Subsequently, on 21 April, the Security Council decided to reduce the size of the UNAMIR force from 2,500 to 270 troops, with a mandate restricted to helping to negotiate an end to the killing (UN Secretary-General 1994; UN Security Council 1994a).

When it became impossible to deny that the violence in Rwanda constituted genocide, it also became impossible for the Security Council not to take more concerted action—or at least to appear to be taking more concerted action. Oxfam first used the term genocide on 24 April, and sometime after that, UNAMIR also started talking in terms of genocide (Dallaire 2003: 333). On 10 May, José Ayala Lasso, the newly appointed UN High Commissioner for Human Rights, arrived in Rwanda on a fact-finding tour and subsequently declared what he had seen to be a genocide (Dallaire 2003: 363). A week later, on 17 May, the UN authorized an additional 5,500 troops for Rwanda in the form of UNAMIR II, with a stronger mandate for the use of force (UN Security Council 1994b). However, no member state was willing to send its troops into such a violent situation, and they were not deployed until three months later, after the RPF had defeated the government (Barnett 2011: 182; Landgren 1995: 448; Prunier 1998: 276).

In the meantime, on 22 June, the Security Council authorized the deployment of a multilateral intervention force with a mandate limited to a two-month period—or less if an expanded UNAMIR force was deployed sooner—to contribute to the security and protection of civilians and to provide security and support to humanitarian relief operations (UN Security Council 1994c, d). France launched Opération Turquoise the next day, and, in the event, only Senegal sent troops to join the French effort (Prunier 1998:

291). On 2 July, Opération Turquoise established a so-called safe humanitarian zone in south west Rwanda (Landgren 1995: 449). This 'safe humanitarian zone' served not only to protect Hutu civilians from retaliation by the RPF but also facilitated the escape of much of the FAR and the interim government—those responsible for organizing the genocide—into Zaire (Hammerstad 2014: 215; Prunier 1998: 308; Terry 2002: 163).

Not only did the international community fail to prevent the genocide, but also the international humanitarian response during the three months of the genocide was severely limited. Water and power were cut off in Kigali and elsewhere, and people lacked access to food, but the more overwhelming need was for security and physical protection. Despite some impressive accomplishments, neither humanitarian agencies nor the much-reduced and ill-equipped UNAMIR force were able to provide assistance or protection at the scale required. Between them, the International Committee of the Red Cross (ICRC) and UNAMIR succeeded in protecting tens of thousands of threatened civilians who had sought refuge in ICRC hospitals, UN compounds, and places such as Amahoro Stadium and the Mille Collines hotel (Dallaire 2003: 263; Gaillard 2002; ICRC 1995: 55).

The ICRC managed to keep its hospitals in Kigali and Kabgayi (around 40 km south west of Kigali) running and providing emergency surgical care for the survivors of massacres as well as doing limited protection work (Eriksson et al. 1996: 23; ICRC 1995: 54). However, most international agencies withdrew from Rwanda when the genocide started (Hammerstad 2014: 218). In Kigali, only the ICRC (with Médecins Sans Frontières (MSF) support) and the UN Advance Humanitarian Team were operating, and their operations were constrained by high levels of violence (Eriksson et al. 1996: 24). Even UNAMIR was unable to make full use of the material and foods placed at its disposal by the UN Department of Humanitarian Affairs, as it came under fire when it tried to take control of the UN humanitarian warehouses (Dallaire 2003: 297).

Until the French Opération Turquoise established the so-called safe humanitarian zone in south west Rwanda, only the ICRC, Catholic Relief Services, and the World Food Programme (WFP) were able to operate in the south and west of the country, and the volume of assistance they could provide was severely limited (Eriksson et al. 1996: 24). From 20 April, the ICRC was providing assistance to 8,000 people in Nyarushishi camp for displaced persons near Cyangugu in the south west. At the end of the genocide, these 8,000 were the only survivors in the prefecture of Cyangugu (Gaillard 2002).

International agencies had greater access in the RPF-controlled areas in the north and east of Rwanda and were able to deliver significant amounts of aid

(Eriksson et al. 1996: 24). Once the Tanzanian border was open to humanitarian convoys in mid-June, for example, the ICRC began transporting food from its Ngara office in Tanzania (ICRC 1995: 56). However, humanitarian agencies were subject to significant restrictions and control (Eriksson et al. 1996: 24). The RPF limited their freedom of operation to ensure they did not witness the massacres it was perpetrating as it took territory (Des Forges 1999: 493). Furthermore, the RPF maintained significant control over the distribution of aid, diverting a share for its own forces (Dallaire 2003: 299, 390).

Mass exodus to neighbouring countries

As the RPF advanced across Rwanda, hundreds of thousands of Hutus fled the country. In late April 1994, around 170,000 fled to Tanzania in the space of twenty-four hours, with thousands more following in subsequent days and months (Terry 2002: 171). By the end of July, around 2 million—including both those responsible for the genocide and civilians fearful of reprisals from the RPF or pressured to leave by the near-defeated state and the FAR—had fled the country, most going to neighbouring Zaire and also to Tanzania and Burundi (Barnett 2011: 182; Terry 2002: 155).

In just four days from 14 to 17 July 1994, between 500,000 and 800,000 Rwandans crossed into North Kivu in Zaire, overwhelming the capacity of aid organizations to mount an adequate response (Terry 2002: 171). Polluted water sources and a lack of sanitation facilities provoked cholera and dysentery epidemics which were the cause of extremely high death rates in Goma, with some 50,000 people dying in just 4 weeks (Eriksson et al. 1996: 25; Terry 2002: 171). Malnutrition rates were also high in Goma, and there was additionally a meningitis outbreak (Terry 2002: 193).

Believing that neither local integration in neighbouring countries nor resettlement in third countries were feasible options for such a large number of refugees, in August 1994 the United Nations Refugee Agency (UNHCR) sent a three-person mission, led by Robert Gersony, to Rwanda and to refugee camps in neighbouring countries to find ways to speed repatriation (Des Forges 1999: 495; Ogata 2005: 188–190). Although it did not set out to investigate abuses by the RPF, it was presented with detailed and convincing witness accounts of arbitrary arrests, disappearances, house-to-house killings, and massacres perpetrated by the RPF both during and after the genocide, including of Hutus seeking to flee Rwanda or returning from neighbouring countries (Des Forges 1999: 496; Ogata 2005: 191–192).

UNHCR concluded that the conditions were not conducive to refugee return and initially suspended its role in repatriation, but the UN Secretary-General, Boutros Boutros-Ghali, supressed the Gersony Report (Des Forges 1999: 496–497; Terry 2002: 173–174, 199). The findings of Gersony's team were embarrassing for UNAMIR and other UN agencies that had been operating in Rwanda apparently unaware of the documented atrocities, and there was a strong sense of guilt among the international community for having done nothing to stop the genocide as well as a desire to delineate the 'good side' from the 'bad side', to avoid weakening the new government, and to have refugees return (Des Forges 1999: 496–497; Ogata 2005: 194; Terry 2002: 174).

Meanwhile, security in the refugee camps was worsening. The FAR fled to Zaire almost intact, and the former regime imposed governance structures and authority within the camps there, replicating the neighbourhood 'self-defence' system that had proved an effective mechanism with which to perpetrate genocide just months earlier (Prunier 1998: 267; Terry 2002: 156–157, 178). Around 22,000 soldiers of the former FAR as well as somewhere between 10,000 and 50,000 militia were present in the camps, and they had tens of millions of dollars, vehicles, machinery, and considerable quantities of military hardware at their disposal (Terry 2002: 160–161). They used the camps to regroup, restrengthen, and re-arm, and they began to launch attacks across the border into Rwanda soon after their arrival in Zaire (Barnett 2011: 183; Terry 2002: 164). The militarized camps contributed to tensions within Rwanda, between Rwanda and Zaire, and within Zaire itself. The ex-FAR and the *Interahamwe* not only launched attacks into Rwanda but also massacred and intimidated local Tutsis and created an atmosphere of generalized violence in eastern Zaire (Hammerstad 2014: 215).

Humanitarianization of the crisis

The international focus followed the exiles into neighbouring countries. Media coverage of the refugee camps provided the first sustained images most Western audiences had had of the genocide (Barnett 2011: 182). With the exodus of Rwandans to Tanzania at the end of April 1994, the focus of the international media, NGOs, and Western governments switched from the genocide inside the country to the refugee crisis, and the international debate switched from being largely at the political level to the humanitarian level (African Rights 1994: 34). Indeed, a study of coverage by the *New York Times* found that there were more stories about the Rwandan genocide in each of

May and June 1994 than in April, when the largest number of killings took place, suggesting that the refugee crisis got more coverage than the genocide itself (Chari 2010: 337). The humanitarianization of the crisis thus accelerated with the mass exodus to Zaire in July, and the international response exploded with the arrival of refugees in Goma (Barnett 2011: 182).

The scale of the response in Goma in particular was extraordinary, prompted by guilt about the failure of the international community to prevent genocide, intense media coverage of the influx, and the subsequent cholera outbreak (Barnett 2011: 182–183; Eriksson et al. 1996: 27; Terry 2002: 203). The epidemics enabled the international community to frame the crisis as humanitarian and to provide a large-scale and highly visible response (Prunier 1998: 302–304; Terry 2002: 171). Much as they were doing in Bosnia, the main donor states opted to provide relatively generous funding for the humanitarian effort in the Kivus—and the cholera outbreak even prompted the United States and the United Kingdom to mobilize their armed forces to provide logistical support for that humanitarian effort—but to avoid committing serious political or military resources to address the security situation in the region (Hammerstad 2014: 217). Providing generous funding allowed donor governments to show that they were doing something to respond, but 'there was an unwillingness to actually understand, to engage or to analyse' (Vaux 2001: 187). This unwillingness, in turn, likely contributed to some of the unintended consequences of the humanitarian response.

International humanitarian response

If success is measured in terms of technical achievements and abstracted from the political context, a positive picture of the humanitarian response can be painted. For example, Oxfam alone provided clean water to over 2.5 million refugees and internally displaced persons (IDPs) in the Great Lakes region of Africa during 1994 (Stockton 1998: 352). WFP and, to a lesser extent, the ICRC, supplied large quantities of food aid (Eriksson et al. 1996: 23). After the meningitis outbreak surfaced in Goma in late July 1994, a mass vaccination campaign against meningitis was carried out in addition to the routine measles vaccination campaign (Terry 2002: 193).

Such achievements undoubtedly saved lives and reduced suffering. There were tens of thousands of preventable deaths—a major inter-agency evaluation estimated that 80,000 people died in the refugee camps in Zaire and Tanzania, and in the IDP camps inside Rwanda, during 1994, mainly from cholera and dysentery (Eriksson et al. 1996: 23). However, given the scale of the influx, it is a credit to the agencies involved in the response that the number was not much higher (Eriksson et al. 1996: 25). The dysentery and cholera epidemics were eventually contained by the provision of clean water and intensive rehydration of thousands of victims (Stockton 1998: 352). Likewise, the fact that there was not widespread starvation can be attributed in part to the effective performance of the food aid supply systems (Eriksson et al. 1996: 23).

However, the efforts of humanitarian agencies were not always focused on those with greatest need, as required by the principle of impartiality.[5] Relative to that in the refugee camps, the humanitarian response inside Rwanda after the genocide was on a much smaller scale. Certainly, many of the people in the camps in neighbouring countries had urgent and significant needs, but so did people inside Rwanda (Storey 1997). Within Rwanda, the focus of international aid was, to a large degree, determined by ease of access rather than the level of humanitarian need. Humanitarian agencies had easier and safer access to IDPs in camps in the so-called safe zone established by Opération Turquoise, which led to the relative neglect of those in the north west of the country (Eriksson et al. 1996: 24). Outside Rwanda, media attention was a major driver of humanitarian focus, and the shift in international attention to Goma from mid-July resulted in the transfer of resources and personnel away from Tanzania (Eriksson et al. 1996: 25).

Moreover, a focus on technical measures or outputs obscures the limits and unintended consequences of the humanitarian response. The level of violence within the camps was extremely high, with an estimated 4,000 camp inhabitants killed in Zaire (Eriksson et al. 1996: 28). Refugees were killed because they were suspected of spying for the RPF, of being against the former regime, or simply for saying they wanted to return to Rwanda (Prunier 1998: 310). Several international and many more local aid workers were also killed in the camps, and intimidation and threats were common (Hammerstad 2014: 224). The high levels of insecurity directly affected the effectiveness of the relief efforts as most international aid workers were unable to remain in the camps overnight, and medical personnel were not always able to maintain continuous care of patients (Eriksson et al. 1996: 28). Moreover, aid was deliberately diverted and manipulated in such a way that it facilitated the violence within (and beyond) the camps.

[5] See 'Principles of humanity, impartiality, neutrality, and independence' in Chapter 13 on Politics, Principles, and Humanitarian Action.

Diversion, manipulation, and unintended consequences of aid

The perpetrators of the genocide gained control of the aid distribution systems within the camps and taxed, diverted, and stole aid (African Rights 1994: 34–35; Storey 1997: 387; Terry 2002, 186, 188). As is often the case in emergency response, international aid agencies employed members of the affected population who had previously held leadership positions within society to assist with camp management and to oversee food distribution. People with leadership experience are often well qualified to organize relief operations and distributions, but employing them to do so can also serve to reinforce pre-existing power structures. In the case of the Rwandan refugee camps, this meant that the perpetrators of the genocide gained legitimacy as representatives of the refugees (Storey 1997: 388), and were able to divert and manipulate aid.

The political and military authorities of the former regime inflated population numbers in the camps and sabotaged efforts to conduct refugee registration exercises. Exaggerating population numbers served not only to generate economic benefits from the provision of additional international aid but also to challenge the new Rwandan regime's right to power (Terry 2002: 188). For the same reason, they also sought to stop refugees from repatriating to Rwanda (Hammerstad 2014: 219). The more Rwandan people outside Rwanda, the less legitimacy the regime inside Rwanda could claim. Furthermore, as the former regime controlled food distribution in the camps, it was able to divert rations from refugees to feed its own forces, with the consequence that, among certain groups, particularly the elderly and female-headed households, there were high rates of malnutrition (Borton 1996: 29; Terry 2002: 188).

In principle, the host state is responsible for ensuring that refugee camps are exclusively civilian and humanitarian, but the states neighbouring Rwanda did not live up to this responsibility. The Zairean government vacillated between reluctant cooperation with UNHCR and maintaining good relations with the former Rwandan regime (Hammerstad 2014: 217; Terry 2002: 183). The camps in Tanzania were also militarized, and although the Tanzanian government was somewhat better able to resist the attempts of the *génocidaires* to control them, the destabilizing presence of militants among the refugee population ultimately led to Tanzania forcibly repatriating all Rwandan refugees in December 1996 (Hammerstad 2014: 216). The record of the wider international community in addressing the violence was equally unimpressive. As in Bosnia, powerful states offered a humanitarian response as a substitute for political action.[6] Between 1994 and 1996, international donors provided funding of USD 1.3 billion for humanitarian assistance in the refugee camps, while refusing funds to disarm the militants (Lischer 2003: 80). The United States, France, Israel, Japan, and the Netherlands even despatched military contingents to fight 'a war against cholera' but did not mandate them to deal with the *génocidaires* or address the insecurity in the camps (Terry 2002: 193).

There was little that humanitarian agencies could do to resist the demands of what was effectively a well-armed military. Aid workers were physically threatened when they sought to remove the *génocidaires* from the distribution network and to reduce the amount of aid being diverted (Barnett 2011: 183; Terry 2002: 201–202). In June 1994, an attempt to remove a known *génocidaire* from Benaco refugee camp in Tanzania almost caused a riot when his followers surrounded the UNHCR compound (Terry 2002: 193). When the Canadian branch of CARE established an independent security structure in Katale refugee camp in Goma, CARE staff received death threats, and thirty-five of the Rwandan 'Scouts' tasked with controlling camp access and directing camp traffic were murdered and replaced by armed militia members, prompting CARE Canada to withdraw (Terry 2002: 176). UNHCR ended up paying and equipping Mobutu's presidential guard to police the camps, but they were not tasked with disarming the ex-FAR and militias, and their deployment only served to improve security until they allied themselves with the Hutu militants in response to shifts in political and security dynamics in eastern Zaire (Hammerstad 2014: 219).

Even if resisting the ex-FAR and the militias was beyond the capacity of the agencies in charge of the camps, they could have taken better precautions to avoid the militarization of the camps in the first place. The diversion of aid to support the restrengthening of the former regime and the militarization of the camps were not only foreseeable but also foreseen by some experts (Terry 2002: 170). Nonetheless, it was arguably understandable that UNHCR and other agencies failed to analyse the history and nature of the refugees arriving very suddenly in Tanzania in April 1994 (Terry 2002: 192–193). Some have also argued that the power of those who had perpetrated the genocide was not immediately clear to UNHCR or other international agencies in the camps in Zaire in July (Hammerstad 2014: 218). That may the case, but it certainly should have been clear. By June 1994, aid agencies were well aware of the nature of the refugee leadership in Tanzania, and yet the same mistakes were made in Zaire in July (Terry 2002: 193).

[6] See 'Humanitarian alibi' in Chapter 4 on the Bosnian War, 1992–1995.

Moreover, there is at least one way in which humanitarian agencies can resist in an already militarized context—they can invoke what MSF calls an ethic of refusal. The question of whether to remain or withdraw from the camps in Zaire created a great deal of division not only between but also within different humanitarian agencies. The International Rescue Committee (IRC) and the French section of MSF decided to withdraw once the emergency phase of the relief operation had subsided, but the vast majority of international agencies remained (Terry 2002: 195). Other MSF sections and agencies such as Care UK and Oxfam considered withdrawing but ultimately decided to stay (Terry 2002: 197, 201). Of the many NGOs that remained, most did so because they prized the humanitarian imperative above concerns for the consequences of their work, or considered consequences only in terms of technical service delivery, focusing on outputs rather than outcomes (Terry 2002: 200). Many also followed an institutional or economic logic. Aid agencies depend on—and compete for—voluntary donations, and media coverage of their work offers free advertising. Many could not afford to miss the intense media coverage of the refugee exodus and cholera outbreak in Goma, or calculated that doing so would not be worth it when a competitor agency would surely take their place and the camps would continue to function in the same way regardless (Cooley and Ron 2002: 25–31; Terry 2002: 202–203).

UNHCR faced calls (from both within and outside of the organization) to withdraw from the camps, but Sadako Ogata, the High Commissioner, never seriously considered this. Within UNHCR, there were fears that withdrawing could make the agency irrelevant in future refugee situations, especially in the global context of increasingly restrictive asylum policies and erosion of refugee protection principles (Terry 2002: 198). Ogata claimed that the institutional mandate obliged UNHCR to help refugees even though some—or even much—of the aid was being diverted, and UNHCR was unable to fulfil its primary responsibility to provide protection to the refugees (Hammerstad 2014: 223; Terry 2002: 197–198). Ogata frequently repeated that the fault lay with political actors and called on the international community to move the camps away from the border, to remove the extremists from them, and to monitor the borders (Hammerstad 2014: 221; Ogata 2005: 197–199). None of these measures were taken, and UNHCR continued to run the camps anyway. Perhaps burned by Ogata's failed attempt at stopping operations in Bosnia,[7] UNHCR arguably became complicit in that its own presence and actions were having negative consequences for security.

[7] See 'Humanitarian alibi' in Chapter 4 on the Bosnian War, 1992–1995.

While aid inevitably has unintended impacts on political and conflict dynamics, three factors set the Rwandan case apart from the norm. First, genocide is a particularly heinous crime, making complicity in strengthening the perpetrators of genocide particularly unpalatable, though this is arguably more a question of justice rather than impact on conflict dynamics. Second, unusually, or even uniquely, the responsibility of the interim government for genocide was fully established before significant humanitarian operations began (African Rights 1994: 28). Third, *only* international aid was sustaining the viability of the old regime, and the aid mechanisms themselves enabled the former government to maintain control over the refugee population (Terry 2002: 196).

Closing the camps

The new regime in Rwanda was never going to allow the IDP camps in Rwanda and the refugee camps in neighbouring countries to sustain the old regime indefinitely, so one way or another they would be closed down. From the perspective of the RPF-led government in Kigali, UNHCR and the other international agencies running the camps had not simply been aiding refugees but also supporting the military struggle of the *génocidaires* (Hammerstad 2014: 217; Terry 2002: 192).

Ultimately, the camps were closed by force, and hundreds of thousands of refugees died or disappeared. In April 1995, perhaps 4,000 IDPs were killed in Kibeho, the largest IDP camp in the zone formerly controlled by the French Opération Turquoise (Eriksson et al. 1996: 24; Rever 2018: 51; Terry 2002: 210). Then, in October and November 1996, after repeated warnings from Kigali, the new Rwandan army (the RPF forces), together with Zairean rebel forces, attacked and destroyed the camps in Zaire—apparently with at least the tacit support of the US government, given that efforts to encourage voluntary repatriation had been largely unsuccessful and the camps were expensive to maintain and served to destabilize the region (OHCHR 2010: 80–82, 85–86, 87, 92–97; Terry 2002: 185, 191). The refugees from the Goma camps in North Kivu and from the camps in Bukavu and Uvira in South Kivu fled in several directions, pursued and, in many cases, forcibly repatriated or massacred by RPF forces and allies (OHCHR 2010: 82–92, 97–118). There is evidence to suggest that RPF forces had infiltrated UNHCR and other humanitarian organizations and intercepted their communications to identify the location of escaping refugees in order to kill them (Rever 2018: 36–37).

The massacres in eastern Zaire following the closure of the camps may have constituted genocide (OHCHR 2010: 278–284). For broadly the same reasons that the Gersony Report was supressed, international observers resisted acknowledging the scale of abuses perpetrated by the new Rwandan regime, and, for several years, the massacres were not widely acknowledged. International agencies were frequently prevented from accessing camps and makeshift camps until the refugees inhabiting them had been killed and their bodies removed (see, e.g. OHCHR 2010: 105, 108–110). Nonetheless, humanitarian workers cannot have been completely ignorant of the massacres. On occasion, they encountered survivors or pre-emptively evacuated refugees when they had warnings that RPF and Zairean rebel forces were advancing (OHCHR 2010: 105, 116). The violence perpetrated by the ex-FAR in and around the camps from 1994 to 1996 and by the RPF during and after the closure of the camps also affected political and military dynamics in eastern Zaire, exacerbating inter-ethnic tensions and sparking the First Congo War (September 1996–May 1997), destabilizing the region, and ultimately leading to many more deaths in the Second Congo War (August 1998–July 2003) and beyond—over 5 million, by some estimates (Paddon 2010).

Had UNHCR and the other aid agencies supporting the camps closed them and terminated the aid operation in 1995 or early 1996, the refugees would have had to return to Rwanda, which would likely have been a better outcome for the tens or hundreds of thousands who were killed during the eventual dispersal in 1996–1997 and also for regional stability (Hammerstad 2014: 21). However, the militarization of the camps posed genuine dilemmas to the agencies operating them. Closing the camps risked punishing the civilian majority for the crimes of the *génocidaire* minority, sacrificing lives today for only the possibility of saving lives tomorrow.[8] While it may have been impossible to predict all the long-term consequences of sustaining the camps, there was some warning that the RPF would attack—Kagame had repeated the threat on several occasions, and, after the massacre at Kibeho in April 1995, it should have been clear that the threat to close the camps in neighbouring countries was very real (Terry 2002: 184–185). Furthermore, it was not just longer-term consequences that were at stake; aid was being used against those for whom it was intended (Terry 2002: 195). Nonetheless, withholding aid to the camps could have meant sacrificing lives, and most agencies were not prepared to take that risk.

[8] For a discussion of this issue in more general terms, see 'Desirability of apolitical action' in Chapter 13 on Politics, Principles, and Humanitarian Action.

Conclusions

Much like the experience of the Bosnian war, that of the Rwandan genocide highlighted the severe limits to what unarmed humanitarians can achieve in contexts of major violence and in the absence of a concerted political effort to stop that violence. It exemplified the risks of aiding combatants and prolonging conflict when combatants and civilians are mixed up (in this case, in refugee camps). Operational agencies justifiably lament the way that powerful states use humanitarian action to mask or excuse a lack of more concerted political or military action. At the same time, however, those agencies become complicit in the so-called humanitarian alibi when they accept funding from donor governments who are paying for humanitarian action as a way to avoid foreign policy engagement (Terry 2002: 214). In sustaining militarized refugee camps, aid agencies contributed—albeit unintentionally—to very harmful consequences for the refugees, and for the region, stimulating a debate in the sector on how to combine the humanitarian imperative with the need to minimize the harmful consequences of aid, and on the merits and demerits of instrumentalizing aid in pursuit of wider goals.[9]

[9] For more on these debates, see 'Political or apolitical humanitarian action' in Chapter 13 on Politics, Principles, and Humanitarian Action.

6
Afghanistan, 2001–2014

In October 2001, the United States launched Operation Enduring Freedom, with UK and US airstrikes targeting al-Qaeda and the ruling Taliban in Afghanistan. Although the Taliban was quickly overthrown, conflict continued and international forces involving more than fifty countries were engaged in combat operations for just over thirteen years. The conflict had a range of humanitarian consequences, including direct consequences, such as civilian casualties and internal displacement, and indirect consequences such as poor health outcomes due to violence limiting the reach of vaccination campaigns and health-care provision. Humanitarian needs did not disappear—or even diminish—in 2014, but the International Security Assistance Force (ISAF) in Afghanistan, a North Atlantic Treaty Organization (NATO)-led military mission, was disbanded, and responsibility for internal security was handed to the Afghan government, marking the end of Operation Enduring Freedom.

Throughout the 2001–2014 period, aid was politicized in a variety of ways, with significant consequences for how and where assistance was provided. The United States and its military partners labelled the context as 'post-conflict' from 2002 and emphasized chronic poverty and underdevelopment rather than humanitarian needs. Data on humanitarian needs was limited and of poor quality, making it difficult for aid agencies to push back against this conceptualization and emphasis. The discourse of post-conflict development was reflected in donor government funding decisions as well as in the efforts of ISAF. Together with the political branches of the United Nations (UN), ISAF sought to instrumentalize assistance for the purpose of stabilization and state-building and to co-opt international non-governmental organizations (INGOs) and UN humanitarian agencies in pursuit of these wider goals.

Many aid agencies adopted a development and reconstruction agenda after 2002, in line with the objectives of their donors, many of whom were also participating in ISAF. Whereas humanitarian agencies usually seek to keep their work separate from political and military actors, in this context, some worked closely with intervening forces, especially to begin with. In doing

The Politics and Everyday Practice of International Humanitarianism. Miriam Bradley, Oxford University Press.
 DOI: 10.1093/oso/9780198840695.003.0006

so, they frequently compromised on the humanitarian principles of independence, impartiality, and neutrality.[1] After the 2009 troop 'surge', many increasingly sought to distance themselves from the UN and from donors with troops on the ground. The efforts of ISAF and the Afghan government to win 'hearts and minds' were not obviously successful, but the perception that aid agencies were associated with foreign military forces contributed to Afghanistan becoming one of the most dangerous places for aid workers worldwide. For the most part, aid agencies responded to high levels of insecurity through recourse to remote management and hard security measures, modus operandi that had negative consequences for the reach and quality of the humanitarian response and, hence, for the Afghan population.

Background

From 1996 to 2001, Afghanistan was governed by the Taliban, and al-Qaeda was operating inside the country. When al-Qaeda members launched a series of coordinated terror attacks on the United States on 11 September 2001, killing close to 3,000 people, the US government responded with an ultimatum requiring the Taliban, among other things, to turn over the al-Qaeda leadership based in Afghanistan. The Taliban quickly rejected the ultimatum, and although it subsequently offered some compromise options, none of these were acceptable to the United States.

With authorization from the UN Security Council, the United States launched Operation Enduring Freedom, beginning with airstrikes on al-Qaeda and Taliban targets on 7 October 2001. The subsequent war in Afghanistan was both the launchpad and the centrepiece of the so-called global war on terror.

The Taliban was quickly overthrown, with the US-backed Northern Alliance seizing Kabul in November 2001. Representatives of the United States and its international coalition gathered in Bonn with their key Afghan allies in December and established the Afghan Interim Authority, headed by Hamid Karzai—and to be replaced by the Afghan Transitional Authority six months later. The Bonn Agreement also set out the role of a UN-mandated military force, and as soon as the Interim Authority was in place, the UN Security Council established ISAF with a mandate to assist Karzai in Kabul, providing basic security, and training the Afghan forces (UN Security Council 2001). ISAF initially had around 4,500 soldiers, and there was a smaller

[1] See 'Principles of humanity, impartiality, neutrality, and independence' in Chapter 13 on Politics, Principles, and Humanitarian Action.

number of US-led forces hunting for Osama bin Laden and other senior Taliban members in the south-eastern provinces bordering Pakistan (Suhrke 2008: 214).

Keen to publicize and consolidate their apparent quick win, the Karzai government and its international backers designated the context as 'post-conflict', despite continuing hostilities on the ground. In fact, the years immediately after the installation of the transitional government and the establishment of ISAF saw a gradual restrengthening of the Taliban, ongoing conflict, and frequently increasing rather than decreasing levels of violence (Crombé and Hofman 2011: 55; Suhrke 2008). In August 2003, NATO took over leadership of ISAF, and in October, the Security Council expanded ISAF's mandate beyond Kabul to cover the whole of Afghanistan (NATO 2015). From 2004, ISAF deployed provincial reconstruction teams (PRTs), small civil-military teams which sought to 'stabilize' the relatively calm areas in which they operated by contributing to reconstruction and generating support for the central government (Suhrke 2008: 223). However, the international forces also relied heavily on aerial bombing in their response to increasing Taliban influence, resulting in civilian casualties which undermined ISAF's legitimacy and increased support for the insurgents (Crombé and Hofman 2011: 55; Suhrke 2008: 221–222, 230). By 2008, six years after the supposed end of the war, the size of the international forces in Afghanistan had increased almost ten-fold to nearly 50,000 troops, and the country was increasingly insecure (Suhrke 2008).

To address growing insecurity, the United States authorized a troop 'surge', which began at the end of 2009 and involved the deployment of an additional 40,000 troops, almost doubling the size of ISAF (Haysom and Jackson 2013: 9; NATO 2015). The surge was accompanied by an increase in violence from armed opposition groups, with insurgent attacks increasing by 50% between June 2010 and 2011 (Featherstone 2012: 5). As conflict intensified following the surge, so too did the number of civilian casualties and the level of internal displacement (Poole 2014: 7).

At conferences in London and Kabul in January and May 2010, representatives of the Afghan government, ISAF troop-contributing nations, donors, and the wider international community elaborated a plan for a transition aimed at transferring security, governance, and development responsibilities to the Afghan government through a gradual withdrawal of international forces (Featherstone 2012: 5). This transition was launched in July 2011 and completed at the end of 2014, by which time all PRTs had been phased out and the ISAF mission was completed. In January 2015, NATO launched the follow-on, the Resolute Support Mission, to train, advise, and support

Afghan security forces and institutions. By July 2020, that non-combat mission numbered 17,000 troops from 39 NATO allies and partner countries (NATO 2020). The Mission was terminated in 2021, and Taliban forces quickly regained control of the country.

Humanitarian needs and vulnerability

Assessing humanitarian needs in Afghanistan during the years of Operation Enduring Freedom was extremely difficult due to a lack of good-quality data—and sometimes of any data at all.[2] The paucity of data was, in part, a consequence of the same factors that were contributing to humanitarian vulnerability, including state fragility and high levels of insecurity, and was exacerbated by the failure of the humanitarian sector to prioritize data collection and analysis (Benelli et al. 2012: 5). Furthermore, the ways that many aid agencies responded to insecurity—with bunkerization and remote management and without dialogue and relationship-building with the Taliban and other armed opposition groups—exacerbated the paucity of information (Donini 2011: 154 fn. 30).

While there is broad consensus that levels of vulnerability and need were high throughout the period covered in this chapter, the picture is complicated by the question of whether (and which) needs were properly 'humanitarian' and the consequence of 'humanitarian crisis', or better conceptualized as the consequence of underdevelopment and other structural factors. In Afghanistan, this was a highly political question. From 2002, because the US-led coalition of international forces and the Afghan government wanted to portray the context as 'post-conflict', and hence not of humanitarian concern, they sought to minimize humanitarian need in their discourse or to rebrand it as chronic underdevelopment (Benelli et al. 2012: 5; Donini 2011: 151; Featherstone 2012: 10). The 'post-conflict' designation was accepted by all but a handful of analysts and served to warp the analysis because, in fact, conflict was ongoing and humanitarian needs did not disappear (Donini 2011: 148).

Several factors suggest that many of the needs and vulnerabilities in Afghanistan even after 2002 could be described as 'humanitarian', resulting from the conflict and other disasters, and requiring a humanitarian rather than—or as well as—a development response. First, the patchy data available reveal acute needs typical of 'emergency' definitions (see, e.g. Benelli

[2] For a general discussion of the importance and challenges of humanitarian needs assessments, see 'Assessing needs, contexts, and capacities' in Chapter 21 on Needs Assessment, Evaluation, and Response Decisions.

et al. 2012: 6–10). Second, much of the vulnerability and many of the needs identified through the 2001–2014 period were clearly a consequence of ongoing armed conflict, including: civilians killed by parties to conflict (Haysom and Jackson 2013: 8; UNAMA 2015: 23); civilians displaced or inaccessible to humanitarian agencies due to counter-insurgency operations and other conflict-related violence (Featherstone 2012: 9; Poole 2014: 7); and access to basic health care limited due to attacks against health-care facilities and personnel (Benelli et al. 2012: 7–8). Third, regardless of what was driving needs, the fact remained that there was ongoing conflict across the country. In many respects, therefore, a principled humanitarian response, mindful of the sensitivities required for working in conflict contexts, was more appropriate than a developmental response, implicitly premised on broadly peaceful operating conditions.[3]

The many good reasons to emphasize a humanitarian rather than development response clashed with the interests of the many donors who were committed to the post-conflict narrative. In this context, the failure of the humanitarian sector to prioritize quality information and analysis was particularly serious. Without a credible picture of how the conflict was affecting health and other essential services across the large parts of the country that were not controlled by the government, operational humanitarian agencies were not in a strong position to push donors to acknowledge that a robust humanitarian response was necessary and to provide funding for such a response (Donini 2011: 154; Featherstone 2012: 10–11).

Funding for humanitarian response

Despite the lack of accurate data on humanitarian needs, it is clear that funding for humanitarian action in Afghanistan was not driven by need and need alone.[4] This is evident from examining how funding changed over time, who was providing the funding, and what conditions they attached to it.

In the 2001–2014 period, international humanitarian funding for Afghanistan was relatively plentiful in some years and scarce in others. Figure 6.1 shows how it fluctuated across that period. Some fluctuations reflected changes in the level of need. For example, the large increase in

[3] For a discussion of what is meant by a principled humanitarian response and why many agencies consider it important, especially in conflict contexts, see 'Principles of humanity, impartiality, neutrality, and independence' in Chapter 13 on Politics, Principles, and Humanitarian Action.

[4] On the determinants of humanitarian funding decisions more generally, see 'In accordance with need?' and 'Other factors shaping funding decisions' in Chapter 18 on Donors and the Funding of Humanitarian Action.

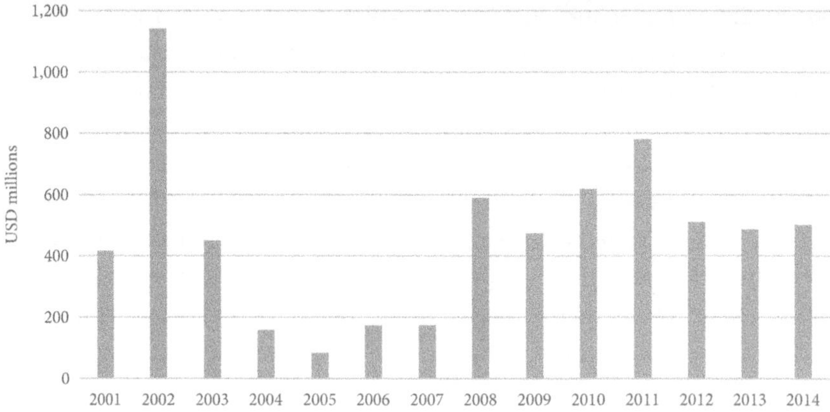

Figure 6.1 International funding for humanitarian action in Afghanistan, 2001–2014

Source: data taken from ReliefWeb Financial Tracking Service.

humanitarian funding in 2008 was a response to an increase in humanitarian needs caused by food shortages and insecurity (Featherstone 2012: 7). However, other fluctuations reflected political and military interests. The big drop in funding between 2002 and 2003, for example, did not reflect a corresponding drop in the level of need. It was accompanied by a large increase in development funding and seemed to be a consequence of donors deprioritizing humanitarian needs in favour of support for development, reconstruction, and stabilization (Featherstone 2012: 6, 7; Poole 2014: 10).

Most humanitarian funding for Afghanistan came from the governments of countries with troops participating in ISAF (that is to say, parties to the conflict), raising concerns about the ability of humanitarian agencies to work independently and impartially. Between 2003 and 2012, for example, 78% of humanitarian funds were provided by governments who also contributed troops to ISAF, and of the top ten donors, only the European Union (EU) institutions and Japan were not also participating in ISAF (Poole 2014: 9). Moreover, this funding was often skewed to the provinces where international forces were present (Featherstone 2012: 15–16; Haysom and Jackson 2013: 10; Jackson and Giustozzi 2012: 3).

Beyond geographical targeting, funding was also directed at activities aimed at bolstering the Afghan government. In 2003, for example, a multi-million-dollar programme was established to fund non-governmental organizations (NGOs) selected by the Afghan Ministry of Public Health and international donors to provide basic health care—an important locus of legitimacy for the main donors, including the United States, the European Community (EC) and the World Bank—in rural areas (Crombé and Hofman 2011: 52). Under the rubric of counter-terror policies,[5] some funders also prohibited activities that risked bolstering the armed opposition in any way—and the prohibitions attached to United States Agency for International Development (USAID) funds in particular were deemed to limit the ability of recipient agencies to operate impartially (Featherstone 2012: 14).

Most agencies accepted funds from donor governments who also contributed troops, although some refused to utilize that money in the same provinces in which the respective donor's troops were present or had red lines as to particular donors and conditions (Featherstone 2012: 17–18; Haysom and Jackson 2013: 7). Others accepted funding directly from PRTs or highly conditional funding from the donor agencies of the PRTs (Featherstone 2012: 14–16; Haysom and Jackson 2013). However much their programming may have responded to needs, by working in areas determined by donor belligerents, operational agencies knowingly furthered the political and military interests of one side in the conflict, undermining their ability to advocate for a truly impartial and neutral humanitarian response (Haysom and Jackson 2013: 12).

International humanitarian response

Many humanitarian agencies adopted a development and reconstruction agenda after 2002, in line with the objectives of their donors, ISAF and the Afghan government—and sometimes working closely with the latter (Donini 2011: 147–148; Jackson and Giustozzi 2012: 3). Whether operational agencies aligned themselves this way because the funding on offer was too hard to resist or because they were optimistic about the prospects for peace and reconstruction, it led to them being viewed with suspicion by the armed opposition, limiting acceptance of their presence and their work in areas not controlled by the government, and hence restricting their access to those they sought to protect and assist. Thus, while the injection of billions of dollars of aid money contributed to modest development gains between 2001 and 2014, basic services were still lacking, inconsistent, or unsustainable in much of the country, and many Afghans remained acutely vulnerable to shocks (Poole 2014: 6–7).

[5] On the impact of counter-terror policies on humanitarian action more generally, see 'Counter-terror legislation' in Chapter 14 on Law and International Humanitarianism.

Integration and coordination

The UN pursued an integrated approach in Afghanistan, which subordinated humanitarian action to a partisan political agenda. The United Nations Assistance Mission in Afghanistan (UNAMA) was established by the Security Council on 28 March 2002 and brought together all UN political, assistance, and human rights functions under a single official in what was the most integrated UN mission to that point (Donini 2011: 148; UN Security Council 2002). As part of the integrated approach, the Office for the Coordination of Humanitarian Affairs (OCHA) closed its office in 2003, and humanitarian affairs were subsumed under the wider UN mission (Haysom and Jackson 2013: 7). Only in 2007 was a humanitarian coordination unit established within UNAMA, however, and even then humanitarian action was subordinated to the political goal of supporting the Kazai government (Donini 2011: 148; Jackson and Giustozzi 2012: 3).

The integrated approach placed aid agencies that wanted to avoid being seen to take sides in a difficult position. Immediately after the installation of the Karzai government, many believed that peace was within grasp, and several agencies were not too concerned about working with UNAMA or as part of an integrated approach that subordinated humanitarian action to reconstruction and development. However, as the intensity of conflict increased and it became apparent that the situation was not one of post-war reconstruction, especially after the 2009 troop 'surge', NGOs increasingly sought to distance themselves from donors with troops on the ground and from the UN (Donini 2011: 148; Haysom and Jackson 2013: 10; Jackson and Giustozzi 2012: 3). UNAMA was seen to be politicized by its close association with the Afghan government and ISAF, and hence risked compromising the independence and neutrality of associated operational humanitarian agencies.[6]

In response to requests by Afghan and international NGOs who believed that humanitarian coordination needed to be separate from UNAMA, OCHA returned to Afghanistan in 2009 (Haysom and Jackson 2013: 10; Jackson and Giustozzi 2012: 3). However, even when re-established, OCHA was unable to fulfil the strong and independent coordination role that many agencies had hoped for—and, indeed, that it is mandated to fulfil—notably in negotiating access and acceptance on behalf of humanitarian actors, both UN agencies and NGOs. In 2011, OCHA did propose a formal collective

[6] On the meaning and importance of the principle of independence, see 'Principles of humanity, impartiality, neutrality, and independence' in Chapter 13 on Politics, Principles, and Humanitarian Action.

access strategy to NGOs, but this was rejected as NGOs sought to distance themselves from the UN and were unwilling to be part of UN-led initiatives (Jackson and Giustozzi 2012: 8). Among UN humanitarian agencies as well as international and national NGOs, it was widely believed that OCHA lacked sufficient capacity and independence from the integrated mission to play a productive role in any collective negotiations (Donini 2011: 153; Jackson and Giustozzi 2012: 3).

Addressing violations of international humanitarian law

International humanitarian law (IHL) was violated by all parties to conflict in Afghanistan, including the foreign military forces, and international humanitarian agencies were slow to challenge these violations.[7] Civilians were targeted by both government and anti-government forces in reprisals against communities thought to be supporting the other side. The use of improvised explosive devices (IEDs) by anti-government forces and of airstrikes by government forces and ISAF posed major threats to civilians. Not all ISAF troops respected their obligations to distinguish themselves from the civilian population, instead dressing in civilian clothing and driving white vehicles, essentially disguising themselves as aid workers as a deliberate force protection strategy, which resulted in increased insecurity for aid workers as some civilian convoys were mistakenly attacked by insurgents (Haysom and Jackson 2013: 8; Terry 2011b, 175). However, most international humanitarian agencies on the ground initially failed to offer a robust challenge to these violations.

In the early years after the Bonn Agreement, the integration of the UN mission—and, more broadly, the philosophy of subordinating humanitarian action to the peace-building agenda of the coalition forces—meant that there was anyway little scope to persuade those forces to comply with IHL. The political side of UNAMA saw the UN humanitarian agencies as antagonistic because they sought to maintain a principled approach and to resist the politicization of humanitarian action, and this made a constructive dialogue on protection difficult or impossible (Donini 2011: 148). Above all, there was a widespread belief among important political and military actors that ISAF was legitimate, the opposition reprehensible, and that, in such a context, humanitarian neutrality was unacceptable and the Geneva Conventions did not fully apply (Terry 2011b: 182–183). Effectively, war against the

[7] See 'International humanitarian law' in Chapter 14 on Law and International Humanitarianism.

Taliban had no limits; in the Bush administration's 'war on terror', the ends were seen to justify the means. The distinction between *jus ad bellum* and *jus in bello*, according to which rules regarding the conduct of hostilities (IHL) apply regardless of the legitimacy or illegitimacy of the recourse to force on either side, was challenged not only in the *practice* of belligerents (which is relatively common) but also in the *rhetoric* of senior UN and US officials.

From 2008 onwards, the UN and international humanitarian agencies upped their efforts to persuade ISAF to comply with IHL, and the foreign military forces began to conduct hostilities with more concern for the civilian population. UNAMA and the Office of the High Commissioner for Human Rights (OHCHR) played an important role investigating and documenting harm to civilians and became more vocal on issues of civilian protection, likely contributing to the subsequent reduction in civilian fatalities attributed to ISAF—from 828 in 2008 down to 316 in 2012 (Donini 2011: 153; Haysom and Jackson 2013: 11). Following sustained lobbying and advocacy, in 2009, the ISAF Commander issued a tactical directive on air strikes, after which airstrikes dropped dramatically—from accounting for nearly two-thirds of civilian fatalities in 2008 to only 4% in 2012—and new directives on entry to, and the use of force in, medical facilities, after which health staff and structures saw far fewer incidents (Haysom and Jackson 2013: 11; Terry 2011b: 184).

While the dialogue and advocacy efforts of UNAMA/OHCHR and international humanitarian agencies appear to have played an important role in improving the conduct of ISAF and the Afghan government forces, other factors may have been equally important. On the one hand, evidence was collected systematically and used effectively to highlight the impact of conflict civilians. Importantly, 'UNAMA Human Rights/OHCHR cultivated relationships with key stakeholders, received significant amounts of high-level UN support, and developed a neutral position' (Haysom and Jackson 2013: 12). On the other hand, it is far from clear that these efforts would have yielded significant results in the absence of fundamental shifts in thinking on military strategy among top officials in major ISAF troop-contributing countries, including the United Kingdom and the United States. By late 2008, there was increasing acceptance of the idea that a negotiated settlement might be the best way forward and that the military strategy to that point had failed to curb support for the insurgency (Terry 2011b: 184). With increasing civilian casualties following the 'surge', aid agencies were able to leverage ISAF's own new counter-insurgency doctrine, with its rhetoric of protecting the population, to exert pressure on the military forces regarding civilian protection issues (Haysom and Jackson 2013: 11).

In general, international agencies focused their dialogue and advocacy efforts on ISAF, but the International Committee of the Red Cross (ICRC) was also engaged in protection dialogue with the armed opposition, trying to increase compliance with IHL. While the Taliban was receptive on paper, increasing the number of IHL provisions in its 2009 Code of Conduct as compared with the 2006 version, attacks that harmed and killed civilians continued on a regular basis, as did the targeting of medical staff and facilities (Terry 2011b: 185). In early 2010, the ICRC took the unusual step of publicly denouncing the planting of IEDs during Operation Moshtarak in Helmand Province (Terry 2011b: 185). By 2012, the head of the ICRC delegation in Afghanistan reported that the ICRC was able to raise its concerns about IHL violations—with the various parties to conflict—more directly and candidly than it was able to earlier in the conflict, and with better results (quoted in Benelli et al. 2012: 18).

Instrumentalization of aid by military actors

International military actors in Afghanistan emphasized the provision of aid as a means of 'winning hearts and minds' and as part of a broader strategy to 'stabilize' the country. Especially in UK and US military circles, service delivery was seen as fundamental to generating support among local populations, enhancing force protection through increased access to intelligence, and changing attitudes towards governments (Gordon 2010: S369). ISAF troops sought to build support for themselves and for the Afghan government by providing material assistance, delivering services, and working to build or rebuild infrastructure themselves, though the ways and the extent to which they did so varied significantly. For example, the United States in eastern Afghanistan emphasized service delivery through PRTs, with military planes dropping both bombs and food packages, while the United Kingdom in Helmand was mostly not providing material assistance directly, emphasizing instead governance programmes and medical aid (Gordon 2010: S736–S737; Lischer 2007: 99, 104). Like those led by the United States, Australian PRTs implemented stabilization projects focused on visible short-term service delivery and aimed at securing support for their troops (Haysom and Jackson 2013: 6). By contrast, Dutch PRTs focused on addressing local grievances and conflicts and only providing aid discreetly, fearing that if it were traced back to international forces, it would undermine the legitimacy of the local actors whose power and popularity they sought to bolster (Fishstein and Wilder 2012: 24; Haysom and Jackson 2013: 5–6).

The services delivered by military forces were not always good quality and ran into various problems as a direct consequence of being delivered by a party to conflict with other priorities. For example, NATO forces built too many schools, without sufficient teachers or funding to sustain them in the long run (Hopkins 2012). Furthermore, they were often left unused or underused because Afghans feared they would be targeted by anti-government forces if they made use of facilities provided by ISAF (Jackson and Haysom 2013: 20). Similar criticisms were levelled at the UK military's medical programmes (Gordon 2010: S379).

Whether such assistance should be labelled humanitarian is debatable since its end goal was not the well-being of the population but stabilization. The new military strategy that accompanied the 2009 'surge' focused on counter-insurgency, and intelligence gathering was an explicit and publicly stated goal of ISAF, with elders asked to provide assistance in tracking down insurgents in return for the 'generosity' of the army and coalition partners providing relief to the Afghan people (Donini 2011: 150; Haysom and Jackson 2013: 9). International military forces dropped pamphlets over southern Afghanistan telling residents that if they wished to continue receiving aid, they were to provide information on the Taliban and al-Qaeda (Terry 2011b: 175). Likewise, the UK military saw the provision of basic health care as a means to gain influence and collect intelligence (Gordon 2010: S378). In many cases, the civilian population paid a high price for these strategies—insurgents attacked villages that accepted this kind of aid, in retaliation for supposed collaboration with ISAF, while NATO forces bombed or raided villages thought to be harbouring insurgents on the basis of intelligence gained in exchange for aid (Terry 2011b: 175).

In addition to providing assistance directly, the international forces in Afghanistan sought to forge alliances with international humanitarian agencies, including through the PRTs. As early as October 2001, and much to the horror of many aid agencies, US Secretary of State Colin Powell described NGOs as a force multiplier and an important part of the US combat team (Powell 2001). Then, in April 2009, the US special envoy for Afghanistan, Richard Holbrooke, claimed that '90% of US knowledge about Afghanistan lies with aid groups' (quoted in Crombé and Hofman 2011: 60). Following the 'surge', there was increased pressure on aid agencies to support the development and government dimensions of the military strategy (Haysom and Jackson 2013: 10). For example, donors sometimes offered funds—albeit not necessarily earmarked as humanitarian—for NGOs to operate alongside PRT officials in the aftermath of battle or made funding conditional on the exchange of information with the army (Featherstone 2012: 14).

Different humanitarian agencies reacted in different ways to efforts by ISAF to integrate their work into the stabilization agenda, and many changed the way they interacted with the Afghan government and the international forces over time. Donor pressure and principles pulled agencies in opposing directions. Some agencies opted to prioritize presence and funding over principles, probably in part due to an arguably naïve expectation after the overthrow of the Taliban that the conflict was over and that post-conflict reconstruction would be relatively straightforward (Donini 2011: 147; Haysom and Jackson 2013: 11). On the other hand, some were reluctant to participate in stabilization operations from the outset and sought to resist donor pressure to do so, avoiding direct engagement with ISAF and the PRTs in particular (Donini 2011: 151; Jackson and Giustozzi 2012: 3). After the 'surge', many agencies resisted pressure from ISAF to participate in what they saw as a battlefield clean-up operation, and increasingly avoided interaction with the military, but some were prepared to work alongside PRT officials and even to provide information to the military as a condition of funding (Featherstone 2012: 14; Haysom and Jackson 2013: 10). Moreover, private contractors and for-profit 'quasi-NGOs' were 'much more ready, willing, and able' (Donini 2011: 151).

There is not much to suggest that the 'hearts-and-minds' strategy was effective in stabilizing Afghanistan. In the short term, limited benefits may have accrued to the international military forces and their counter-insurgency operations. For example, there is some evidence that—at least, in parts of the country—communities who received aid projects were more willing to report IEDs, provide other useful information, or interact with ISAF (Fishstein and Wilder 2012: 54). In the longer term, however, other factors were more important. The continued corruption of the Afghan government, for example, created the perception that despite the long-term presence of foreign troops, nothing changed. Worse, international aid was seen to be fuelling government corruption and benefitting those in power at the expense of the wider population (Fishstein and Wilder 2012: 61–62). Furthermore, ISAF's combat mission undermined the 'hearts-and-minds' activities and, indeed, the stabilization efforts more broadly (Haysom and Jackson 2013: 6; Suhrke 2008: 230). The fact that the same militaries that were inflicting civilian casualties were also handing out aid generated grievance and resentment among the population, which the Taliban was able to exploit.

Even if the hearts-and-minds strategy largely failed to achieve its objectives, it did have consequences for the security of aid workers and agencies. The roles taken by international military forces in Afghanistan, and the relationships between those forces and international civilian actors, are seen to have had significant implications in terms of aid worker security and aid agency access to populations in need of their assistance. Attributing direct causality may be impossible, but it is clear that Afghanistan became more dangerous for aid workers and agencies in the years following the US-led invasion, and that humanitarian access suffered as a result.

Insecurity and aid agency access

According to data in the Aid Worker Security Database, in every year from 2002 to 2014, Afghanistan was one of the top three countries worldwide in terms of numbers of major security incidents involving aid workers and agencies. From 2001 to 2013, there was a fairly steady increase in the number of major security incidents involving aid workers in Afghanistan (see Figure 6.2). However, some of the year-on-year increases may reflect better reporting of incidents and an increase in the number of aid workers in Afghanistan rather than an increase in the level of danger they faced.

While we lack good data on the *rate* of security incidents, an upward trend in the absolute numbers of incidents combined with some high-profile attacks to engender a sense of insecurity, limiting operational access for many humanitarian agencies in large parts of Afghanistan. In March 2003, an ICRC water engineer was deliberately killed by the Taliban as he travelled from Kandahar to Tirin Kut in southern Afghanistan. In June of the following year, five Médecins Sans Frontières (MSF) staff were ambushed and killed in northern Afghanistan. By the end of 2007, nearly half of the districts in Afghanistan

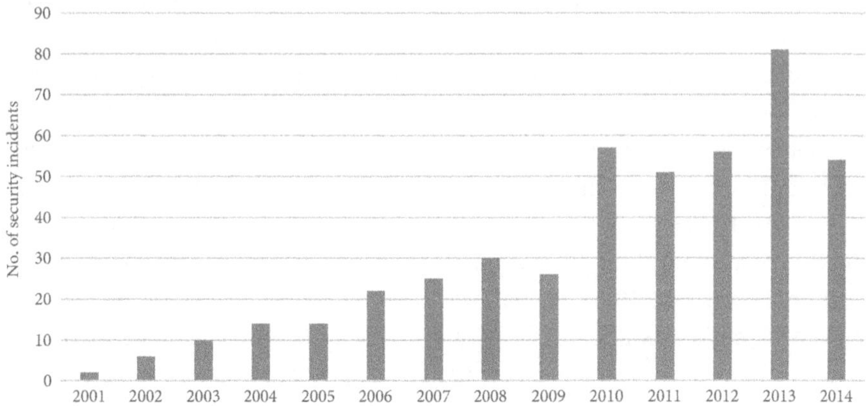

Figure 6.2 Major security incidents involving aid workers in Afghanistan, 2001–2014
Source: data taken from the Aid Worker Security Database.

were out of bounds for UN staff because they were considered too dangerous (Haysom and Jackson 2013: 9). As troop presence increased with the 'surge', insecurity increased and spread to previously stable provinces, making it nearly impossible for many aid agencies to access large parts of the country (Haysom and Jackson 2013: 10).

Different explanations are proposed to explain attacks on aid workers and aid agencies in Afghanistan. They are not mutually exclusive, and the evidence for each is mixed, suggesting that some combination of these factors was at play. Part of the explanation surely lies in general insecurity, and not all attacks were intentional or targeted (see, e.g. Jackson and Giustozzi 2012: 11). At the same time, agencies could be deliberately targeted because they were seen to be taking sides, working with ISAF, or in some other way breaking the rules or conditions set by the Taliban or other armed groups (Jackson and Giustozzi 2012: 15, 16). Many members of the armed opposition could not easily distinguish between different humanitarian agencies or between humanitarian and political actors such that even those agencies that refused to work with ISAF may have been tarred with the same brush as those that did (Donini 2011: 154; Featherstone 2012: 16–17; Jackson and Giustozzi 2012: 4). Confusion about the role and goals of humanitarian agencies was exacerbated by the involvement of international military forces in providing assistance (Haysom and Jackson 2013: 7). Beyond any such confusion, international humanitarian agencies were often seen by opposition forces to represent a fundamentally Western, liberal project that ran counter to their own principles and interests (Jackson and Giustozzi 2012: 18; Terry 2011b: 176).

Aid agency responses to these challenges

Despite some efforts to achieve access through so-called acceptance strategies, according to which aid agencies seek to negotiate consent for their operations from local communities, authorities, and armed actors, operational security strategies in Afghanistan have frequently focused on remote management and hard security measures. Many agencies sought to distance themselves from international military forces, which can be seen as a passive acceptance strategy, but given the high insecurity and low tolerance of humanitarian agencies by the armed opposition, working outside of areas controlled by the government required more proactive approaches.

Some agencies—notably the ICRC and MSF—sought to engage in more active acceptance strategies centred on maintaining dialogue with the armed opposition and demonstrating the value of neutral humanitarian actors to both sides (Donini 2011: 156; Jackson and Giustozzi 2012: 6; Terry 2011b: 181–184). Other agencies either did not engage in direct, structured negotiations with the Taliban and other armed groups or opted not to speak publicly about such engagement. In part, the lack of dialogue stemmed from initial optimism that the conflict was over and there was no longer any need to maintain dialogue with the Taliban (Donini 2011: 147). Some agencies, particularly those accepting funding from USAID, avoided engagement for fear of the repercussions under counter-terror legislation and donor policies (Featherstone 2012: 14). Responsibility and risk were often transferred to junior Afghan employees, and some international staff apparently expected that national staff would engage with local-level Taliban to gain and maintain acceptance without official support or authorization from their respective agencies, whose senior managers preferred not to know (Jackson and Giustozzi 2012: 5). OCHA, whose mandate includes engaging with all belligerents to negotiate access on behalf of the humanitarian community as a whole, was first absent due to the structure of the integrated mission and then failed to fulfil this part of its role (Donini 2011: 152, 156).

Engagement with the Taliban and other opposition groups may have been a necessary condition for operational access and aid agency security in much of Afghanistan, but it was not sufficient. The ICRC and MSF both appear to have been more committed to dialogue than those other international aid agencies, and they frequently achieved greater access than other organizations. By 2010, for example, both were operating health-care programmes in Helmand, a notoriously insecure province (Gordon 2010: S377). The World Health Organization (WHO) had to rely on ICRC contacts for its vaccination drives, demonstrating the relatively higher level of trust developed by the ICRC with the anti-government forces (Donini 2011: 152; Terry 2011b: 183). However, the ICRC and MSF are also the two most purely 'humanitarian' of the major international aid agencies, as the others have mandates not only for emergency response but also for more development-focused or human rights work.[8] It may be that their ability to secure access depended as much on their more limited mandates as on their approach to dialogue. Moreover, neither organization ever achieved anything like complete acceptance.

[8] See 'Perspectives of the largest international humanitarian agencies' in Chapter 13 on Politics, Principles, and Humanitarian Action.

Those international humanitarian agencies in Afghanistan that sought to engage with anti-government forces faced a number of obstacles to productive dialogue. There were multiple armed opposition groups and so acceptance and access required engagement with—and the consent of—more than just the Taliban (Crombé and Hofman 2011: 61, 65; Jackson and Giustozzi 2012: 13–14, 22). Highly decentralized, there was sometimes tension and competition between different networks within the Taliban, often along tribal lines, and attitudes towards international humanitarian agencies varied across the different networks (Jackson and Giustozzi 2012: 2, 22). Aid agencies found it difficult even to establish contact with the midranking and high-level commanders who may have been able to secure access, particularly as ISAF's strategy of targeting Taliban leaders had eliminated many such commanders (Jackson and Giustozzi 2012: 6; Terry 2011b: 179). Where they were able to negotiate security guarantees, internal divisions within the Taliban meant that such guarantees were not necessarily reliable (Jackson and Giustozzi 2012: 9, 10, 16, 18, 29; Terry 2011b: 178–179). Agencies also risked repercussions from other actors if they engaged with the Taliban. In late 2007, for example, the Afghan government expelled two Western diplomats for allegedly engaging in political talks with the Taliban, and although none of the major Taliban groups were listed by the United States as terrorist organizations until September 2012, some agencies were concerned about the impact of counter-terror measures (Jackson and Giustozzi 2012: 6).[9]

Attacks on aid workers and operations pushed many operational agencies to adopt remote management and hard security measures, which allowed them to reduce some kinds of risk but negatively affected the quality and reach of their programming—and generated other kinds of risk. International staff, especially in UN agencies, were often bunkerized in fortified compounds, restricted to travelling in armoured vehicles, and even then, much of the country was off limits to them. Programmes were managed from a distance, with only the bare minimum of staff—usually predominantly local staff—remaining in the area of operations, resulting in a transfer of responsibility and risk to those staff who remained operational. Senior staff were unable to visit programme activities, which was detrimental to aid agency relationships with the communities they sought to assist and limited the scope for monitoring (Donini 2011: 154; Terry 2011b: 180–181). Moreover, the use of hard security measures meant that there was little to distinguish UN compounds in major Afghan cities from those of ISAF or private security

[9] For a wider discussion of the impacts of counter-terror measures on humanitarian response, see 'Counter-terror legislation' in Chapter 14 on Law and International Humanitarianism.

companies, reinforcing perceptions that the UN was part of the same enterprise as the foreign military forces (Donini 2011: 154). These strategies may have reduced aid agency vulnerability to any given attack but may also have increased the likelihood of attacks against them, as their would-be attackers came to see them as (more of) a legitimate target.

Conclusions

The depiction of the Afghan context as a post-conflict situation from 2002 was a political move by the Afghan government and its international backers that did not reflect the reality on the ground and had problematic implications for the humanitarian response. Initially optimistic, many aid agencies worked more closely with political and military actors than they usually would in armed conflict. After 2002, funding for humanitarian response decreased, while funding for development and stabilization—which often came with conditions attached—increased, further incentivizing some humanitarian actors to compromise their commitments to the principles of impartiality, neutrality, and independence. While many agencies later sought to distance themselves from the international military forces and the Afghan government, others complied with demands from donor governments and troop-contributing countries, and where they resisted, for-profit actors were ready to fill the gaps. The instrumentalization of aid for stabilization purposes does not seem to have been very effective in Afghanistan but may have further undermined already weak relationships between aid agencies and the armed opposition, exacerbating aid worker insecurity and access difficulties and limiting protection dialogue with the Taliban and other insurgent groups.

7

Indian Ocean Tsunami, 2004

The 2004 Indian Ocean tsunami was one of the deadliest disasters in living memory, and the response from the global public was unprecedented. For the first time, a humanitarian emergency was not underfunded but overfunded, with many humanitarian agencies receiving more donations than they knew what to do with. Established agencies rushed to respond, as did many inexperienced individuals and organizations, private companies, and international financial institutions. Despite all the interest and goodwill, much of the international assistance was inappropriate or inadequate, the relationships between different actors were often competitive rather than cooperative or coordinated, and in some cases, aid may have served to reinforce pre-existing inequalities. Overall, there is significant consensus that 'the response did not achieve the potential offered by the generous funding' (Cosgrave 2007: 3).

While the level of funding makes the tsunami case unique, it provides useful insights into the functioning of international humanitarianism more generally precisely because it attracted unprecedented donations. First, as an 'extreme case' in terms of funding, it helps to shed light on what drives humanitarian funding decisions. Second, because the budgetary constraints normally associated with humanitarian action did not exist in this case, any faults in the tsunami response cannot be explained away as a consequence of insufficient funding. This helps to expose other fundamental issues in the international humanitarian system. Third, with so much money available, aid agencies were able to try out new approaches and—both individually and collectively—to invest significant resources in learning and evaluation, which facilitates drawing lessons to inform future responses.

Both the needs arising from the tsunami and the capacities of different actors to meet those needs were misrepresented by international media and humanitarian agencies. The capacities of local actors were underestimated, while the capacities of international aid agencies were overestimated. International humanitarian agencies anticipated greater need for emergency response—that is, their core expertise—than there was. At the same time, they also sought to work on longer-term recovery and rehabilitation, tasks which are beyond their core expertise and experience, raising questions about

The Politics and Everyday Practice of International Humanitarianism. Miriam Bradley, Oxford University Press.
 DOI: 10.1093/oso/9780198840695.003.0007

whether humanitarian actors are best placed for this kind of work and about where the boundaries of 'humanitarian' action should lie.

The tsunami and its immediate aftermath

On 26 December 2004, a massive 9.1 magnitude earthquake off the Sumatra coast led to a series of tsunamis[1] in the Indian Ocean. The tsunami claimed lives across fourteen countries, of which India, Indonesia, Sri Lanka, and Thailand were the hardest hit. It is estimated that around 230,000 people were killed, of whom 170,000 died in Indonesia and 30,000 in Sri Lanka, as well as considerable numbers in Thailand and India and smaller numbers in Myanmar and the Maldives (Brauman 2009: 108). Economic losses and damages were (very roughly) estimated at USD 10 billion, nearly half of which was suffered by Indonesia (Cosgrave 2007: 7). The scale of destruction and loss of life placed the 2004 tsunami among the very worst 'natural' disasters of the past century (Cosgrave 2007: 8).

Not everyone was affected equally. Different communities and individuals had different levels of vulnerability, and vulnerability is the product of—among other things—economic, political, and social factors.[2] For example, some groups in the Nicobar and Andaman islands had traditional knowledge of tsunami risk and moved inland before the tsunami struck, whereas a lack of tsunami knowledge in Sri Lanka meant that some people went to collect fish left stranded on the shore when the tide withdrew after the first waves and were then drowned in subsequent waves (Cosgrave 2007: 13). Across the tsunami-affected areas, as a consequence of different levels of strength, knowledge of swimming, locations, childcare duties, clothing, and other factors, death rates were higher for women than for men, and for those under fifteen and those over fifty (Cosgrave 2007: 6). In Sri Lanka, already marginalized Muslim and Tamil populations along the coastal belt were disproportionately affected, relative to the Sinhalese majority (Silva 2009: 65). The Muslim communities in Ampara District, the hardest-hit region in Sri Lanka, most of whom were involved in the fishing industry and lived in congested housing settlements along the coast, suffered extremely high rates of mortality, morbidity, and property damage relative to the Tamil and Sinhalese communities (de Silva 2009: 254).

[1] Although it is accurate to refer to the Indian Ocean tsunamis in plural, in this book, I adopt the more commonly used singular, writing of the series of tsunamis collectively as 'the tsunami'.

[2] See 'Crises in the contemporary world' in Chapter 1 on Humanitarian Emergencies.

The tsunami struck mainly middle-income countries, many of which had significant domestic capacity for emergency response and longer-term recovery and rehabilitation. The two worst-affected countries—Sri Lanka and Indonesia—were additionally characterized by decades-long armed conflicts when the tsunami struck.[3] In both countries, the tsunami disproportionately affected conflict areas—the northern and eastern parts of Sri Lanka, where the civil war between the Liberation Tigers of Tamil Eelam (LTTE, or Tamil Tigers) and the government of Sri Lanka was concentrated, and the province of Aceh on the Indonesian island of Sumatra, where the Free Aceh Movement (Gerakan Aceh Merdeka, or GAM) was fighting the Indonesian government.

Across the tsunami-affected region, the first response was from local communities, and national and regional actors were then also quickly mobilized (Cosgrave 2007: 10). Because rapid-onset disasters tend to impact limited geographic areas, there is often national capacity to respond from relatively unaffected areas of the same country. In Sri Lanka, for example, the tsunami swept over a strip of coastline which varied from 100 to 300 metres wide. The rest of the country was functioning normally, and Sri Lankan doctors and nurses, together with members of civil society, got to the tsunami-affected areas almost immediately, bringing food, shelter, and medical equipment and able to cope with the numbers of patients arriving at hospitals without international assistance (Brauman 2009: 111; Stirrat 2006: 12). The Indian army mobilized more than 8,300 troops for rescue and relief operations in both India itself and neighbouring Sri Lanka, with the first troops deployed within six hours of the tsunami (Price and Bhatt 2009: 20). In Aceh, the response began local and then went national, with thousands of Indonesian citizens contributing, although local government and military capacity was severely reduced by the tsunami, and damage to transport and communications infrastructure slowed the arrival of outside help (Cosgrave 2007: 10; Fan 2013: 5; IFRC 2005: 86).

The national governments of the different tsunami-affected countries responded to offers of international assistance in quite different ways—in part according to differences in needs and capacity. India and Thailand largely rejected international assistance for relief operations, while Indonesia and Sri Lanka both gave international agencies relatively free access, resulting in a so-called 'second tsunami' of aid agencies, with more than 200 arriving in Aceh and around 500 in Sri Lanka (Cosgrave 2007: 16; IFRC 2005: 81; Silva 2009: 67). The rebel groups in Indonesia and Sri Lanka adopted very different attitudes to international aid. In Aceh, GAM—a group also largely dependent on,

[3] On the conflict in Sri Lanka, see 'Background' in Chapter 8 on Sri Lanka, 2008–2009.

and responsive to, the local population—welcomed international assistance, while in Sri Lanka, the LTTE—a group whose funding derived principally from international sources and which frequently instrumentalized the local population—denied international non-governmental organizations (NGOs) access to the North and East provinces, insisting on controlling all aid itself, even if NGOs refused to provide resources under such conditions (Beardsley and McQuinn 2009: 638–640).

Given the significant domestic capacity in the affected countries and the nature of the tsunami, there was more need for international assistance with recovery and reconstruction than with emergency relief. In contrast to armed conflicts, sudden-onset disasters like earthquakes and tsunamis generally lead to more deaths than injuries and do not usually lead to outbreaks of infectious diseases (Brauman 2009: 210; Guha-Sapir and van Panhuis 2009). The 2004 tsunami was no exception, although the ratio of dead to injured survivors varied from place to place, 6:1 in Aceh, 1.5:1 in Sri Lanka and 0.3:1 in India (Cosgrave 2007: 6). Even in Aceh and Sri Lanka, where international agencies arrived in large numbers, they mostly arrived too late to participate in rescue missions, and the immediate medical and subsistence needs of the survivors were met by domestic actors (IFRC 2005: 86; Silva 2009: 66).

Recovery and reconstruction needs were significant, with hundreds of thousands of people having lost their homes, livelihoods, or both. Industries based at, or close to, the coast, especially fishing and tourism but also farming, were the worst affected, and in Aceh, ports and harbours were also destroyed (Cosgrave 2007:, 8). Of a pre-tsunami population of 4.25 million in Aceh, almost 500,000 survivors lost their homes and around 750,000 lost their livelihoods (Fan 2013: 5). In Sri Lanka, 98,000 houses were damaged or destroyed and 150,000 livelihoods affected, with almost 17,000 fishing boats (75% of the total fishing fleet) damaged (Government of Sri Lanka data, cited in Khasalamwa 2009: 76).

Media coverage

The media played an important role in shaping how Western audiences understood and responded to the tsunami through a framing which, in several important respects, obscured realities on the ground.[4]

[4] On the ways in which media portrayals contribute to the construction of humanitarian emergencies, see 'Framing the agenda' in Chapter 17 on Media and Celebrities.

Disproportionate attention was—somewhat unsurprisingly—paid to Western victims and survivors of the tsunami. While the island of Sumatra in Indonesia was the hardest hit, initial coverage of the tsunami in the global North focused on Thailand and Sri Lanka, with video footage by Western tourists caught up in the disaster during their Christmas holidays (Brauman 2009: 108). Moreover, while less than 1% of those who died were tourists, much of the media coverage in donor countries—40%, according to a study of print coverage on the effects of the tsunami in eight Western countries—focused on them (Cosgrave 2007: 6; Franks 2006b: 283).

Western media coverage of non-Western survivors of the tsunami emphasized their victimhood. Photographs on the front pages of the *New York Times* depicted non-Western survivors of the tsunami as helpless, passive victims enduring appalling conditions and suffering visible physical or emotional hardship—and, to the extent that Westerners appeared in the images, they were in the role of capable, brave, rescuers (Hutchison 2014). The choice of images that depict Westerners and non-Westerners and the relationships between them in this way played to—and reinforced—preconceived notions in the West and obscured the importance and capacities of the individuals, civil society, and governments in affected countries. While Indonesian coverage emphasized the courage and solidarity of the Acehnese, of sixty-seven BBC reports on Aceh in the two months after the tsunami, not one report focused on self-reliance and only a handful even mentioned it (IFRC 2005: 87).

Had Western media instead sought to challenge those preconceived notions with, for example, images that reflected the fact that the first responders were overwhelmingly local or national, the coverage would almost certainly have packed less emotional punch. The images favoured by the *New York Times* (and other Western media) employed colonial tropes which 'feed into established, historically constituted Western emotional regimes' to generate a 'politics of pity', which clearly distinguishes between those who suffer and those who do not, and in which the suffering is perceived as an inconceivable event that has happened to the unfortunate (Hutchison 2014: 3, 8). In depicting the non-Western survivors of the tsunami as passive victims, dependent on Western assistance for their survival, the media coverage thus elicited pity in the viewers—pity which helped both to constitute hierarchical relations between viewers and victims and to generate unprecedented levels of financial donations.

Funding for humanitarian response

The tsunami response was extraordinarily generously funded, breaking many records. An estimated USD 13.5 billion came from international sources, and at least USD 190 million in private donations and USD 2.7 billion in public funding from affected countries themselves (Cosgrave 2007: 19). Total funding for the tsunami response translated to a conservative estimate of USD 7,100 per affected person, compared with just USD 3 per head pledged to assist the 36 million people affected by the Bangladesh floods earlier in the same year (Telford and Cosgrave 2007: 7). A few months after the tsunami, an earthquake struck Kashmir, killing 75,000 people and seriously injuring tens of thousands more in Pakistan without attracting any great outpouring of public sympathy (Brauman 2009: 109). The tsunami was, in this respect, quite unique.

Private donors were especially generous. Official funding normally dwarfs private giving, but in response to the tsunami, international funding from private sources almost equalled official government pledges (Cosgrave 2007: 19). Linked to this, international non-governmental organizations (INGOs) and the Red Cross & Red Crescent (RC) Movement received a higher proportion of the funding than in most emergencies (Cosgrave 2007: 18). A range of factors likely contributed to the outpouring of empathy and compassion around the world which, in turn, contributed to the enormous magnitude of financial donations—from both public and private sources.

Surveys of the public in Germany and Spain identified a number of motivations behind the generosity of private donors (see Cosgrave 2007: 20). Respondents mentioned the scale of death and destruction and the fact it was a 'natural' disaster and those affected were 'innocent'. This is consistent with what we know about private donors' general preference for funding disasters linked to natural phenomena over armed conflicts.[5] Social proximity to Westerners was also cited as a factor motivating giving. Many individuals in the main donor countries had visited one or more of the affected countries, and some reported a sense that 'it could have been me'. Although Indonesia, Sri Lanka, India, and Thailand suffered by far the greatest loss of life, Germany and Sweden were the fifth and sixth worst affected nationalities, albeit with less than one in four hundred of total fatalities each (Cosgrave 2007, 7). The time of year was also cited as important, with charitable giving in many of the main donor countries at its peak in the December holiday season, and some survey respondents also mentioned a feeling of peer pressure

[5] See 'Donors' in Chapter 18 on Donors and the Funding of Humanitarian Action.

as everybody was giving. All of this was underpinned by the high level of media attention which, as we have seen, emphasized both the social proximity of Western audiences to some of the victims and the innocence of the 'helpless' non-Western survivors. Thus, both the scale and the nature of media coverage played a role, and funding decisions by private donors were based on an emotional reaction rather than a reasoned assessment of how much was needed and where.

Arguably, it is precisely because mass solidarity is not based on rational reasoning alone that institutional donors and operational humanitarian aid agencies should provide a more level-headed perspective, but for the most part they did not (Brauman 2009: 109). Donor government funding was largely driven by media coverage and political or institutional factors rather than needs assessments, with the scale of public concern leading to what the European Union (EU) aid Commissioner termed a 'donor beauty contest' (Cosgrave 2007: 20). While individual agencies carried out needs assessments to inform their own programming, an atmosphere of competition ensured that these were mainly not shared, and the lack of comprehensive and authoritative assessments meant that the mass media—and not humanitarian professionals—was the main influential source of information, not only for individual but also for institutional donors (de Ville de Goyet and Morinière 2006: 11). For most operational agencies, 'mobilizing goodwill and transforming emotion into donations became ends in themselves, and the question of how the funds raised would be used was regarded in practice as meaningless' (Brauman 2009: 114). Either the agencies themselves were carried away by the outpouring of compassion, or they put institutional interests above their social missions. They could have stopped accepting donations and/or requested permission from donors to reallocate some of the funds to other crises, but for the most part, they did not (Telford and Cosgrave 2007: 8). Just a week after the tsunami, Médecins Sans Frontières (MSF) did, in fact, announce that they would not be accepting any more donations for this emergency, but their decision was deemed controversial and met with protests from other big international humanitarian agencies, most of whom continued to accept and to solicit donations quite actively until the end of January or later (Brauman 2009: 110; Cosgrave 2007: 18).

Compared with most humanitarian emergencies, donations were faster and more flexible, but they were not based on formal assessments of needs and there were some important imbalances. Funds were provided unusually quickly (often official donors take much longer to actually disburse the money they have pledged) and, largely due to the high proportion of private donors, the operational agencies receiving the funding had significant flexibility as to what they could spend it on (Cosgrave 2007: 19; Telford and Cosgrave 2007: 4). Although the response in the Maldives was severely underfunded, the worst affected countries did, for the most part, receive the largest flow of funds (Telford and Cosgrave 2007: 5). However, nearly half the total official pledges were for emergency relief, despite it being clear that reconstruction would be the biggest need (Cosgrave 2007: 20–21). Although some governments—notably the United States, Spain, Germany, and Sweden—strongly favoured reconstruction, others—notably Ireland, Canada, Japan, and the Netherlands—funded mainly relief (Telford and Cosgrave 2007: 5).

International humanitarian response

The huge amounts of funding were matched by huge numbers of organizations and relief workers arriving in the affected countries. The need for such a large-scale international response was not self-evident, and it instead reflected—or was justified by—incorrect preconceptions among international aid agencies and media about humanitarian needs and response (Brauman 2009: 112). Moreover, international agencies lacked the capacity to use the vast funds they received effectively, overestimating what they could achieve with the money, underestimating how long it would take to achieve those things, and without enough experienced staff globally to make good use of such large quantities of money (Cosgrave 2007: 14, 15).

Emergency response

The 'emergency' element of the international response was, in many ways, inappropriate and poor quality, characterized by misplaced priorities and insufficient coordination. In some respects, the problems with the emergency response were mitigated by its scale—with enough funding, the response can be effective even while it is inefficient (de Ville de Goyet and Morinière 2006: 13). In other respects, however, the scale of the funding and the resulting operational response exacerbated problems.

Just as donors did not base their funding decisions on formal assessments of needs, operational agencies acted only selectively on their many needs assessments.[6] International agencies relied heavily on data collected by local authorities in their assessments, but failed to invest in national capacity to

[6] For a discussion of this issue in more general terms, see 'Allocating resources, choosing response modalities, and designing programmes' in Chapter 21 on Needs Assessment, Evaluation, and Response Decisions.

improve the quality of that data in order to gain a consolidated picture of overall needs (de Ville de Goyet and Morinière 2006: 13). The resulting assessments were piecemeal and uncoordinated, and, in many cases, agencies went ahead with responses based either on preconceived notions of what was needed or on what they could offer, given their expertise and experience, ignoring lessons from previous 'natural disasters' which suggested that emergency relief efforts should be limited (Brauman 2009: 115; de Ville de Goyet and Morinière 2006: 13).

A focus on emergency relief and on epidemics reflected what international agencies were most able to offer but was out of step with what communities directly affected by the tsunami were in most need of.[7] These issues were not unique to the tsunami response but were exacerbated by the scale of public support, media interest, and financial donations. The high levels of funding and media coverage created organizational incentives for agencies to be present and active, irrespective of how useful their work was in practice. For the sake of their brands (and hence their capacity to work in future emergencies), it was important for them to be visibly operational in response to such a high-profile disaster. Indeed, findings of needs assessment exercises were ignored when they would have discouraged such self-serving forms of assistance (de Ville de Goyet and Morinière 2006: 13).

Even if emergency relief was not the priority need for most affected populations, it did give recipients the security that they needed to begin planning what to do next. Perhaps the most positive dimension of the response was the provision of cash transfers at scale.[8] National governments and international humanitarian agencies provided different tsunami-affected populations with cash or vouchers of different values and with different goals, in some cases unconditional and elsewhere with conditions attached (Adams 2007; Cosgrave 2007: 13). This allowed communities and individuals greater choice than does the provision of in-kind aid, enabling them to buy what they needed and wanted rather than what aid agencies thought they needed and wanted—or what donors and operational agencies had available or considered media-friendly. The flexibility and will to adopt cash programming at scale were undoubtedly linked to the large amounts of funding available and, while the scale of the cash response to tsunami-affected populations may look timid with hindsight, at the time it was ground-breaking, and it provided evidence which paved the way for cash-based responses to future emergencies (see, e.g. Adams 2007).

[7] See 'Assessing needs, contexts, and capacities' in Chapter 21 on Needs Assessment, Evaluation, and Response Decisions.

[8] For a wider discussion of the use of cash transfers in humanitarian response, see 'Cash and voucher assistance' in Chapter 22 on Material Assistance and Direct Service Provision.

Other aspects of emergency response were less successful, with widespread reports of unsolicited and inappropriate in-kind donations, significant weaknesses in coordination, and a tendency to sideline local capacity. Local ownership of the tsunami response declined over time, and international agencies sometimes pushed local and national actors aside, setting up parallel mechanisms and eroding local capacity (Cosgrave 2007: 11–12; Silva 2009: 67–68).[9] A huge number of international actors, including completely inexperienced organizations (and individuals), were operating, and this exacerbated endemic coordination problems in the humanitarian sector. Coordination problems in turn led to gaps and overlaps in the response (Cosgrave 2007: 17). Furthermore, as a result of the extensive media coverage and global public interest, some agencies received huge quantities of unsolicited in-kind donations, including inappropriate items, such as tinned pork provided for Muslims in Aceh, and unwanted items such as large quantities of used and sometimes unsuitable clothing (Cosgrave 2007: 17; Fritz Institute 2005: 3–4).

Recovery and rehabilitation

There was a rapid transition from relief to recovery and rehabilitation, which was widely seen as the right thing in principle but as not working so well in practice, raising questions about whether humanitarian agencies are the right actors beyond emergency response. The generous and flexible funding for the tsunami response facilitated efforts to 'build back better' (BBB), and enabled initial recovery activities to start almost immediately, but it also limited those efforts as agencies were under pressure from donors to spend money quickly, with no time for deeper analysis of the political, social, and economic contexts and relations (Fan 2013: 9; Stirrat 2006: 13; Telford and Cosgrave 2007: 15). Even without such pressures, many of the international agencies involved lacked the expertise to handle complex development processes and treated recovery and rehabilitation as technocratic questions of replacing lost assets (particularly, homes and fishing boats) and rebuilding infrastructure, rather than about facilitating societal and political change. Such projects—with their emphasis on large, physical items—were appealing to aid agencies because they provided visibility and demonstrable results (Silva 2009: 68; Stirrat 2006: 13). However, assets were often replaced without due consideration for market conditions, let alone efforts to shape those conditions, and thus sometimes served to reinforce rather than redress pre-existing inequalities.

[9] The issue of eroding local capacity was not unique to the tsunami response and is discussed in more detail in Chapter 19 on Government and Civil Society in Affected States.

The replacement assets were not always targeted to those most in need. In Sri Lanka, for example, there were more new houses than tsunami-affected households in some Sinhalese communities, while hundreds of families from the already marginalized Muslim community in Ampara District had no prospect of a new home more than five years after the tsunami (Khasalamwa 2009: 80; Mulligan and Nadarajah 2011: 365). Likewise, with many agencies involved in boat distribution, more were provided than had been destroyed, with reports of indiscriminate distribution whereby some who had lost their boats did not receive any replacement while other households managed to acquire several and non-fisher folk entrepreneurs also received boats (de Silva 2009: 268; Khasalamwa 2009: 83; Stirrat 2006: 14).

In some cases, the assets provided were inappropriate to local conditions. For example, international aid agencies rushed to give affected people equipment to help them start or re-start self-employment or small businesses, without due consideration of the economic or environmental context. Little thought (or funding) was given to how to develop the fishing sector, with agencies instead focusing on replacing lost boats and gear. In Sri Lanka, some of the boats were unsuitable for local waters (Stirrat 2006: 14). Boats were oversupplied in areas where there were already concerns about overfishing (Cosgrave 2007: 15, 18). Fishing, retail, and three-wheeler microenterprises failed because there was not sufficient market for them (Mulligan and Nadarajah 2011: 364). In Sri Lanka, there was insufficient meaningful consultation—and sometimes no participation at all—of intended beneficiaries on housing design, with negative consequences for the completion rates and quality of houses constructed (Silva 2009: 68). Houses were designed without proper ventilation for the use of biomass fuels for cooking, on the assumption that gas or electricity would be used, but their cost was prohibitive for many rural households (Khasalamwa 2009: 82).

Land tenure and ownership systems were of central importance for housing, and often also for livelihoods, yet many humanitarian agencies never fully grasped the implications of these systems and worked without questioning the legal and policy framework. In Sri Lanka, for example, the government introduced a coastal buffer zone, which extended to 65 metres from the sea in Ampara District, in which building was prohibited (de Silva 2009: 263; Silva 2009, 69). This buffer zone policy had enormous economic and social impacts, relocating people to marginal lands such that many, particularly those who depended on the sea, lost their livelihoods (Khasalamwa 2009: 81; Mulligan and Nadarajah 2011: 361). In addition, many livelihoods strategies—most obviously agricultural production—depend on the land itself, and for some tsunami-affected communities in Sri Lanka, the land tenure and ownership system was an obstacle to secure livelihoods (Khasalamwa 2009: 84).

In accepting the legal and policy frameworks as given, aid agencies limited the impact of their response. At least one new settlement in Ampara was abandoned by those who had been rehoused there because it was too far from important facilities, and elsewhere in the district, tsunami survivors defied the new restrictions, refusing to relocate to houses offered to them in a new village more than 20 km away (Mulligan and Nadarajah 2011: 361). Agencies also risked reinforcing or exacerbating inequalities as they worked to build houses for already marginalized populations on the even more marginal lands they had been allocated, instead of protesting that allocation. At the same time, it is far from clear that humanitarian agencies are the right actors to be advocating for land reform. Analysis of land ownership and tenure systems is beyond the skills and expertise of most humanitarian workers, and working to reform such systems implies a more transformative goal than fits comfortably with the principle of neutrality, to which most of the major international humanitarian agencies subscribe.[10]

Moreover, in line with their largely technocratic approach to reconstruction and rehabilitation, the recovery efforts of international agencies were often insensitive to gender relations, social inequality, and poverty and to the conflicts in Aceh and Sri Lanka. The BBB mantra encapsulated the goal 'not just to reinstate what the tsunami had destroyed, but to leave the communities it had affected better, fairer, stronger and more peaceful than they had been before the disaster struck' (Fan 2013: 1; see also, Khasalamwa 2009). However, the reality fell short of this for several reasons. A focus on replacing assets led to the neglect of the poorest, who lacked assets to begin with. In some places, it also had a gendered impact as women with few assets received less in compensation than men who already had many assets (Scheper et al. 2006: 9). Furthermore, the international response was largely focused on the 'tsunami-affected', creating issues of inequity where the very poor but not directly-affected were neglected, and undermining the goal of constructing fairer, stronger, and more peaceful communities (Fan 2013: 8; Khasalamwa 2009: 84).

The potential for recovery activities to promote equality and a fairer society was demonstrated by the Rehabilitation and Reconstruction Agency (BRR) of Aceh-Nias, a ministerial-level agency established in March 2005, which pursued much more transformative goals under the rubric of recovery. For

[10] See 'Principles of humanity, impartiality, neutrality, and independence' in Chapter 13 on Politics, Principles, and Humanitarian Action.

example, taking advantage of the mass resettlement to change property rights for women and children, the BRR persuaded the central government to implement a land titling initiative, providing land titles with joint ownership between husband and wife to those being resettled (Fan 2013: 9). While this kind of initiative fits squarely within the remit of national and local governments, arguably redressing inequalities falls outside the proper scope of humanitarian ambition, at least insofar as it implies political choices about how society should be organized and structural change to reorganize society in that way.[11] At the same time, it is important to recognize that replacing assets and physical infrastructure without incorporating analysis of distribution and equity is likely to reinforce the status quo ante, complete with whatever inequalities that entailed.

Coordination, competition, ownership, and accountability

International humanitarian agencies largely failed to coordinate their response to the tsunami—either among themselves, between themselves and local and national civilian actors in affected states, or between themselves and national and international military forces.

Poor coordination among international humanitarian agencies was exacerbated by the huge numbers of international NGOs in Aceh and Sri Lanka, itself a consequence of the high levels of global public concern and funding. The multiplicity of actors worked against effective coordination both because the enormous number of actors made coordination enormously difficult and because increased competition—whether for resources, visibility, or even beneficiaries—reduced the incentives for cooperation (Cosgrave 2007: 16; Silva 2009: 68; Stirrat 2006: 13–16). The huge levels of funding for the tsunami response meant that humanitarian agencies were not competing with one another for funding. Instead, they were competing for ways of spending that funding that were visible and that conformed to the ideas of their donors about what aid looks like (Stirrat 2006: 14, 16). The failure to coordinate effectively led to duplication and waste (Khasalamwa 2009: 84; Stirrat 2006: 14).

Local capacities were, in many ways, undermined rather than bolstered by the international response, with negative consequences for the quality of the tsunami response itself but also for future response capacity in affected countries. While interaction with national governments was deemed relatively

[11] On these issues, see 'Political or apolitical humanitarian action' in Chapter 13 on Politics, Principles, and Humanitarian Action, and Chapter 16 on the Nexus Concept.

constructive, and national capacities for future disaster preparedness were mostly strengthened by international agencies, insufficient early engagement with community-based organizations and local NGOs undermined local ownership of the response (Scheper et al. 2006: 9). 'Poaching' staff from local organizations and imposing burdensome reporting requirements on implementing partners served to undermine local capacity, while (much more expensive) expatriate staff were often placed in roles that would have been better filled by people with an understanding of the social structures in local communities (Scheper et al. 2006: 35, 37). None of these issues were unique to the post-tsunami context[12] but were particularly notable in this case because of the size of the international response and the fact that there was significant local capacity to begin with. Pressures to spend money rapidly may have contributed to international agencies behaving as if they were saving lives long after the emergency response was over, when they could have invested more in supporting local organizations, something that would likely have improved the effectiveness and efficiency of the international response in the longer term (Scheper et al. 2006: 36–37).

The international humanitarian response to the tsunami emphasized 'upward' accountability to governments, the media, and the public in donor countries at the expense of 'downward' accountability to the public in affected countries and 'lateral' accountability to other agencies and the governments of affected countries (Cosgrave 2007: 11; Stirrat 2006: 3). The lack of downward accountability to the affected populations can be traced back to the needs assessments. People affected by the tsunami 'felt over-assessed but not consulted', and the resulting reports too often 'served the interests or mandate of the assessing agency more than those of the potential beneficiaries' (de Ville de Goyet and Morinière 2006: 11, 12). While it is common that international humanitarian agencies are more accountable to their donors than to those they seek to assist, the huge amounts of funding exacerbated this imbalance in the tsunami response.

Conclusions

The international humanitarian response to the 2004 Indian Ocean tsunami was extraordinarily well funded but failed to live up to the potential that funding offered. International responders overestimated their own capacities at the same time as they underestimated the capacities of local actors. Their

[12] See Chapter 19 on Government and Civil Society in Affected States.

responses were shaped more by their expertise and by the potential for visibility and media coverage than by the needs of the affected populations. As a result, while the extraordinarily high levels of funding facilitated significant success if measured in outputs or technical achievements, outcomes were less impressive. Efforts to engage in recovery and rehabilitation focused on the technical tasks of asset replacement without sufficient analysis or understanding of the economic, political, and social contexts in which assets were being replaced—and with the consequence that housing and livelihoods interventions sometimes failed and, further, sometimes served to reinforce pre-existing inequalities.

8
Sri Lanka, 2008–2009

After decades of sporadic fighting between the Liberation Tigers of Tamil Eelam (LTTE—or Tamil Tigers) and the Sri Lankan state, in 2009, the government opted for an all-out military solution to the civil war. This chapter focuses on the events of 2008–2009 specifically, covering the end of the war and its immediate aftermath, because this is the period that created the most serious humanitarian consequences and presented international humanitarian agencies with the greatest challenges and dilemmas.

Both the Sri Lankan army and the Tamil Tigers committed serious violations of international humanitarian law (IHL), including the targeted and wilful killing of civilians. They also sought to manipulate international humanitarian actors, and the government in particular was very successful at this, with severe consequences for the civilian population. The government prevented aid agencies from operating in the worst-affected areas and restricted the delivery of food and essential medicines. After the end of the conflict, hundreds of thousands of civilians were interned in closed and militarized government camps, which were sustained by international humanitarian assistance. Throughout, the government obstructed humanitarian action, restricting access to affected civilians and, hence, limiting the range and quality of protection and assistance those civilians could access.

In pursuit of access to populations in need, aid agencies frequently compromised their principles and turned a blind eye to abuses of the civilian population by the government. They never achieved the access they sought, and in their pursuit of it, they arguably became complicit in the government's brutal campaign, not least because each time the international community failed to react in any meaningful way, the government appeared emboldened to commit further abuses.

Background

In 1948, Sri Lanka gained independence after 450 years of colonial rule, first by the Portuguese and then, from 1815, by the British, which had accentuated ethnic divisions and Sinhala nationalism. A large majority of

The Politics and Everyday Practice of International Humanitarianism. Miriam Bradley, Oxford University Press.
 DOI: 10.1093/oso/9780198840695.003.0008

the population—74% according to the 2001 census—is (mainly Buddhist) Sinhalese, and there are two main minorities: (mainly Tamil-speaking) Muslims—7.4%; and (mainly Hindu) Tamils—18.2% (Kumar 2019: 71).

Political violence and conflict in the 1970s and 1980s killed tens of thousands of people, and from 1983 to 2009, there was civil war, with fighting interspersed by three relatively ineffective ceasefire agreements (Internal Review Panel 2012: 5; Niland 2014: 130). The LTTE launched in 1976, after two decades of state-sponsored marginalization of the Tamil community, demanding greater autonomy and seeking to establish the state of Tamil Eelam in the north and east of Sri Lanka, where most of the Tamil population lived (Internal Review Panel 2012: 5; Niland 2014: 131). The LTTE strategy was brutal, involving suicide attacks against military, police, and civilian targets as well as indiscriminate violence in the south of the country (Internal Review Panel 2012: 40). The Tigers also regularly proved willing to sacrifice the safety and well-being of the Tamil population to their military and political goals. Intimidation of the population and the forcible recruitment of both adults and children were widespread, and the LTTE used violence to silence rival Tamil groups (Internal Review Panel 2012: 40; Keen 2009: 50, 2014: 6). The hard-line strategies of the Tigers were mirrored by the two main Sinhala political parties, and the state was responsible for widespread extrajudicial killings and disappearances (Internal Review Panel 2012: 5; Niland 2014: 131).

Several factors, both domestic and international, contributed to the strengthening of the government position in the latter years of the conflict. In the 1990s, India switched its allegiance—and its military and political support—from the Tigers to Colombo, and in 2004, the Tamil Makkal Viduthalai Pulikal split from the LTTE and sided with the government (Internal Review Panel 2012: 41; Niland 2014: 132). Management of reconstruction aid after the 2004 tsunami reignited the conflict over sovereignty between the Sri Lankan government and the LTTE (Weissman 2011b: 16). In 2005, the Tigers encouraged the Tamil population—with force, according to some reports—to boycott the presidential election, contributing to the victory of hardliner Mahinda Rajapaksa, who, backed by a nationalist coalition, narrowly defeated a more moderate candidate (Beardsley and McQuinn 2009: 635; Internal Review Panel 2012: 41; Weissman 2011b: 16, 17).

Rajapaksa pursued a military solution to the conflict without serious challenge from international political actors. His government was adept at portraying the fight against the Tamil Tigers as part of the global war on terror at the same time as appealing to domestic and other Asian constituencies by portraying itself as standing up to Western colonialism (HPG 2010: 1).

By 2006, the Cease Fire Agreement of 2002—the latest major initiative of the Western-sponsored liberal peace-building agenda in Sri Lanka—had largely fallen apart (Internal Review Panel 2012: 5). Domestically, this helped to create the political space for an ultra-nationalist and anti-Western backlash, while internationally, Sri Lanka increasingly aligned itself with Eastern powers, with China proving an important ally, and Western power dissipating (HPG 2010: 15; Niland 2014: 32). Post-9/11, there was growing international consensus against engagement with non-state armed groups using terror tactics, and in May 2006, the European Union (EU) declared the LTTE a terrorist organization, which served to ensure that the Tigers were politically isolated and ostracized internationally (Internal Review Panel 2012: 41).

Violence escalated in the north east of the country in late 2005, with the government pushing the LTTE into an area in the north of Sri Lanka called the Vanni and regaining control of the entire eastern region by July 2007 (Internal Review Panel 2012: 42; Weissman 2011b: 16). The government formally launched its military campaign in the Vanni in 2007, and the violence intensified over the following eighteen months until the LTTE was defeated in May 2009 (Internal Review Panel 2012: 8).

Final stages and immediate aftermath of the war

In the end stages of the conflict, both sides adopted harsh military strategies with little concern for the civilian population, and the war was particularly inhumane in the Vanni. As the government launched its final offensive there after September 2008, the Tigers dramatically increased their use of forcible recruitment and employed increasingly violent means to hold the largely Tamil population against its will, using civilians as a human shield in the vain hope that this would slow the advance of the Sri Lankan army (Internal Review Panel 2012: 9; Keen 2009: 50, 2014: 6; Weissman 2011b: 17). The LTTE leadership mistakenly judged that mass civilian deaths would compel the international community to respond with a military intervention, which it hoped would provide some respite to the Tamil fighters (Niland 2014: 134). At the same time, the government—which had the upper hand militarily—unilaterally declared no-fire zones, and told civilians to move into them (Internal Review Panel 2012: 9). Civilians concentrated in these supposedly safe areas, but the LTTE installed its offices and military equipment nearby and shot civilians who tried to escape, while government forces carried out intense shelling, killing and injuring many civilians (Internal Review Panel 2012: 10, 11).

From October 2008, there was significant migration out of the Vanni to government-controlled territory, with an estimated 36,000 internally displaced persons (IDPs) in government-held Vavuniya in early March and some 180,000 registered there by mid-April 2009 (Internal Review Panel 2012: 66, 79). They fled for a variety of reasons, seeking to escape LTTE intimidation as well as prolonged shelling and aerial bombardment by government forces from November 2008 to April 2009, which posed a direct threat to civilian lives and additionally destroyed essential infrastructure in the area (Keen 2009: 69). Government restrictions on aid into the Vanni over a significant period were also a factor (Keen 2009: 70). Many more fled the violence of the final assault, in which government forces shelled the no-fire zones where civilians had congregated as per government instructions. The Sri Lankan army broke through the LTTE's defence lines on 20 April 2009, cutting its territory in two and triggering the evacuation of 100,000 civilians in just a few days (Weissman 2011b: 27). Many were desperate to leave, but many were also forced onto government buses and misled about what to expect in Vavuniya, where conditions were poor and they would be denied freedom of movement (Keen 2009: 70, 2014: 7).

The government declared victory on 19 May. It is likely that tens of thousands of civilians were killed in the final stages of the war, with estimates based on credible sources and information ranging from 40,000 (according to the UN Panel of Experts) to 70,000 (the number unaccounted for, based on estimates by the Office for the Coordination of Humanitarian Affairs (OCHA) and a local government agent from Mullaithivu), while some sources within the central government claimed that the number was well below 10,000 (Internal Review Panel 2012: 14, 38). Furthermore, a large proportion of the Tamil population from the Vanni was interned in overcrowded and underprepared camps (ICG 2010: 6; Keen 2014: 10). As people left the Vanni, the government screened them to identify LTTE cadres among the civilians, interning close to 280,000 civilians in IDP camps and more than 10,000 alleged LTTE members in separate 'surrendee' camps (ICG 2010: 6; Internal Review Panel 2012: 20). The government called them 'welfare villages', but the IDP camps were surrounded by barbed wire and controlled by the military, with complete restriction on movement in or out (Keen 2009: 71). Following two huge waves of arrivals and internments in April and May, the camps were overcrowded and conditions were life-threateningly poor, with inadequate or non-existent water and sanitation facilities (Keen 2009: 71; Weissman 2011b: 29).

Although the war in Sri Lanka generated a humanitarian emergency of an intensely and obviously political nature, it did not attract a robust response from international political actors (HPG 2010: 6; Keen 2009: 51). The Sri Lankan government framed its advance into the Vanni as a humanitarian mission and a hostage rescue operation, and the international community broadly accepted this, with international bodies such as the UN Human Rights Council echoing this language (Keen 2009: 49; Weissman 2011b: 17). Despite the severe violence of the final stages of the conflict, the UN Secretary-General was not willing to take on the government, and none of the Security Council, Human Rights Council, or General Assembly held any formal meeting on Sri Lanka during those months (HPG 2010: 14; Internal Review Panel 2012: 14). The Security Council did hold several unofficial meetings, at which Secretariat officials presented statements focused mainly on the humanitarian situation without calling attention to the responsibilities and obstructions of the government and the LTTE or providing full information on civilian casualties, and the Council only issued a press statement three days before the end of the conflict (Internal Review Panel 2012: 4; Keen 2009: 78).

In addition to the international political failure, there was a collective weakness on the part of humanitarian actors, with the response and discourse from humanitarian agencies largely focused on establishing the numbers of the displaced and shipping food to them (HPG 2010: 3; Keen 2009: 51). UN agencies in particular sought to distinguish humanitarian from political action and classified as political anything which would offend the government, with the consequence that they failed to fulfil basic humanitarian responsibilities and allowed themselves to become complicit in the government's counter-insurgency strategy (Internal Review Panel 2012: 19; Weissman 2011b: 15).

International humanitarian response

In the final stages of the war, there were high levels of humanitarian vulnerability and need in the Vanni. Given the military strategies of both sides and the scarcity of respect for IHL, protection was a major concern, and by the end of 2008, half the inpatients in Puthukkudiyiruppu hospital had been admitted for war wounds (Internal Review Panel 2012: 56). Food and medicines were in short supply. The escalation of violence caused huge disruption to food production in both the north and the east of Sri Lanka such that important rice-producing areas in the Vanni probably had no significant production in 2009, and doctors in the Vanni reported thousands of deaths due to malnutrition and lack of antibiotics (Internal Review Panel 2012: 18;

Keen 2009: 61). There was also significant need for shelter, water, and sanitation, in part due to repeated displacements, and the need for shelter materials and medicines was all the more urgent after the start of the monsoon in October 2008 (Internal Review Panel 2012: 56; Keen 2009: 63).

Despite these high levels of need, most international aid agencies left the Vanni in September 2008, with severe consequences for the delivery of assistance and for the potential to protect civilians through presence and monitoring (Internal Review Panel 2012: 8). This relocation was prompted by repeated shelling close to UN compounds combined with the government's announcement that it could no longer guarantee the safety of aid workers (Internal Review Panel 2012: 8). The UN withdrew all international staff and some national staff (many national staff chose to stay behind because the LTTE prevented staff dependents from leaving), despite some inhabitants of the region pleading with the UN to stay, fearing that with no one to bear witness, the government would be less restrained and the civilian population exposed to increased violence (Internal Review Panel 2012: 8; Keen 2009: 60, 2014: 8).

In October 2008, the UN reached an agreement with the government and the LTTE for weekly convoys of humanitarian assistance to the Vanni—convoys on which the government imposed significant restrictions (Internal Review Panel 2012: 8; Keen 2014: 8). The government engaged in endless discussions on access to aid, signing up to international standards like the Guiding Principles on Internal Displacement, with such engagement aimed at convincing aid agencies that its intentions were good, even if the discussions did not lead to improvements in access (HPG 2010: 2, 7; Niland 2014: 134). It allowed small amounts of food to get through, as apparent evidence of its good intentions, even though the quantity of food allowed through was far from sufficient to meet the nutritional needs of the population in the Vanni, and other items, including essential medicines and materials for shelter and sanitation, were largely prohibited, ostensibly because they could be used by the LTTE (Internal Review Panel 2012: 56; Keen 2014: 8–9). Starvation would have generated dramatic media coverage, whereas blocking medicines and water tanks was not so visible, even if it may have been a deliberate attempt to make the population sick (Keen 2009: 63). Artillery shelling by government forces, and sometimes from unidentified sources, exploded close to several of the convoys (Internal Review Panel 2012: 54).

In many ways, the World Food Programme (WFP) responded flexibly to overcome government obstruction, but even so, the food delivered was inadequate. Flexible practices included purchasing food locally in the north of Sri Lanka to circumvent official obstructions to transportation and to provide some support for livelihoods in local agricultural communities, but it was insufficient to make up for the shortages in deliveries from elsewhere (Keen 2009: 53–54). WFP also used a logistics hub at Vavuniya, deployed fourteen mobile storage units to areas close to conflict-affected populations, and had its own fleet of thirty trucks, all of which facilitated some degree of operational autonomy (Keen 2009: 54). For all WFP's efforts, however, the tonnage successfully delivered from October to mid-December 2008 was only around 40% of the planning amount and less than 20% of the requirements for actual population numbers (Internal Review Panel 2012: 55). In addition to UN convoys, India sent a consignment, the Sri Lankan government organized convoys with food for sale at low prices, and the International Committee of the Red Cross (ICRC) delivered some food by ship, but all this was insufficient to plug the gaps (Internal Review Panel 2012: 55; Keen 2009: 62). Moreover, the government refused to allow UN staff to monitor the distribution of food in the Vanni, and one estimate suggested that the LTTE may have taken 20% of the assistance delivered by the convoys (Internal Review Panel 2012: 18, 55).

The eleventh UN relief convoy, which arrived in the Vanni on 16 January 2009, ended up being the last to make the journey. As with previous convoys, it had clearance from both the Sri Lankan Army and the LTTE, and pre-approval to return the following day, but in the event, the government refused clearance for its return journey until 22 January, apparently due to ongoing military operations (Internal Review Panel 2012: 57–58). Suspecting that this would be the last convoy, and deeply concerned for the safety of UN national staff and their dependents in the Vanni, the convoy leader sought to evacuate them in the returning trucks, but the LTTE prevented this (Internal Review Panel 2012: 57–59). He and one other international UN staff member remained with the seventeen national staff and their eighty-six dependents, and the government instructed them to go to Udaiyaarkaddu, a village in Mullaitivu District located within an area the government had unilaterally declared a no-fire zone (Internal Review Panel 2012: 59).

By this time, civilians were being squeezed into a shrinking zone in Mullaitivu, where conditions were dangerous and squalid (Keen 2009: 66). To protect themselves from shelling, IDPs were sleeping in muddy wet trenches, and by March, thousands were reported dead from inadequate food and medical care (Internal Review Panel 2012: 9). Hoping that the UN flag would offer some kind of protection, hundreds set up tarpaulin shelters close to the site occupied by those parts of the UN convoy that had remained (Internal Review Panel 2012: 60). However, they encountered heavy shelling and came under intense artillery fire, mainly from government forces (Internal

Review Panel 2012: 9, 59–62). In just forty-eight hours, before the convoy left to escape the shelling, dozens of civilians were killed or injured within metres of the UN bunkers. Many more were killed or injured in other makeshift IDP sites in the government-designated 'safety zones' as well as in hospitals. The wounded had access only to rudimentary care within the LTTE-controlled zones, provided by eight Sri Lankan doctors who had refused to abandon their posts and worked with medical supplies delivered by the ICRC, which also managed to evacuate 6,600 wounded and seriously ill people to government-controlled areas (Weissman 2011b: 24).

After the end of the war, more than fifty non-governmental organizations (NGOs) and UN agencies operated in the internment camps set up by the Sri Lankan government—sustaining them through the provision of shelter, food, water, and sanitation, etc. (Weissman 2011b: 28–29). However, government obstructions limited the ability of the aid agencies supporting them to take a principled approach. The government initially planned to house people in these camps long term—for up to three to five years (Keen 2009: 76–77; Weissman 2011b: 25). Many aid workers had serious reservations about supporting camps that were closed and militarized, where IDPs were denied freedom of movement and the access of international agencies to the camps, as well as their communication with the IDPs within them, were restricted (Internal Review Panel 2012: 20). International agencies were concerned about weaknesses in the government screening processes, particularly given the history of disappearances of people the government suspected or accused of being linked to the LTTE (Keen 2009: 72). Neither the UN nor the ICRC, despite its mandate for monitoring the treatment of people detained in relation to armed conflict, were allowed to be present where the Sri Lankan security services screened those leaving the Vanni (Internal Review Panel 2012: 20). With restrictions on access, communications, and activities, systematic needs assessment and a range of important interventions were impossible (HPG 2010: 4; Weissman 2011b: 28–30). Unable to carry out needs assessments or properly monitor distribution, aid agencies all but jettisoned minimum standards and relied on government lists of who needed different kinds of assistance (HPG 2010: 5). Poor distribution systems and government refusal to share registration data meant that aid was not always getting to the most vulnerable within the camps (Keen 2009: 74–75).

Donors and operational agencies alike faced a dilemma. Should they provide for people's basic needs in what were dire conditions or refuse to legitimize (and make just about survivable) what were effectively concentration camps, in the hope that this would force the government to close them and let people leave? Reluctant to support indefinite internment, they sought to underline the temporary nature of the camps by refusing major improvements (permanent shelters, cement floors, latrines with septic tanks, extension of health and water infrastructure), decisions which may have contributed to the early closure of the camps, but additionally made for uncomfortable and insanitary living conditions in the meantime (Holmes 2014: 146; Weissman 2011b: 31). Following concerted pressure from donors, as well as shifts in Sri Lankan domestic politics, the camps were opened, and some freedom of movement for IDPs was introduced by the end of 2009 (HPG 2010: 5; Weissman 2011b: 31–32).

Manipulation of aid and aid agencies

The Sri Lankan government—likened to a highly skilled chess master and a poker player—was masterful in manipulating humanitarian agencies and the international community more broadly (HPG 2010: 3; Niland 2014: 132). Committing to international standards and engaging in endless discussions about humanitarian access and assistance, the government sought to create the impression that it was changing in a positive direction, while the humanitarian community 'lacked the skill, strategy, resolve and political backing needed to manoeuvre successfully against the government to protect or secure any kind of effective humanitarian space' (HPG 2010: 3; see also, Niland 2014: 134).

One of the ways in which the government manipulated humanitarian operations was through limiting access to information about what was going on in Sri Lanka and particularly about abuses perpetrated by the state. For example, although Colombo defended the evacuation of humanitarian agencies from the Vanni as necessary for the security of aid workers, it had a clear ulterior motive in terms of manipulating information flows. Since recent shelling incidents and the main threat to aid agencies came from the government side, to say that the same government could not guarantee their safety was disingenuous—it could have ordered the army not to attack them. However, the withdrawal of security assurances served to remove potential witnesses and hence to facilitate the government's final offensive (Internal Review Panel 2012: 17; Keen 2014: 8). This was obvious to many aid workers and to much of the civilian population and the LTTE, who wanted the aid agencies to remain (Internal Review Panel 2012: 17, 50, 52; Keen 2014: 8). Apart from the ICRC and Caritas, however, aid agencies left the Vanni, apparently without questioning the logic of relocating operations because of a government security warning, when most of the immediate security threat came from government forces themselves (Internal Review Panel 2012: 17).

As part of its manipulation of information flows, the Sri Lankan government played politics with numbers. While the LTTE had long inflated population numbers in order to claim a larger constituency and additional humanitarian assistance, Colombo claimed that the number of people remaining in the Vanni after the relocation of the UN and most other international aid agencies was much lower than it really was. Estimates by local government authorities put the total population of the Vanni at 428,000 in October 2008, while the national government in Colombo insisted that there were no more than around 70,000 people (Internal Review Panel 2012: 18). The government used low population numbers as an excuse to limit relief deliveries and to argue that the quantities being delivered were adequate (Internal Review Panel 2012: 37). Claiming that there were far fewer people in the area further served as a pre-emptive method of reducing casualty figures, and the government used low population estimates to rebut allegations of civilian casualties (Internal Review Panel 2012: 18).

The humanitarian community responded passively to government efforts to restrict access and information. OCHA is mandated to negotiate access for the humanitarian community as a whole, and NGOs relied heavily on the UN to raise issues with the government, but the UN was deferential to Colombo. On withdrawing from the Vanni in September 2008, for example, the UN issued a statement saying that it had been forced to relocate because it had assessed the security situation to be too dangerous, without mentioning that it had been forced to leave by the government (Keen 2014: 8). The UN did not explain the full circumstances of the evacuation either to member states or the general public, and the Sri Lankan government did not face significant criticism for pushing international agencies out of the main theatre of war (Internal Review Panel 2012: 17; Niland 2014: 134). By allowing limited food aid into the Vanni, the government was able to use the existence of relief operations as evidence of its humanitarian intentions and to inhibit collective pressure from international agencies (Keen 2014: 14). At the same time, aid agencies were competing for visibility and for space on the convoys, and negotiations to allow non-food items delayed departure, all of which reduced the chances of developing a common advocacy position (Internal Review Panel 2012: 56; Keen 2014: 9).

A combination of bureaucratic obstacles and threats de-energized aid agencies and de-incentivized them from challenging the government. Colombo was adept at ensuring that donors and operational agencies had to expend enormous effort overcoming bureaucratic obstructions such that they had little capacity left to address higher-level strategic issues (HPG 2010: 11).

Visas, authorizations to travel within the country, and import licences were all frequently delayed and obstructed by the government and were often implicitly—sometimes explicitly—contingent on aid agencies (and individual staff members) being 'cooperative' and not making public statements about the conduct of the government and armed forces (Internal Review Panel 2012: 41; Niland 2014: 133; Weissman 2011b: 20–24). Aid agencies and individual staff members feared being expelled from Sri Lanka, not always for entirely disinterested reasons, and they also feared for their safety (Keen 2009: 87–88). Sri Lanka was a dangerous country for aid workers, and there were good reasons for the sense of fear and insecurity. NGOs in particular had been the subject of a systematic media campaign that depicted them as LTTE sympathizers, and seventeen Sri Lankan staff of Action contre la Faim had been executed in their office in Muttur on the east coast in August 2006, a few hours after government forces recaptured the town (HPG 2010: 5; Weissman 2011b: 18).

Most aid agencies allowed themselves to be manipulated in that they refused to call out the government restrictions on aid and instead responded pragmatically—accepting restrictions if it meant some aid could get through rather than maintaining a principled stand. Thus, they agreed to seal convoy trucks under military supervision, and allowed body searches at checkpoints, and gradually gave in to more and more government demands and harassment (HPG 2010: 13; Keen 2009: 54, 80). This was driven in part by beliefs among UN and humanitarian decision-makers that each accommodation made to the government was the lesser of two evils. Hindsight casts doubt on such beliefs, and even in the moment, many individual aid workers saw that these accommodations were doing more harm than good, but they lacked common coordinated positions which could have strengthened pressure on the government. Competition among agencies combined with government restrictions on inter-agency coordination to reduce the potential for collective pressure (HPG 2010: 9, 10, 13; Keen 2009: 63). The weakness of a competitive and uncoordinated humanitarian sector was exacerbated by the lack of political will in the UN and the lack of unity among donors (HPG 2010: 9, 10, 11).

A failure of protection

Both during the final stages of the war and in its immediate aftermath, the protection of the civilian population was of paramount importance and a resounding failure of the humanitarian response. Although much of the work undertaken by humanitarian agencies came under the rubric of protection, this included activities such as psycho-social care, the provision of food and shelter, recreational activities, and staff training, none of which necessarily help keep people safe from violence (Internal Review Panel 2012: 18). Such a broad definition 'obscured the very limited extent to which the UN's protection actions actually served to protect people from the most serious risks' (Internal Review Panel 2012: 19). As the note of a round-table event on humanitarian space in Sri Lanka put it, 'it is estimated that thousands of people were killed and approximately 300,000 were forced into closed militarised camps and humanitarian agencies responded by handing out water and plastic sheets' (HPG 2010: 6). Where UN action to address issues would have provoked criticism from the government, senior UN staff tended to label those issues as political and, hence, outside the remit of humanitarian response, with the consequence that concerns about how many civilians were being killed and by whom, and even how many civilians were in the Vanni, were not adequately addressed or reported—and, indeed, such issues were often not even monitored (Internal Review Panel 2012: 19).

When the Sri Lankan government began its final offensive in the Vanni, the government did its best to block information about civilian casualties, and the UN was not systematically monitoring civilian deaths and injuries or acting on the evidence it did have. The two international staff members from the last convoy, who had remained in the Vanni for an additional two weeks, returned to Colombo at the end of January 2009 with their own eyewitness accounts and documentation of events that other sources were already reporting (Internal Review Panel 2012: 10, 59–62, 63). The UN then had incontrovertible evidence of the shelling and killing of civilians by government forces but did not use it to pressurize the government. At the beginning of February, the UN finally implemented a system for the collection of data on killings and injuries, which showed that a large majority of civilian killings were the consequence of shelling and aerial bombardment by government forces and a smaller proportion were the result of LTTE actions (Internal Review Panel 2012: 19). UN satellite pictures became public in April 2009, providing the basis for new estimates of how many people remained in the Vanni and confirming that there was ongoing artillery shelling by government forces (Internal Review Panel 2012: 12–13). However, the UN did very little with all this evidence.

The international humanitarian agencies in Sri Lanka in 2008 and 2009 only engaged in limited and one-sided advocacy in the final stages of the war. The UN spoke out about both verified and unverified abuses by the LTTE but not about abuses by the government, even when they had been directly witnessed by UN staff (Internal Review Panel 2012: 20; Keen 2014: 15).

On 14 March 2009, the Office of the High Commissioner for Human Rights (OHCHR) did release a statement that made specific reference to casualty numbers and allegations of violations of international law by the government. However, this statement was an exception to the general position of UN officials, who avoided criticizing the government because they did not want to prejudice humanitarian access, and it also went against the express wishes of the Chef de Cabinet, the Under Secretary-General for Humanitarian Affairs, and the UN Resident Coordinator in Sri Lanka—who also wore several other UN hats, including that of Humanitarian Coordinator (Internal Review Panel 2012: 11–12). Drawing on the experiences of its Dutch section in the IDP camps in Vavuniya District, in the first months of 2009, Médecins Sans Frontières (MSF) spoke publicly about the artillery fire in the Vanni and the lack of freedom of movement for the IDPs in Vavuniya (Weissman 2011b: 26), but for the most part, the international humanitarian community avoided any public criticism of the Sri Lankan government.

Neither humanitarian agencies nor the wider UN engaged in much forceful dialogue privately with the government either, although they moved increasingly to do so after the end of the war. NGOs had almost no direct contact with the government and hence relied heavily on the UN to raise concerns (HPG 2010: 4). For the most part, UN officials went out of their way not to offend the government, and even where they associated government fire with civilian deaths, they did not specify the provisions of IHL that had been violated (Internal Review Panel 2012: 20). For example, the UN Resident Coordinator wrote to the government on 7 February stating UN minimum estimates for civilian casualties in the Vanni over the past weeks and saying it was highly probable that fire from both sides had led to these casualties, 'despite the best efforts' of the government (Internal Review Panel 2012: 10). Given that the UN had irrefutable evidence of government forces shelling and killing civilians, including attacks on UN installations and hospitals, and had data which showed that government forces were responsible for more casualties than the LTTE, the contents of the letter were excessively deferential to the government.

There was also a lack of behind-the-scenes advocacy to mobilize other actors who might be able to influence the Sri Lankan government. In a briefing to members of the diplomatic corps in Colombo on 9 March, members of the UN Country Team, including the Resident Coordinator, presented estimates of casualties between 20 January and 2 March in Mullaitivu District, describing a range of human rights and IHL violations by the LTTE, without explicitly addressing government responsibility for shelling and without presenting data which showed that a large majority of the casualties were reportedly attributable to government forces (Internal Review Panel 2012: 11). In the same briefing, and in subsequent documents, the UN Country Team informed the diplomats of UN efforts to be present at locations where the government was screening those emerging from the conflict zones but neglected to mention reports of human rights violations and disappearances from other screening locations, to which the UN was denied access (Internal Review Panel 2012: 13). In the last days and weeks of the war, some members of the Security Council, as well as senior officials from the Secretary-General's office, complained that they were receiving almost no information from the UN and had begun to rely on reports from international NGOs (Internal Review Panel 2012: 13).

Advocacy vs access and assistance?

As in many cases, humanitarian actors faced a dilemma: should they speak out and risk losing access, in the hope that the perpetrators of IHL and human rights violations would be shamed into improving their conduct, or stay quiet in the hope of saving lives and alleviating suffering through the delivery of relief?[1] Most humanitarian agencies decided that access was more important than advocacy. They favoured (the possibility of) the delivery of food, medicine, and other material assistance over (the possibility of) the protection of the civilian population. While this is a genuine dilemma, access is not always best served by silence, and advocacy does not necessarily entail protection.

Dialogue and public or private advocacy may be explicitly focused on aid agency access, but in Sri Lanka, aid agencies avoided criticizing the government on *both* protection issues *and* questions of access. Although the UN repeatedly lobbied both the Sri Lankan government and the LTTE for improved humanitarian access, it did not speak out about government restrictions on access and humanitarian aid, and 'the UN did not confront the Government directly with the fact that obstructing assistance was counter to its responsibilities in international law' (Internal Review Panel 2012: 18). Moreover, the decision to prioritize access over public advocacy or forceful bilateral dialogue is not a one-off decision, but in Sri Lanka, the consequences of earlier decisions did not appear to inform later decisions. Limiting criticism of the government did not have any positive impact on access, which got worse rather than better, yet, there was no move to more vocal criticism.

[1] For a more general discussion of this dilemma, see 'Neutrality, access, and humanitarian space' in Chapter 23 on Dialogue, Negotiation, and Advocacy.

The lack of strongly voiced objections to government conduct created the impression that massacres, civilian casualties, and the expulsion and exclusion of humanitarian actors was acceptable, and the government may have seen this apparent acceptance as a green light from the international community to go ahead with more and worse abuses of human rights and IHL (HPG 2010: 12; Keen 2014: 15).

Given that repeated deference to the government was not yielding improvements, the prioritization of technical approaches to delivering material assistance over protection and advocacy was particularly problematic and cannot be explained simply as the result of an impossible dilemma. Several analysts and field-level staff suggested that more prosaic considerations played a role, namely, donor preferences and the opportunity for operational agencies—including those such as the United Nations Refugee Agency (UNHCR), which ought to be leading on protection—to raise revenue and capture market share (HPG 2010: 6; Internal Review Panel 2012: 21, 110; Keen 2014: 6). Fear of the government response was undoubtedly part of the reason for avoiding any focus on protection or advocacy—and while fear of being expelled from the country often had more to do with aid worker self-interest than security concerns, the government also posed a very real security threat to aid workers and agencies, as outlined above (HPG 2010: 5; Keen 2009: 85–86, 87–88).

It is difficult to know how much leverage humanitarian agencies and other international actors, including member states and the political bodies of the UN, could have exercised had they sought to do so (Bradley 2014: 150–151). John Holmes, who was the UN Emergency Relief Coordinator at the time, has argued that nothing the humanitarian community could have done would have been likely to make the government listen and change its approach (Holmes 2014). Certainly, aid agencies had limited influence. The reputation of international NGOs in Sri Lanka had been tarnished after the 2004 tsunami by the arrival of huge numbers of international agencies, some with little experience (the so-called second tsunami of NGOs), together with claims of corruption and misuse of funds by some organizations (HPG 2010: 3).[2] Furthermore, more strident advocacy could have had counterproductive results, with talk of human rights playing into the government's discourse that human rights were imperialistic and, hence, consolidating government power domestically (HPG 2010: 8). Moreover, the lack of unity among aid agencies undoubtedly reduced their leverage—if one agency refused to support the camps, for example, another would take its place (Keen 2009: 80).

[2] See Chapter 7 on the 2004 Indian Ocean Tsunami.

For all the limits of its influence, however, there are at least two reasons to think that aid agencies and the wider UN system had more leverage than they sought to exercise. First, they did not provide clear information on human rights in Sri Lanka to powerful states that may have had some influence over the Sri Lankan government (Keen 2014: 18). UN documents referred so prominently to protection that some members of the diplomatic corps assumed that the UN had a system for monitoring and responding to violence against civilians (Internal Review Panel 2012: 19). While powerful states were undoubtedly at fault for the weakness of their response, the humanitarian community and the UN could have provided them with better information. Second, the rare occasions where humanitarian agencies did make strong representations to the government or publicly highlighted government abuses suggest that the government was concerned about its reputation. For example, the government responded robustly to any suggestion that there were civilian casualties (Internal Review Panel 2012: 12). It also improved its conduct in response to one-off interventions by OCHA, UNHCR, and WFP (Keen 2009: 17, 18, 2014: 86, 93–96). Likewise, concerted pressure from donors and other humanitarian actors was instrumental in the opening of the IDP camps at the end of 2009 (HPG 2010: 5).

Conclusions

The UN as a whole—including its humanitarian agencies—was severely criticized for its role in facilitating abuses of the population by the government, not least in the report of its own Internal Review Panel, commissioned by the UN Secretary-General at the time, Ban Ki-moon. The civilian population suffered abuses at the hands of both government forces and the LTTE, and humanitarian actors did little in response, largely ignoring abuses by the government side in particular. Focused almost entirely on technical issues related to the provision of material assistance, they allowed themselves to be manipulated by the Sri Lankan government—manipulation that was itself 'a key factor in the intensification of suffering and death' (Niland 2014: 134). Failing to do anything to address government abuses, restrictions, and obstructions, aid agencies were arguably complicit in government strategies of targeting civilians and restricting access to humanitarian assistance.

9

Haiti Earthquake, 2010

On 12 January 2010, an earthquake registering 7.3 on the Richter scale struck Haiti just 17 km from the capital, Port-au-Prince, with devastating consequences (Fan 2013: 19). Of a population of around 10 million, somewhere between 100,000 and 316,000 people died, approximately 300,000 were injured, and 2 million—one-fifth of the population—were suddenly made homeless (Patrick 2011: 2). Death and destruction were on a scale comparable with that resulting from the 2004 Indian Ocean tsunami,[1] but, in this case, it was concentrated largely in one city, Port-au-Prince (Levine et al. 2012a: 5).

The humanitarian response that followed was, in many ways, highly problematic. Failures and limitations were in abundance, but just as in the case of the tsunami, they cannot be blamed on a lack of funds. Dominant discourses in the United States and elsewhere in the global North after the earthquake constructed the Haitians as in need of saving by (white Western) outsiders. Reflecting and reinforcing these discourses, the international response largely sidelined Haitian contextual knowledge and capacities, which significantly limited the impact of the well-funded response. While the emergency response was generally considered a success, international aid agencies failed to fulfil the promise to 'build back better' (BBB), either in terms of building better physical infrastructure or in terms of building better public institutions.

The international response let Haitians down not only by not living up to its own promises but also by doing harm in two particularly notable ways: UN peacekeepers who arrived as part of the international response inadvertently (but avoidably) brought cholera to Haiti, causing thousands of deaths; and both peacekeepers and international aid workers were involved in the sexual exploitation of Haitians.

[1] See Chapter 7.

The Politics and Everyday Practice of International Humanitarianism. Miriam Bradley, Oxford University Press.
 DOI: 10.1093/oso/9780198840695.003.0009

Background and context

Structural vulnerabilities and a history of fragile governance exacerbated the impact of the earthquake (Levine et al. 2012a). The poorest country in the Caribbean, Haiti suffered from systemic poverty, which exacerbated people's vulnerability to the earthquake and increased obstacles to recovery.[2] At the time of the earthquake, for example, more than half the population lived on less than USD $1.25 a day, child mortality rates were double the regional average, around one-third of the population was considered food-insecure, and only 58% of the population had access to clean water and sanitation (Fan 2013: 19; Patrick 2011: 2). Forty per cent of the population lacked physical access to health care, and many more faced financial barriers to using existing services (Biquet 2013: 130). Several decades of rapid urbanization combined with weak urban planning in Port-au-Prince had resulted in poor-quality buildings not designed to withstand earthquakes and the expansion of *bidonvilles* or 'shanty towns' without sanitation and often in precarious locations (Levine et al. 2012a: 9; Patrick 2011: 2; Versluis 2014: S97). That the earthquake hit a densely populated urban centre in which housing was mostly not quake-resistant largely accounts for the high number of deaths in Haiti,[3] the majority of which were the result of immediate crushing and suffocation from the collapse of buildings (DARA 2011: 4; Schuller 2012: 172).

The most urgent needs of survivors were treatment for those injured in the earthquake and shelter for those whose homes were destroyed or made uninhabitable. Around 300,000 were injured—an unusually high number for a rapid-onset disaster—and thousands required life-saving amputations and psychosocial care (DARA 2011: 4; Patrick 2011: 2). Hundreds of thousands of people already lacked housing in Port-au-Prince, and the earthquake destroyed around 100,000 houses and damaged about another 200,000 (Levine et al. 2012a: 23). Of the 2 million people whose homes were destroyed or made uninhabitable, it is estimated that more than half a million left the capital, at least temporarily (Versluis 2014: S95). Others who lost their homes, or were afraid to sleep indoors in the face of frequent aftershocks, settled spontaneously in the few safe open spaces—parks, roads,

[2] For a discussion of the historical causes of Haiti's vulnerabilities and the role of countries in the global North—as well as bilateral and multilateral aid programmes—in creating and perpetuating them, see Oliver-Smith (2010).

[3] A significantly higher magnitude earthquake—8.8 on the Richter scale—in Chile that same year killed around 500 people, just a fraction of the number killed in Haiti.

a golf course—creating camps with sheets and cardboard, which they gradually replaced with tarpaulins and tents (Levine et al. 2012a: 13; Versluis 2014: S95, S101).

The nature and extent of damage also undermined the already weak capacity of the national government and posed challenges to national and international responders engaged in the relief effort. The earthquake generated 10 million cubic metres of rubble, and the debris restricted road access throughout earthquake hit areas (Levine et al. 2012a, 15; Patrick 2011: 2). Because the epicentre was so close to Port-au-Prince, much national infrastructure and many schools, hospitals, and government buildings were destroyed or badly damaged (Patrick 2011: 2; Schuller 2012: 73). An estimated 60% of the country's economic and administrative infrastructure was damaged or destroyed in the earthquake, and many civil servants—one-third of the government—died, while others were absent through injury or because they were caring for people within their own families (Fan 2013: 19; Levine et al. 2012a: 1, 14; Patrick 2011: 2). Despite the damage and loss of personnel, as well as the personal suffering and trauma experienced by survivors, Haitians were the first to respond—local staff of international non-governmental organizations (INGOs) were extremely effective in the initial response, and within days, the government had made important steps in making fuel available, repairing damaged electric plants, reopening banks, and paying public sector workers (DARA 2011: 6; Patrick 2011: 5).

After more than 100 years in which Haiti had been free of the disease, one of the largest outbreaks of cholera in modern history began in October 2010 (TDC et al. 2013: 8). Lacking prior experience of the disease, Haitians were unfamiliar with the basic treatment (simple, oral rehydration), and the fatality rate was unusually high for a cholera outbreak, particularly among the rural poor (DARA 2011: 5). Within three months, more than 200,000 cases and around 4,000 deaths were recorded (Biquet 2013: 133). In the first year of the epidemic, over 470,000 cases had been identified, with over 6,600 deaths (TDC et al. 2013: 12). Cholera continues in Haiti today, although cases and deaths have been in steady decline since 2016 (CDC 2021). The first cases were clustered around a UN base housing Nepalese peacekeepers and the strain of cholera was a perfect match for a strain found in Nepal, from which it could be inferred that cholera was brought to Haiti by the peacekeepers who had deployed as part of the earthquake response, and that poor sanitation at the UN camp sent sewage into local waterways, from which members of the wider population were infected (Pilkington 2020; TDC et al. 2013: 1, 8, 13–14).

Western discourses, media coverage, and celebrity engagement

The earthquake in Haiti attracted significant media coverage in the global north, much of which was racialized and served to evoke pity, reproducing the dynamics of colonialism.[4] In the US media, for example, repetition of images that underscored the pitiful state of Haiti, and of words such as chaos, dysfunction, violence, and hopelessness, made the victims of the earthquake objects of pity in a process facilitated by their racial otherness (Balaji 2011: 50, 65). In some coverage, traumatized Haitians were depicted as indifferent and the nation as desperate and lawless, unable to function without assistance (Balaji 2011: 51; Ulysse 2010: 38). Countless images of white Western aid workers helping to rescue black Haitians from the rubble and an emphasis on the benevolence of the West and the efforts of US or European aid workers served to reinforce this idea (Balaji 2011: 51, 55). The dynamics of imperialism were evident in the way the plight of victims was highlighted in contrast to the heroism of the white Western rescuers, and it has been argued that the dominant Western media representations of Haiti after the earthquake reproduced narratives and stereotypes that could be traced back as far as the nineteenth and early twentieth centuries, including the period of US occupation and its aftermath (Balaji 2011: 57; Ulysse 2010: 37).

Extensive celebrity involvement told a similar story and called on audiences in the global North to 'save Haiti' by donating funds and buying related products. Celebrity engagement within and beyond the United States included talk show appearances, telethons, and other 'charitainment' events (see, e.g. Cloud 2014; Driessens et al. 2012; Duvall 2015). Hope for Haiti Now, spearheaded by George Clooney and featuring over 100 celebrities, included a telethon broadcast live by MTV, CNN, Fox, ABC, and many more US and other broadcasters worldwide and a website which prompted visitors to donate money to organizations including Oxfam, the American Red Cross, and the United Nations Children's Fund (UNICEF) and to buy an album of the music performed during the telethon (Driessens et al. 2012: 711, 714; Duvall 2015: 583). Much celebrity engagement shared three important characteristics. First, the celebrities involved tended either to pay little attention to the question of the factors that made the earthquake so catastrophic or to blame ill fortune and irresponsibility on the part of Haitians rather than international economic relations (Cloud 2014: 48). Second, they portrayed

[4] For a more general discussion of the roles and impacts of media and celebrities in humanitarian crises and responses, see Chapter 17 on Media and Celebrities.

the Haitians as helpless and in need of saving by superior (mainly white) donors in the global North, whose humanity was emphasized rather than that of the Haitian victims (Balaji 2011: 62; Cloud 2014: 53). Third, they proposed 'saving' Haitians through the privatizing logic of neoliberalism, according to which charity efforts were commodified and audiences called on to respond as individual consumers rather than as political actors (Balaji 2011: 62; Cloud 2014: 45; Duvall 2015: 583).

These kinds of representations both reflected and reinforced the views of donors and operational humanitarian agencies and the kinds of response that were seen as possible and desirable. According to these views, vulnerability was primarily a product of a domestic political environment that 'had been characterised by dictatorship, political violence, fraudulent elections, exclusion from political processes, stark socio-economic inequality and the denial of basic rights to the majority of the population' (Fan 2013: 20). General pessimism about Haiti's prospects underpinned the expectations of donor governments that external events and domestic vulnerabilities would keep the country trapped in a chronic humanitarian crisis over the medium-to-long term (Muggah 2010: S446). Moreover, global media appear to have massively overstated levels of crime and violence in the aftermath of the earthquake (Muggah 2010: S460).

There are other ways to represent and understand Haiti. For instance, a range of scholarship highlights the role of the international community, and in particular the United States, in generating, sustaining, or exacerbating vulnerabilities in Haiti. Several authors point to two centuries of US intervention as an underlying cause of governmental weakness (Cloud 2014: 43; Dupuy 2010: 15; Oliver-Smith 2010: 34–35). From the 1970s, the Haitian government followed structural adjustment policies prescribed by the international financial institutions in return for international military and economic support, mainly from the United States, Canada, and France—policies which turned Haiti into a supplier of the cheapest labour in the Western hemisphere and a major importer of US food, exacerbating underdevelopment, poverty, and inequality (Dupuy 2010: 15–17). More recently, the Clinton, George W. Bush, and Obama administrations arguably all used crises in Haiti 'to expand tourism, support textile sweatshops, and weaken state economic control through privatization and deregulation', creating or exacerbating the vulnerabilities that made the impacts of the earthquake so catastrophic (Cloud 2014: 43). Alternatively, Haitians' resilience, innovation, and self-reliance could be highlighted instead of vulnerability, suffering, and impoverishment, and state weakness seen not as a failure but as resulting from the refusal of a population descended from slaves to be controlled by the state (Fan 2013: 20).

Explaining the complexity of the Haitian context, apportioning blame internationally, and emphasizing the resilience of Haitians would, however, likely not have been so effective in pulling at the heartstrings of Western audiences or at prompting wide interest and large-scale financial donations.

Funding for humanitarian response

Humanitarian funding reached USD 3.5 billion in 2010 alone, making the international humanitarian response in Haiti the highest funded since the 2004 Tsunami (Levine et al. 2012a: 5). Western donor governments pledged massive amounts of aid, with the United States the biggest by a long, long way. Contributions from 'non-traditional' donors, including Brazil, Cuba, and Venezuela, were also significant (DARA 2011: 3). One-third of that USD 3.5 billion came from private donors, compared with less than 4% of all international humanitarian funding elsewhere in the world in 2010 (Levine et al. 2012a: 13). Within eight days of the earthquake, 18% of US households had reportedly donated and another 30% said they planned to (Pew Research Center 2010: 2).

Many of the same factors that contributed to the extraordinarily high levels of public interest and donations for the humanitarian response to the 2004 Indian Ocean tsunami were also at play here.[5] As a so-called 'natural' disaster, the earthquake could be understood as a crisis for which no one was to blame, even if the economic, social, and political vulnerabilities that made the consequences of the earthquake so disastrous for so many people were anything but natural. The spectacular imagery of the devastation caused by the earthquake attracted significant media attention, and high visibility inspired both private and public donors (DARA 2011: 7). The fact that several significant non-traditional donors were Latin American governments suggests that the location of Haiti in their neighbourhood in the Western hemisphere may have played a role in driving their donations. Just as the social proximity of the tsunami-affected zones in 2004, some of which were destinations popular with Western tourists, created a sense of 'it could have been me' among Western publics, the geographical proximity of Haiti to the United States, less than 1,500 km from Florida, is likely to have been important in generating US public interest and concern. For the US government in particular, a long history of interventions in Haiti and a desire to contain would-be refugees in their country of origin may also have motivated high levels of funding.

[5] See 'Funding for humanitarian response' in Chapter 7 on the 2004 Indian Ocean Tsunami.

Official donors were generous in their pledges but slow to convert those pledges into contributions. In March 2010, the Haitian government presented an ambitious Action Plan for National Recovery and Development at a donor conference in New York, and more than USD 5 billion was pledged for the 2010–2012 period, around another USD 3 billion for programmes up until 2020 and almost USD 1 billion in debt relief (Fan 2013: 19–20). Eighteen months after the March pledging conference, however, only 43% of the promised funds had been disbursed, and the percentage was much lower for some of the largest donors—the United States, for example, had disbursed only 18.8% (Schuller 2012: 171). By November 2012, 53.2% of pledges from official donors (USD 6.04 billion) had been disbursed, and international assistance from private donors amounted to an estimated USD 3 billion (OSE 2012: 8, 15).

Official donors provided significant quantities for recovery and development, and they favoured funding international agencies rather than the Haitian government or Haitian non-governmental organizations (NGOs).[6] Some donors preferred to support only the emergency phase and provided only short-term funding, but by November 2012, bilateral and multilateral donors had contributed USD 2.41 billion in humanitarian funding and USD 3.63 billion in recovery funding (DARA 2011: 10; OSE 2012: 15). However, despite—or perhaps because of—the widely shared perspective that weak national governance was at the root of many of Haiti's vulnerabilities, the vast majority of that funding went to international, not national, actors. Of the USD 6.04 billion from official donors that had been disbursed in humanitarian and recovery funding by November 2012, less than 10% was estimated to have gone to the government and less than 0.6% to Haitian organizations and businesses (OSE 2012: 15–16). Even when funds for recovery are separated out of the total, only 15.4% was channelled directly to the government (OSE 2012: 15). This is consistent with a narrative in which Haitians are helpless and Haiti's problems are self-made and need solving by (white, Western) outsiders, and it also implies a limited understanding of recovery and development, focused on assets and infrastructure rather than building up the government and other Haitian actors.

A final important source of funding for the earthquake response was the many Haitians living abroad. The Haitian diaspora is estimated at 1–2 million people—87% of them living in the United States, the Dominican Republic, and Canada—a very high number relative to the population of Haiti itself,

[6] On the preferred channels of different types of donors, see 'Donors' in Chapter 18 on Donors and the Funding of Humanitarian Action.

around 40% of whom have family living abroad (Versluis 2014: S96–S97). The diaspora is a major source of income, with one in five Haitians receiving remittances from abroad, collectively estimated to contribute at least 19% of Haiti's gross domestic product (GDP), far exceeding international aid in a typical year (Versluis 2014: S96, S98). It is impossible to know exactly how much support the diaspora provided in the aftermath of the 2010 earthquake, but in general, remittances are expected to increase after a crisis, and one study found that those who had received remittances before the earthquake tended also to have received them afterwards (Versluis 2014: S103). Such 'informal' aid tended to be in the form of cash transfers direct to affected individuals and households and was deemed more timely and flexible than most formal aid (Versluis 2014: S103–S104).

International response

There was already a major presence of international agencies prior to the earthquake—indeed, Haiti was nicknamed the 'Republic of NGOs', both because more than 10,000 of them were operating there and because of the way many of them bypassed the government in their operations (Schuller 2012: 171). Many of these NGOs scaled up their operations after the earthquake, and many more arrived on the scene (DARA 2011: 5). A large number and diversity of private organizations went to Haiti wanting to help, despite damage to the main airport and seaport rendering both inoperative, limiting access to Haiti for international responders in the immediate aftermath and slowing the arrival of international aid and aid workers (Levine et al. 2012a: 13; Patrick 2011: 2; Versluis 2014: S95). Many of the individuals arriving in Haiti were young and inexperienced, and many of the NGOs were established specifically to respond to the Haiti earthquake (DARA 2011: 5; Schuller 2012: 173–174).

A number of weaknesses in needs assessments by international agencies limited their understanding of how best to respond.[7] The international humanitarian community was quick to implement the Rapid Initial Needs Assessment for Haiti but slow to publish it such that many of its findings were out of date by the time they were available to many who might have made use of them (Patrick 2011: 3). Individual agencies mostly conducted their own needs assessments, but the lack of common standards, methodologies, and focus limited their usefulness for overall analysis or strategic

[7] On common challenges to, and weaknesses of, humanitarian needs assessments, see 'Assessing needs, contexts, and capacities' in Chapter 21 on Needs Assessment, Evaluation, and Response Decisions.

planning (Patrick 2011: 3). Moreover, these assessments largely lacked contextual analysis and assessments of the capacity of the Haitian government and other Haitian stakeholders, Haitians were poorly consulted, and valuable studies and assessments carried out by Haitians themselves were largely ignored (Patrick 2011: 3, 5). As a consequence, international responders had poor understanding of social and political dynamics and of the capacities of Haitian government and civil society, and many in the humanitarian community wrongly assumed that there was little or no residual local capacity (Patrick 2011: 3, 4).

Partly as a consequence of these weaknesses in needs and capacity assessments, the international humanitarian response to the earthquake was not well adapted to the social and political context in Haiti. The responses and coping strategies of Haitians were largely overlooked (Patrick 2011: 5). The fact that much of the destruction—and hence much of the need for international assistance—was concentrated in urban areas posed an additional challenge for international humanitarian agencies, most of which are more familiar with operating in rural settings (Levine et al. 2012a: 1; Patrick 2011: 4; Schuller 2012: 173).[8] The resulting humanitarian response was 'self-contained, working outside government systems and reliant on imported material and personnel, supporting displaced individuals in internally displaced persons camps with food and non-food assistance' (Patrick 2011: 4). In many respects, this response was designed to replace, rather than support, local actors and served to disempower, rather than strengthen, Haitian government and civil society (Patrick 2011: 4).

Fortification and militarization

Haiti was perceived as politically unstable and Port-au-Prince in particular as highly insecure (Levine et al. 2012a: 5). Insecurity—or perceptions of insecurity—affected the response.

Due to the designation of high security levels by the UN, international staff members of most agencies were prevented from interacting freely with Haitians (Fan 2013: 22). International aid workers in some organizations were not allowed outside their residences at night, for example, and were forbidden from going to certain 'no-go' zones (Beerli 2018: 80–81). International aid was further 'fortified' or 'bunkerized' as coordination meetings were often held at the UN logistics/military base, where security measures

[8] See 'Urbanization' in Chapter 15 on the Shifting Parameters of Humanitarian Emergencies.

made it difficult for Haitian nationals to enter (Fan 2013: 22). The resulting distance between international aid workers and their intended beneficiaries exacerbated issues of poor contextual knowledge on the part of aid agencies and insufficient inclusion of, and accountability to, local communities in the design and implementation of the international response.

The international response to the earthquake also included a military element. The United States sent 14,000 troops, who got Haiti's seaports and airport functional, setting up an air traffic control system and taking control of the airport in Port-au-Prince (de Waal 2014; Versluis 2014: S95). The US military eventually also played a useful role clearing roads and key areas of rubble (de Waal 2014; Patrick 2011: 7). In the weeks immediately following the earthquake, however, the military was poorly used, and beyond managing the airport and airspace, it mainly provided armed escorts for humanitarian agencies (Patrick 2011: 7). The army later moved into other relief activities, for which it was neither cost efficient nor highly skilled (de Waal 2014). The United Nations Stabilization Mission in Haiti (MINUSTAH), which had been operating since 2004, was reinforced after the earthquake, with additional troops and a mandate extended to include assisting the Haitian government with post-disaster relief and recovery (TDC et al. 2013: 8, 9–10). MINUSTAH troops assisted with clearing debris, distributing food, and rebuilding local infrastructure (TDC et al. 2013: 10).

Emergency assistance

The initial humanitarian response was widely deemed adequate, and short term targets were mostly achieved and sometimes exceeded (Biquet 2013: 130; Patrick 2011: 2). Large quantities of material assistance were provided, with 2.1 million household non-food item kits distributed in the first 6 months, and 4 million people receiving food aid (Patrick 2011, 2). 1.2 million people had access to safe water daily, and 90% of displaced people in the capital had access to health clinics (Patrick 2011: 2). However, the response was limited in important ways. Aid delivery was mostly restricted to the displaced in camps in Port-au-Prince, neglecting those who continued to live among the rubble and those outside the capital (Levine et al. 2012a: 13). Furthermore, the response focused on meeting immediate needs and was not designed in such a way as to support longer-term recovery and livelihoods. For example, goods and services that had previously been provided by Haiti's private sector were procured overseas and distributed for free, undermining business recovery in a country where 43% of the workforce was self-employed (Patrick 2011: 5). While 1 million people participated in cash-for-work projects in the first six months, these projects were administratively burdensome, typically paid minimum wage (USD 5 per day), and provided a maximum of twenty-five days of employment per person, meaning that the money participants earned was insignificant compared with their needs (Levine et al. 2012a: 15; Patrick 2011: 2).

With around two million people displaced from their homes, shelter was an obvious priority—a need that was met much more effectively in the immediate term than in the medium or long term. Within 6 months of the earthquake, 1.5 million people received emergency shelter materials, in the form of tents and sheeting, and this initial shelter response was largely viewed as a success (Fan 2013: 22–23; Patrick 2011: 2). Shelter requires not only materials but also land on which to construct the shelter, and herein lay a significant challenge in Haiti. In rural settings, it is usually easier to find land onto which to relocate people, but most of those displaced by the earthquake in Haiti were in a densely populated urban area, and international agencies participating in the shelter sector saw land rights as a major obstacle (Levine et al. 2012a: 1; Schuller 2012: 173). Both local councils and private landowners sometimes sought to reclaim land by evicting camp occupants, whom they perceived to be squatting on their land (Levine et al. 2012a: 1, 17). Humanitarian agencies did not consider it their responsibility to pay rent to landowners and did not engage in any coordinated or proactive way with the landlords themselves, with UN agencies in particular focusing their efforts on state authorities and seeking—unsuccessfully—to persuade the government to put a moratorium on all evictions (Levine et al. 2012a: 1, 20).

In the context of the cholera epidemic, the provision of water, sanitation, and health services was absolutely crucial, and here results were mixed. Despite the achievements of the initial emergency response, many internally displaced person (IDP) camps lacked water and sanitation services when the epidemic began (Schuller 2012: 172). Médecins Sans Frontières (MSF) reported that water, hygiene, and sanitation response 'was cruelly lacking', forcing the agency to take charge of meeting these needs as well as providing treatment to cholera patients (Biquet 2013: 133). Despite the activation of the Health Cluster after the earthquake, it did not react sufficiently quickly or effectively to the cholera outbreak, with 80% of patients in the first three months taken in by MSF and the Cuban health brigades, neither of which was working within the cluster system (Biquet 2013: 133; see also TDC et al. 2013: 14–15). While much of it may have taken place outside the formal humanitarian coordination system, the cholera response in the camps was relatively effective. Whereas camp-based populations often suffer the highest death rates from disease outbreaks, there was close to zero mortality from cholera in the IDP camps in Haiti (DARA 2011: 5). However, other vulnerable parts of the population were neglected, and many agencies were outraged by the failure to contain the outbreak more effectively, given the massive international presence and resources (DARA 2011: 5).

A lack of alternatives meant that people stayed in camps long term, and over time the services provided were reduced, and insecurity appeared to increase.[9] Humanitarian agencies felt an urgent need to get people out of camps, but the high standards set for longer-term housing solutions. and wide acceptance that the humanitarian response could not limit itself to those who had lost their homes due to the earthquake but also had to deal with the effects of chronic poverty and homelessness, made this very difficult (Levine et al. 2012a: 21). Camp residents, who accounted for one-fifth of the population of Port-au-Prince, were thus stuck in temporary settlements, with no viable alternative, long after emergency funds for life-saving food, water, and sanitation services ran out (Schuller 2012: 172). Furthermore, reports suggest that the incidence of rape had increased several-fold in some Port-au-Prince camps by the end of 2011 (DARA 2011: 4).

Recovery and rehabilitation

As was the case after the 2004 Indian Ocean tsunami,[10] many international responders saw the earthquake as an opportunity for building back better and even catalysing wider structural change in Haiti (Fan 2013: 19, 22). As a minimum, BBB meant physically reconstructing buildings and infrastructure in such a way as to make Haiti more resistant to future disasters, particularly earthquakes and hurricanes, and implied an emphasis on the physical and technical aspects of rebuilding (Fan 2013: 22). In addition, many humanitarian agencies saw the injection of international funding and expertise as an opportunity for direct community involvement in recovery and development and hoped this would lead to more fundamental structural changes in Haitian society and politics (Fan 2013: 22). There were heated debates as to what a sustainable approach to shelter would look like, and some questioned the extent to which international responders should be involved in reconstruction at all (Fan 2013: 23).

[9] For a more general discussion of camps for displaced persons, see 'Politics of camps' in Chapter 22 on Material Assistance and Direct Service Provision.

[10] See Chapter 7.

In practice, despite large amounts of funding, the international response failed to BBB, even in a limited way, as very little reconstruction actually took place. Both donors and humanitarian agencies shied away from housing projects, which frequently posed complex problems and required acceptance of imperfect solutions—in terms of both construction quality and the legalities of land tenure. Donor reticence meant that much reconstruction money went to expensive non-shelter projects, and housing was left to humanitarian funds and agencies such that, by the end of 2012, USD 6 billion of international aid had been disbursed, but more than half had gone to humanitarian relief and only USD 215 million to housing reconstruction (Fan 2013: 22). However, humanitarian agencies are not well placed to provide long-term housing solutions—their timescales and skill sets are not suited to resolving deep-seated land tenure issues, much less achieving structural change in the area of land rights (DARA 2011: 7; Levine et al. 2012a: 23). Moreover, some of the largest official donors were not willing to pay the costs of the kind of housing that would have met the high standards they were talking about (DARA 2011: 7).

Operational agencies were reluctant to compromise and to work with the messy reality of Haiti's system of land administration, and this severely restricted the building of new homes (Levine et al. 2012a: 2, 15). In Haiti, land and economic power are concentrated in the hands of a few large families (Levine et al. 2012a: 5, 7). An antiquated and complex system of land law and administration—according to which it is often unclear who is the legal owner of a piece of land—coexists with the practice of occupying land without formal title and arrangements between landlords and tenants whereby tenants build houses which they own, but on rented land (Levine et al. 2012a: 7–9). Such pragmatic arrangements sat uncomfortably with international agencies, who sought to establish proof of ownership before building new structures and to micro-manage the terms of rental agreements between landlords and tenants. Their inability to relinquish control meant that they failed to address the biggest problem facing those who had been tenants, which was finding enough money for a twelve-month deposit for rent on a new property (Levine et al. 2012a: 2).

At the same time, agencies were reluctant to support people's efforts to repair damaged houses, and no assistance was given for repairs for over a year, despite the fact that repairs cost a fraction of what was spent to provide people with a prefabricated temporary structure (Levine et al. 2012a: 1–2). This reluctance can be attributed to concern about their reputational, legal, and moral responsibilities should repairs prove fatally inadequate in a future disaster and also to a more general reticence to relinquish control, with repairs by their nature custom-designed and hence difficult for aid agencies to budget for and monitor (Fan 2013: 23; Levine et al. 2012a: 2). In the 20 months after the earthquake, only 5,000 homes in the Port-au-Prince metropolitan area were repaired or newly constructed, a pitiful number considering the high levels of funding and the fact that 99,000 homes had been assessed by engineers to be in need of serious repair and another 78,000 marked for demolition (Schuller 2012: 171–172).

Instead of supporting people to rebuild, repair, and relocate in the ways they had previously done, and for which they mostly only lacked the necessary funds, the international response focused on providing its own solutions, which it was unable to do at the required scale (Levine et al. 2012a: 1). This approach reflected humanitarian 'business as usual' and meant that engineers could design shelter solutions to meet technical standards without concerning themselves too much with social, legal, and economic challenges and actors (Levine et al. 2012a: 21). Such an approach also fitted with the dominant representations of Haiti and Haitians in the global North. Had alternative understandings of agency and resilience been dominant, BBB efforts might have looked very different, focused on finding ways to increase people's agency and support their own efforts to rebuild and repair housing rather than providing (an inadequate number of) externally designed shelter solutions to a population portrayed as exploited, impoverished, and disempowered (Fan 2013: 20).

Having abandoned a strategic focus for a technical solution, the international humanitarian shelter response effectively amounted to BBB by T-shelter (the primary transitional shelter response), with the 'better' coming from innovations to construction such as using reinforced steel bars to increase hurricane resistance (Fan 2013: 22–23). This focus on technical solutions and project implementation rather than seeking to address institutional challenges also meant that the humanitarian response lacked a clear exit strategy and became protracted, with international agencies using short-term tools to address chronic problems (Levine et al. 2012a: 21). Paying camp occupants to leave camps was a controversial policy, even more than a year after the earthquake, with agencies divided between viewing it as a way to help people find their own solutions by helping with rental payments, for example, and seeing such an approach as in contravention of international law, in as far as it sought to entice people to leave the camps even in the absence of a durable solution and sometimes without informing them of their rights (DARA 2011: 7; Levine et al. 2012a: 20). Almost 400,000 people remained in temporary camps and settlements two years after the earthquake (Levine et al. 2012a: 5).

Coordination and ownership

The quality of the humanitarian response to the earthquake and subsequent cholera outbreak was negatively affected by poor coordination—both among international actors and between international and domestic actors. A large number of UN staff, including the Special Representative of the Secretary-General, died in the earthquake, and the Humanitarian Country Team was only reconvened three weeks later (Patrick 2011: 6). At the same time, hundreds of new, inexperienced donors and organizations arrived in the country, causing huge challenges in coordination. Many of the amateur, or spontaneous, NGOs did not register with the government, effectively working illegally and demonstrating a lack of respect for Haitian authorities. The large number of NGOs additionally posed a challenge to the coherence of the overall aid effort, with many actors pulling in different directions, resulting in gaps and overlaps in the delivery of assistance (Levine et al. 2012a: 13; Patrick 2011: 7).

The cluster system[11] was quickly activated, with thirteen clusters and multiple working groups established in Port-au-Prince in a mostly unsuccessful attempt to ensure some measure of coordination and organization (Levine et al. 2012a: 13). Certain characteristics inherent in the cluster approach limited its effectiveness in Haiti. Notably, with respect to shelter and housing, the division of labour across at least four different clusters (with the camp coordination and camp management cluster responsible for camps and spontaneous settlements; the shelter and non-food items cluster for the provision of transitional shelter; the early recovery cluster for debris removal, house repairs, permanent housing, and settlement planning; and the protection cluster for advocacy concerning forced evictions, protection for renters, and resettlement) fragmented the response and arguably made it more difficult to develop a common strategy or a consistent, coordinated, and cohesive approach (Fan 2013: 23; Levine et al. 2012a: 13).

The way in which the cluster approach was implemented in Haiti further limited its effectiveness. Lengthy and numerous cluster meetings made for a heavy coordination system (Levine et al. 2012a: 13). Furthermore, cluster meetings in Port-au-Prince were initially held in English at the UN logistics base, which was convenient for multilateral agencies and INGOs—and

[11] The 'cluster system' is a coordination structure involving UN and non-UN humanitarian organizations at both the global and the country level. Each cluster focuses on a particular aspect or sector of emergency response (e.g. food security, camp coordination/management, early recovery, emergency shelter, health, etc.) and has a clearly defined lead agency and a provider of last resort.

facilitated coordination among them—but served to exclude and marginalize Haitian actors, who could not easily enter the base and were not all able to participate in English (DARA 2011: 6; Fan 2013: 22; Levine et al. 2012a: 13). Later on, French became the lingua franca and some meetings were moved to municipal hubs, but some continued to be held in English, and the UN base remained the hub of the international aid community (Levine et al. 2012a: 13).

The Haitian government failed to take an institutional lead on emergency relief, recovery, and reconstruction, or land issues (Levine et al. 2012a: 14), and international aid agencies failed to integrate Haitian authorities and civil society organizations into the response and recovery. Prior to the earthquake, aid had often marginalized and undermined state institutions as it tended to be channelled through parallel institutions, responding to priorities determined by INGOs, without significant input from the government or local communities (Fan 2013: 21). Unsurprisingly, then, government and state representatives lacked good understanding of the institutional architecture of the international humanitarian system, and conversely, aid agencies reported difficulties in understanding the roles and responsibilities of different government roles, ministries, and commissions (Levine et al. 2012a: 14). Cooperation between the most senior level of the Haitian government and top political, humanitarian, military, and diplomatic representatives was relatively good, but this was not extended to the lower tiers of national or local government (Patrick 2011: 5).

Despite wide agreement that BBB should not only mean building physical infrastructure but also stronger public institutions that would be better able to manage public resources and deliver basic services, most international aid did not serve to empower or build the capacity of Haitian government and civil society, instead sidelining them (Fan 2013: 21; Patrick 2011: 5).

Accountability

Running through the whole chapter is the question of accountability, or lack thereof, a perennial problem with humanitarian action. There exist mechanisms of 'upward accountability' from international humanitarian agencies to states (as the main donors to those organizations and, in the case of UN agencies, as member states). Such mechanisms include strict reporting requirements, and states have the power to restrict the mandates of UN agencies and to restrict the funding of both UN and non-UN agencies. However, those who receive humanitarian aid are not those who pay for it, and the people who are intended to benefit from humanitarian action have very little power to hold humanitarian actors accountable.

While the humanitarian sector has been criticized for the unfulfilled BBB promises in Haiti, affected populations were mostly unable to demand accountability from the sector or the agencies within it. Efforts to improve accountability to affected people since the 1990s have encompassed standard-setting initiatives and platforms focused on communication and dialogue between beneficiaries and aid agencies (including complaints procedures). However, it is not clear that either of these kinds of accountability mechanism worked to improve accountability to affected populations in Haiti. Standard-setting initiatives, the most widespread of which is the Sphere Humanitarian Charter and Minimum Standards, first published in 1999, have been criticized for focusing on technical standards and quantitative measures in such a way that they risk undermining the humanitarian ethic and often come at the expense of addressing more difficult ethical dilemmas (Terry 2000). In response to the earthquake in Haiti, such standards at best made little sense (many Haitians were living in conditions that did not meet Sphere standards prior to the earthquake), and at worst contributed to the reticence of international humanitarian actors to support people's own diverse and imperfect solutions to land and housing issues, for example (Levine et al. 2012a: 21). Although some agencies, including the International Federation of Red Cross and Red Crescent Societies, the International Organization for Migration, and Oxfam, developed platforms for dialogue with beneficiaries and community-based groups (Fan 2013: 22), Haitians had little or no meaningful input into the design of programmes managed by international responders and little or no recourse for unmet promises.

Even when international responders did very clear damage to Haiti and Haitians, those affected were (and continue to be) unable to hold them to account for their actions. Two incidents in particular bear highlighting. First, for several years, the UN denied that cholera had been introduced to Haiti by peacekeepers, despite overwhelming evidence (TDC et al. 2013). The UN eventually made some kind of apology for the cholera outbreak in December 2016 but stopped short of accepting formal responsibility or any kind of legal accountability (Pilkington 2020). Those affected have received no compensation, and international funding for initiatives aimed at eradicating cholera in Haiti have been severely underfunded. Second, Oxfam staff responding to the earthquake were involved in sexual abuse, using prostitutes in Haiti, some of whom were reportedly underage (Cooper 2019). The scandal made headlines in 2018, and Oxfam was accused of having covered up its own 2011 investigation, which resulted in four staff members being dismissed, but another three had been allowed to resign before the investigation was complete (Cooper 2019). Senior figures resigned when the scandal broke, and funding from both public and private donors was cut drastically as those who hold the purse strings exercised their power to hold Oxfam accountable (Gayle 2018). However, it is far from clear that this strategy leads to greater accountability or better outcomes for affected populations. Apart from the fact some beneficiaries will likely suffer from programme cutbacks, punishing Oxfam—an agency known to have some of the best policies on sexual violence in the sector and which had not entirely covered up its 2011 report—in this way likely serves to disincentivize transparency about abuse in other agencies (Crack 2018).

Conclusions

The international response to the earthquake in Haiti suffered from insufficient involvement of Haitian actors. In the immediate aftermath of the quake, the perceived need for haste was understandable, and indeed, the emergency response was broadly deemed effective. In the longer term, however, bypassing the Haitian government and local people served to disconnect the response from its context and to undermine recovery efforts (Patrick 2011: 5). This sidelining of Haitian actors was consistent with dominant discourses about Haiti in the global North, which tended to depict the country and its people as in perpetual need of international assistance. It also reflected a humanitarian aid system which is risk averse and emphasizes technical expertise, viewing the people affected by crisis as passive recipients rather than active participants, and as such failed to make good use of local contextual knowledge and failed to support people's imperfect solutions even as it failed to deliver on its own promises (Levine et al. 2012a: 2). The case of Haiti also highlights the power that international intervenors wield, the scope for them to abuse this, and the inadequacy of mechanisms through which affected populations can hold them to account—either for failing to live up to their promises (in this case, to BBB) or for actions that are damaging (the introduction of cholera to Haiti and sexual exploitation of Haitians).

10
Famine in Somalia, 2011–2012

The famine in Somalia in 2011–2012 was triggered by drought in late 2010 and early 2011, but it was not the drought alone that caused the famine. Rather, it was caused by drought combined with rapidly increasing food prices against a background of long-running conflict and an underlying livelihoods crisis, compounded by a weak and delayed international response—a response that was itself hampered by several conflict-related factors.

From August 2010, reliable early warning systems began predicting a food crisis in Somalia, but the international community was slow to respond. Only when the Famine Early Warning System Network (FEWSNET) and the Food and Agriculture Organization (FAO)-operated Food Security and Nutrition Analysis Unit for Somalia (FSNAU) declared a famine in much of the south of the country on 20 July 2011 was the international response scaled up significantly. By October 2011, 750,000 people—17% of the population of south central Somalia—were classified as being in famine conditions (Majid and McDowell 2012: 36). The famine continued into early 2012, and between October 2010 and April 2012 it is estimated that 258,000 people, just over half of whom were children under 5, died as a consequence (Checchi and Robinson 2013: 8). In southern and central Somalia, 4.6% of the total population and 10% of children under five died, with marginalized groups disproportionately affected (Checchi and Robinson 2013: 8, 10; Majid and McDowell 2012).

The international and domestic political context—most notably international counter-terror legislation, restrictions imposed by Harakat Al-Shabaab al-Mujhadeen (Al-Shabaab), and access constraints relating to insecurity—had a significant impact on the international humanitarian response. Decision-makers among humanitarian donors and operational agencies largely saw the risks of action as outweighing the risks of inaction until the declaration of famine, at which point the equation was reversed and a vigorous response mobilized. Even then, however, operations were constrained by insecurity and the long absence of some agencies from the areas most affected by famine. In an effort to overcome these constraints,

The Politics and Everyday Practice of International Humanitarianism. Miriam Bradley, Oxford University Press.
 DOI: 10.1093/oso/9780198840695.003.0010

a consortium of agencies implemented what was then one of the world's largest ever cash transfer programmes, which served to improve food and livelihoods security for much of the famine-affected population—albeit less effectively than could have been expected of an earlier and more extensive response.

Background

Somalia has not had a central government since 1991, when the regime that had been headed by Siad Barre for more than twenty years was toppled following several years of civil war. In the aftermath of the disintegration of the state, severe famine occurred with hyper-inflation, nearly complete crop failure, and more than 50% livestock losses (Checchi and Robinson 2013: 17). Extreme insecurity continued in the early 1990s, and three United Nations (UN)-mandated international military interventions between 1992 and 1995 largely failed to achieve their objectives. The last of these, the UN Operation in Somalia II (UNOSOM II), was widely seen as a party to conflict and had a particularly conflicted relationship with the humanitarian community. On the one hand, many aid agencies depended on UNOSOM for armed escorts, and on the other hand, UNOSOM and major donor governments sought to instrumentalize aid as part of their political and stabilization strategies, which contributed to the perception that aid agencies were taking sides and generated tensions with local militias (Jackson and Aynte 2013: 5).

After the dissolution of UNOSOM II in 1995, the intensity of fighting dropped significantly, and around the same time neighbourhood *sharia* courts emerged as an important source of law and order in the Somali capital, Mogadishu. Working with local businesses and clan militias to provide a degree of security and welfare that made them popular and legitimate in the eyes of much of the population, the courts and associated militias eventually became the basis for the Islamic Courts Union (ICU), a loose umbrella group of Islamist organizations (Abild 2010: 78; Barnes and Hassan 2007: 152). In early 2006, the United States backed a weak and unpopular alliance of warlords to capture or kill individuals in Somalia responsible for US embassy bombings in Nairobi and Dar es Salaam and to stem the influence of the Islamic Courts (Little 2012: 191). Within a few months, the US-backed alliance was defeated by the ICU, who gained control of Mogadishu and most of southern Somalia. Their victory served to increase the popularity and strength of the radical Islamic elements within the ICU, in particular the youthful al Qaeda-affiliated Al-Shabaab (Little 2012: 191–192; Menkhaus 2012: 30).

With US backing, neighbouring Ethiopia invaded Somalia in December 2006, routed the ICU, and occupied Mogadishu for two years, bringing to an end a decade of relative stability, political innovation, and economic dynamism, further fuelling local resentment against outsiders, and increasing support for Al-Shabaab (Abild 2010: 78–79; Hammond and Vaughan-Lee 2012: 3; Little 2012: 192; Majid and McDowell 2012: 38; Menkhaus 2012: 30). With the approval and authorization of the UN Security Council, the African Union Mission in Somalia (AMISOM) was deployed in early 2007, joining and eventually replacing the Ethiopian forces in Somalia (AMISOM n.d.; Menkhaus 2012: 30). AMISOM was mandated to protect the internationally backed Transitional Federal Government (TFG), stabilize the country, and facilitate humanitarian activities, but in practice, Western policies in 2007 and 2008 had the opposite effect, contributing to insecurity and reducing humanitarian space (Menkhaus 2009). Al-Shabaab emerged as the main insurgency fighting the TFG and trying to drive out AMISOM and the Ethiopian forces, turning Mogadishu and parts of south Somalia into a battlefield for several years (Menkhaus 2012: 30). A chronic livelihoods crisis developed after more than 600,000 Somalis were displaced by fighting in Mogadishu during the Ethiopian occupation, and in parts of southern Somalia settled farmers and agro-pastoralists abandoned their farms or lost livestock due to armed conflict (Little 2012: 192). The insecurity additionally led to a significant decrease in the presence and operations of international non-governmental organizations (INGOs) and UN agencies, above all in the areas most in need (Bradbury and Maletta 2012: 117).

The instrumentalization of humanitarian aid by donors since 9/11 further served to reduce humanitarian access. Donors, regional governments, and the UN took sides—and even became belligerents—in the war, with the United States taking unilateral military action against particular targets, other donors funding the training and salaries of TFG forces, and all failing to hold the TFG, Ethiopian forces, or AMISOM to account for violations of international humanitarian law (IHL), even when there was clear evidence of abuse of civilians and despite the fact that they condemned attacks by insurgents (Bradbury and Maletta 2012: 121). Although it was not an integrated mission, donor governments and the UN put pressure on aid agencies to support the TFG, and the UN Department of Political Affairs additionally made some attempts to align assistance with its state-building and stabilization objectives (Bradbury and Maletta 2012: 122–123). Humanitarian agencies largely resisted this pressure, but for-profit contractors provided assistance as part of a stabilization strategy, and this may have affected local perceptions of humanitarian agencies and further compromised acceptance of their presence and work (Bradbury and Maletta 2012: 124).

From late 2009, humanitarian access and funding were massively reduced as a consequence of counter-terror restrictions imposed by the UN and a number of Western countries. The United States had listed Al-Shabaab as a terrorist group in 2008, effectively criminalizing humanitarian engagement with its members.[1] The US Patriot Act and associated counter-terror legislation prohibited providing—even unintentionally—any material benefit to listed groups, and United States Agency for International Development (USAID) officials feared that providing aid to people in territory controlled by Al-Shabaab could make them criminally liable (Menkhaus 2012: 31). In September 2009, the Obama administration suspended all US humanitarian assistance to areas controlled by Al-Shabaab, leading to an 88% decrease in US funding for humanitarian aid to Somalia between 2008 and 2010, with significant implications for food security in southern Somalia, given that the United States is the main source of food aid globally and had been the single largest donor to Somalia (Menkhaus 2012: 31; UN OCHA 2010: 3). The UN and several more countries proscribed Al-Shabaab in early 2010, and Canada suspended humanitarian assistance to Somalia (Bradbury and Maletta 2012: 128). While UN sanctions included the option for a humanitarian exception, this was not adopted by all countries and did not cover all humanitarian agencies (Pantuliano et al. 2011: 9).

Insecurity for aid workers in Somalia and restrictions imposed by Al-Shabaab further restricted humanitarian access and operating space, and by 2010, international humanitarian action in much of Somalia was extremely limited. Since the mid-1990s, insecurity had led most international agencies to manage operations remotely from Kenya, a move which implied a reduction in the number of staff in-country and a transfer of risk to national staff and sub-contracted local organizations (Abild 2010: 83; Fredriksen 2016: 45). After US airstrikes killed Aden Hashi Ayro, a top Al-Shabaab military commander, in May 2008, aid workers were accused of collaborating and officially designated by Al-Shabaab as legitimate targets (Abild 2010: 79–80; Bradbury and Maletta 2012: 118–119). Due to its support for the TFG and the Ethiopians, the UN was denounced by Al-Shabaab as an enemy of Islam, and UN compounds were attacked (Abild 2010: 80; Bradbury and Maletta 2012: 125;

[1] For a discussion of how counter-terror legislation has affected humanitarian action, see 'Counter-terror legislation' in Chapter 14 on Law and International Humanitarianism.

Fredriksen 2016: 45). The UN's official support for one side in the conflict greatly limited international access to the large parts of the country not controlled by the TFG, and in the years immediately prior to the famine, Somalia was one of the most dangerous countries in the world for humanitarian aid workers (Menkhaus 2012: 32; Slim 2012: 8). The leadership of Al-Shabaab was divided with respect to the presence of international aid agencies, but its hardliners prevailed, expelling many agencies—including CARE in early 2009, and subsequently other non-governmental organizations (NGOs)—from areas under its control and imposing tight restrictions on those that remained (Fredriksen 2016: 5; Lautze et al. 2012: 44; Menkhaus 2012: 30). In December 2009, the World Food Programme (WFP) announced that it was suspending operations due to unacceptable levels of insecurity (although the real reason for withdrawal may have had more to do with pressure from the United States), and following the announcement, Al-Shabaab immediately banned WFP (Bradbury and Maletta 2012: 127; Menkhaus 2012: 31).

Proximate causes of famine

Against this background, rapidly rising food prices and rapidly falling rural incomes created a perfect storm in which many people—particularly agriculturalists, agro-pastoralists, labourers, and internally displaced persons (IDPs)—were unable to access the food necessary for survival (Maxwell and Fitzpatrick 2012: 6).[2] The fall in rural incomes was, in large part, the consequence of drought. Somalia has two rainy seasons, and severe drought occurred after the lowest recorded rainfall in fifty years during the late 2010 Deyr season and the subsequent failure of the Gu rains in April–June 2011 (Maxwell and Fitzpatrick 2012: 6). This resulted in the lowest annual cereal crop since the 1991–1994 Civil War and substantial livestock mortality (Hillbruner and Moloney 2012: 20). Reduced harvest and loss of livestock meant that common sources of income—farm labour and livestock sales—were reduced. At the same time, food prices increased significantly, with the price of red sorghum in one part of Somalia rising 240% between June 2010 and June 2011 (Hillbruner and Moloney 2012: 20). Drought-related shortages in the supply of food contributed to price rises, but so too did the massive reduction in food assistance—which had been responsible for an estimated average of 15% of national cereal supply between 2007 and 2010, peaking at 29% during 2009—after WFP withdrew from Somalia in January 2010, itself a

[2] On the meaning and causes of famine in general, see 'Famine' in Chapter 1 on Humanitarian Emergencies.

consequence of the conflict (Hillbruner and Moloney 2012: 25). Even in years of good rainfall, Somalia relies heavily on imported food—accounting for around 55% of national cereal supply between 2007 and 2010 (Hillbruner and Moloney 2012: 25). Much of this takes the form of commercial imports and increases in global prices following the 2007/2008 global food price crisis thus exacerbated the crisis in food access (Maxwell and Fitzpatrick 2012: 5).

In different circumstances, Somalis would have been better able to weather this perfect storm, but years of armed conflict and a consequent livelihoods crisis had eroded their coping strategies—and the safety net of international food aid had largely been pulled out from beneath them. During the civil war, for example, much of the irrigation infrastructure, which facilitated more diverse crop production and provided a buffer against the worst impacts of climatic variability, was looted or fell into disrepair (Majid and McDowell 2012: 38). Livestock losses during the war left agro-pastoralists and pastoralists with fewer, or no, assets to sell, and many of those who had been displaced by armed conflict could not access their land, employment, or other usual livelihood opportunities. International assistance in the years leading to the famine had done little to build up resilience as it had focused more on food aid than strategic livelihood support and largely excluded south central Somalia in favour of those parts of the country that were more easily accessible (Lautze et al. 2012: 46). Remittances from the Somali diaspora[3] had been affected by counter-terror legislation and the UN sanctions regime and had also fallen in the wake of the 2008 financial crisis (Bradbury and Maletta 2012: 117, 129–130). Conflict and restrictions on population movement imposed by Al-Shabaab in 2011 limited the option of migration and mobility, ordinarily a livelihood strategy in pastoral and agricultural contexts across Somalia (Majid and McDowell 2012: 39–40).

By the time famine was declared in July 2011, an estimated 3.2 million people—2.8 million of whom were in the south—were in need of immediate, life-saving assistance (Slim 2012: 8). Certain population groups—predominantly the historically minority and marginalized Reewin and Bantu—were disproportionately affected, as they had been in the famine of the early 1990s (Majid and McDowell 2012: 36). They depended on the agricultural sector for their livelihoods, and their income derived from the same farms from which they bought their food, making them doubly vulnerable to the effects of the drought. Vulnerability peaked in September 2011, with an

[3] It is estimated that the Somali diaspora remits over a billion dollars per year to family members in Somalia, driven by the powerful reciprocal obligations that characterize relations among Somalis of the same lineage (Menkhaus 2012: 34). Indeed, international remittances are believed to be by far the largest contributor to gross domestic product (Majid and McDowell 2012: 39).

estimated 4 million people (53% of Somalia's population) in need of humanitarian assistance, three-quarters of whom were in south and central Somalia, areas largely under the control of Al-Shabaab (Hammond and Vaughan-Lee 2012: 3). Limited provision of health-care services, poor water and sanitation, increased disease transmission due to displacement, and exposure to trauma—all consequences of long-running conflict—exacerbated the effects of food security crisis (Checchi and Robinson 2013: 17). A substantial increase in emergency assistance in September and October, a sharp decline in local cereal prices, and excellent Deyr rains between October and December 2011 brought the famine to an end in early 2012 (Hillbruner and Moloney 2012: 20). It is estimated that more than a quarter of a million people had died (Checchi and Robinson 2013: 8).

Early warning, late response

Reliable warnings with alarming predictions were issued from August 2010, and with eleven months of early warning, international actors knew what was coming and could have responded much sooner than they did. Somalia had suffered a severe famine in the early 1990s, and in response FSNAU and FEWSNET were set up in 1995 to provide regular analysis of the status of food security and nutrition in Somalia. Based on monitoring of a range of indicators, including remote sensing imagery and price and trade data, FSNAU and FEWSNET produce a series of fourteen reports in a typical year, which provide early warning of food security crises (Hillbruner and Moloney 2012: 21). Between August 2010 and the declaration of famine in July 2011, they released an additional sixteen written products focused on the developing crisis as well as providing more than fifty briefings, mainly in Nairobi and Washington DC (Hillbruner and Moloney 2012: 21).

The briefings, alerts, and press releases first described the likelihood of poor rainfall and highlighted the expected impact on food and income sources, then told of a worsening situation, forecasting severe food insecurity and identifying different forms of early livelihoods support that could avert a crisis (Hillbruner and Moloney 2012: 21–23). By early 2011, they were reporting dramatic increases in cereal prices, extreme levels of malnutrition, the collapse of cattle markets in some regions, and a 20% increase in the number of people in need of emergency assistance (Hillbruner and Moloney 2012: 23). In February, March, and May, FEWSNET organized a series of three multi-agency scenario-building workshops, which resulted in consensus alerts that highlighted the urgent need for large-scale humanitarian assistance and stated that even if the Gu rains performed well, significant improvements in food security would take many months, and the current situation warranted an immediate response (Hillbruner and Moloney 2012: 23). In the event, the rains were delayed by a month, performed erratically, and ended early, resulting in the lowest Gu harvest since 1995 (Hillbruner and Moloney 2012: 23).

Despite reliable warnings beginning in August 2010, and sounding loud and clear through the first half of 2011, the international response was not scaled up until famine was declared in two areas of south-central Somalia on 20 July 2011 and subsequently in three more. The Integrated Food Security Phase Classification (IPC), which was first developed in Somalia in 2004 and is now a global tool for classifying levels of food insecurity, consists of a five-phase scale of food emergencies. Phase five (famine) occurs when, in a given geographical area, at least 20% of households face extreme food shortages with limited or no options to cope, the prevalence of Global Acute Malnutrition exceeds 30%, and crude death rates reach or exceed two deaths per 10,000 population per day (Hillbruner and Moloney 2012: 23). By definition, at this stage, extensive damage has already been done and lives and livelihoods lost. Not only could early humanitarian response have achieved more (including saving more lives) but also it would likely have been more cost effective. Restocking livestock herds can cost several times more than keeping the animals alive through supplementary feeding, for example, and it is significantly cheaper to provide support to people in their homes rather than IDP or refugee camps (Save the Children and Oxfam 2012: 19).

Only when famine was declared was there large-scale response, with a substantial increase in funding and emergency assistance. The response was not the result of the repeated warnings but of the declaration—it was a response to an already occurring famine, rather than a concerted effort to prevent that famine from occurring, and only got to scale once the worst was past. In some respects, this is not surprising as worldwide early warning is consistently ignored. The idea of the humanitarian emergency[4] corresponds not to potential future events but to radically present ones, and while sudden-onset disasters and images of famine often provoke significant media attention, early warning systems do not (Fredriksen 2016: 43, 44). However, the more timely provision of humanitarian assistance helped to avert famine in Ethiopia and Kenya, despite their also being affected by drought and food price rises (Slim 2012: 5).

There are three broad reasons why many donors and operational agencies were reluctant to act in Somalia. First, in the decades leading up to 2011,

[4] See Chapter 1 on Humanitarian Emergencies.

conditions that would be considered to have reached crisis levels in other countries had come to be considered normal in Somalia (Bradbury and Maletta 2012: 113). Crisis had been normalized in Somalia to such an extent that the distinction between routine crisis conditions and those requiring a significant scaling up of the humanitarian response was difficult to discern (Hobbs et al. 2012: 50). Second, despite the FSNAU and FEWSNET warnings, access restrictions and remote management practices limited the ability of humanitarian actors to collect information themselves and, hence, to get a 'human sense' of what was happening on the ground (Fredriksen 2016: 46). While the FSNAU data was good quality, it was patchy and was often outdated by the time the Humanitarian Country Team based in Nairobi processed it (Slim 2012: 10). There was no deep situational needs analysis, anecdotal information at the field level was not providing consistent messages, and the country team tended to discount reports by Somali NGOs (Hobbs et al. 2012: 52; Slim 2012: 10).

Third, extreme insecurity, counter-terror legislation, and concerns about the diversion and instrumentalization of assistance made many humanitarian actors cautious. While the diversion and instrumentalization of assistance is by no means unique to Somalia, since the collapse of the Somali state, aid had important impacts on the economy and local politics (Abild 2010: 85; Hammond and Vaughan-Lee 2012: 6–8). The listing by the UN and many countries of Al-Shabaab as a terrorist group exacerbated concerns about providing aid in south central Somalia, making both donors and operational agencies excessively cautious. Extensive risk mitigation measures, including finance checks, tracking and monitoring systems, reduced the flexibility and responsiveness of operations, inevitably slowing any response (Pantuliano et al. 2011: 9). Moreover, the food crisis warnings came at a moment when the question of diversion was particularly sensitive as the UN Monitoring Group on Somalia and Eritrea had published a report in March 2010 alleging that up to half of WFP's food aid in Somalia was being diverted by three of the organization's primary contractors and sold off illegally, possibly benefitting some armed groups (Ali and Gelsdorf 2012: 58).

Funding for humanitarian response

The famine declaration prompted immediate mobilization of funds, with donors—including 'traditional' and 'non-traditional' donor governments and the global public—almost doubling their funding for Somalia virtually overnight (Lautze et al. 2012: 44; Maxwell et al. 2012: 1). This mobilization of funds was crucial as some operational humanitarian agencies had sought to scale up their operations prior to the declaration and had been unable to because they lacked sufficient funding (Hobbs et al. 2012: 52; Lautze et al. 2012: 45). The Consolidated Appeals Process (CAP) launched in December 2010, for example, had been met with a 50% shortfall in committed funds, such that WFP had to reduce rations by around 65% in April/May 2011, and 79% in June 2011 (Fredriksen 2016: 51). While donors can be blamed for underfunding the CAP in the context of a rapidly worsening food security crisis, operational agencies—and above all the Humanitarian Country Team—share responsibility for requesting insufficient funding to begin with. Their pessimism about the generosity and propensity of Western donors to engage in crisis prevention shaped the CAP, appealing for funding in line with what they anticipated risk-averse counter-terror donors would give rather than what the situation might demand (Slim 2012: 10, 16).

Most of the main donors preferred pooled funding mechanisms administered by the UN, but some NGOs questioned whether they should seek funding through such mechanisms, given the UN's political stance and its listing of Al-Shabaab (Bradbury and Maletta 2012: 123). In the end, USD 1.3 billion was raised for Somalia, only USD 800 million of which went through the CAP system (Slim 2012: 8). In part, this was due to significant engagement of non-traditional donors, in particular members of the Organisation of Islamic Cooperation (OIC). Saudi Arabia worked within the CAP system, and in fact, became its third-largest donor with funds of USD 58 million, while some other OIC members did not (Slim 2012: 9, 12). Turkey was especially generous, announcing a total contribution of USD 365 million (Slim 2012: 12). The rapid response of OIC donors was significant not only in terms of its volume but also in its impact on more qualitative aspects—notably the 'how' and the 'where'—of the humanitarian response in Somalia.

International humanitarian response

Even when funding was significantly increased after the declaration of famine, the conflict and political context complicated and hindered the humanitarian response. The larger international aid agencies had extraordinarily limited access to the famine-affected parts of Somalia on account of restrictions imposed by Al-Shabaab, counter-terrorism law and policy in the United States and other countries, and insufficient trust, relationships, and networks on the ground after cutting back operations in previous years. This both limited the scale of activities possible and slowed their implementation. Linked to the significant funds provided by non-traditional donors, the operational response was marked by the involvement of non-traditional actors. Some of the new and inexperienced Islamic NGOs were less concerned about US legislation, welcomed by Al-Shabaab, and broadly accepting of the restrictions the group imposed on them such that they had more room to manoeuvre than Christian and secular agencies (Menkhaus 2012: 31; Slim 2012: 12).

Al-Shabaab controlled almost all the famine-affected areas, and its policies became increasingly restrictive over the course of the famine. It would not allow WFP—usually the main instrument of food aid delivery—to return, so WFP was only able to operate limited food distributions in urban Mogadishu and in accessible border regions (Hobbs et al. 2012: 54; Maxwell et al. 2012: 2). In south Somalia, the International Committee of the Red Cross (ICRC) was the only agency with access and with the capacity to move large quantities of food and so faced the Herculean task of getting food aid to one million people at the same time as negotiating with a nervous, unpredictable, and divided Al-Shabaab leadership (Menkhaus 2012: 30). In late November 2011, Al-Shabaab shut down most of the relief agencies that remained in south Somalia, allowing only the ICRC and a small number of NGOs, including Médecins Sans Frontières (MSF), to continue operating—and then expelled the ICRC at the end of January 2012 (Menkhaus 2012: 31). These restrictions appear to have been driven by non-Somalis in Al-Shabaab's leadership, motivated by personal security fears that aid agencies may include spies of the West, together with disgust at Somali dependence on the West for food aid (Bradbury and Maletta 2012: 120; Menkhaus 2012: 31).

For aid agencies to work safely in territory controlled by Al-Shabaab, they needed its consent—which, in turn, required the agencies to negotiate directly or indirectly with the group and to accept conditions it imposed (Jackson and Aynte 2013: 9). Standard conditions included the payment of registration fees and additional taxes, although this was often denied by more senior aid agency officials in Nairobi (Jackson and Aynte 2013: 9–10, 18). Agencies that had a longstanding presence, strong community support, and a history of structured engagement with Al-Shabaab at all levels were sometimes able to negotiate exceptions to such payments (Jackson and Aynte 2013: 17). Other common conditions included prohibiting aid agencies from proselytizing, from publicly criticizing Al-Shabaab, and from employing Somali women—with the exception of doctors and nurses, whose involvement was encouraged so that women could access health care (Jackson and Aynte 2013: 10, 18). Sometimes, Al-Shabaab insisted on distributing international food aid directly, making it impossible for aid agencies to monitor the distribution or assess the extent to which it was diverted (Jackson and Aynte 2013: 18). Even for those aid agencies officially permitted by Al-Shabaab to operate in southern Somalia, the operational environment continued to be highly insecure.

Western counter-terrorism restrictions were another major impediment to the provision of aid in areas controlled by Al-Shabaab—that is, in almost all the famine-affected areas. The suspension of US humanitarian assistance to territory controlled by Al-Shabaab in late 2009 forced other donor governments and operational agencies to face the fact that counter-terrorism legislation effectively criminalized almost all aid provision within that territory (Menkhaus 2012: 32). US government officials themselves were divided when famine was declared in July 2011, with those prioritizing humanitarian concerns seeking to remove legal obstacles to the provision of food aid, while those prioritizing security or legality viewed the provision of aid as undesirable or unallowable, respectively (Menkhaus 2012: 32). Domestic politics in the United States played a role, with the Obama administration wanting to avoid both blame for the famine and accusations of being soft on terrorism (Menkhaus 2012: 32). Efforts by NGOs to negotiate a waiver yielded a memo and verbal assurances from US government officials that 'incidental benefits' to Al-Shabaab were not the focus of enforcement, but lawyers advised that these assurances did not provide adequate legal protection, and many aid agencies remained wary of prosecution (Burniske et al. 2014: 7; Menkhaus 2012: 32; Pantuliano et al. 2011: 9). Many NGOs were concerned not only about the risk to individual employees, who could face criminal prosecution, but also about the risk to institutional reputation and funding if they were to be publicly accused of aiding a terrorist group (Menkhaus 2012: 32). Ultimately, only a handful of UN agencies and Western NGOs opted to assume the risk and so remained in southern Somalia until they were expelled in late 2011 and early 2012 (Menkhaus 2012: 32).

In addition, a number of newly established Islamic NGOs, many from Turkey, began working in Somalia, and Al-Shabaab welcomed their presence but nonetheless placed tight restrictions on them (Menkhaus 2012: 32). Whereas both Western and local NGOs faced an agonizing dilemma (hand over truckloads of food to Al-Shabaab, in violation of the principle of neutrality and of Western counter-terrorism laws, or cut off aid to famine victims), some of the new and inexperienced Islamic NGOs did not see handing over control of aid distribution to Al-Shabaab as unreasonable or objectionable (Menkhaus 2012: 31). The inexperience of some may have made them more susceptible to co-optation by conflict actors, but Islamic NGOs found more room for manoeuvre and were able to provide assistance for some areas that Western organizations were unable to access (Hammond and Vaughan-Lee 2012: 10; Slim 2012: 12).

Many aid agencies preferred not to acknowledge it in public, but even in the absence of the restrictions imposed by Al-Shabaab and by counter-terrorism legislation, most agencies would have been unable to scale up their operations sufficiently to meet the needs of famine victims. This frustrated donor governments, who were especially keen to increase operations in areas recently captured from Al-Shabaab, but quickly restarting operations in a highly insecure environment after a long absence was deemed by most agencies to be too risky. Having withdrawn altogether from Somalia, suspended operations, reduced them significantly, or subcontracted them to local partner organizations by 2009, international agencies had effectively dismantled or eroded the networks and trust with local communities upon which their operations had depended (Menkhaus 2012: 33). Rebuilding relationships would take time, and until then, delivering large quantities of aid would likely have prompted looting, diversion, and potentially lethal security threats (Menkhaus 2012: 33).

Cash programming

Cash transfers offered a way to overcome some of the political and logistical challenges, given that the reduced presence of food aid actors meant that large-scale food aid was not an option.[5] From March 2011, the Cash Based Response Working Group (CBRWG) argued that donors and major operational agencies should begin planning and coordinating a large-scale cash response (Ali and Gelsdorf 2012: 57). Small-scale cash transfers had been used in Somalia since around 2000, increasing access to basic food and non-food needs and allowing recipient households to make their own decisions about priority needs (Ali and Gelsdorf 2012: 58). Commercial enterprises in Somalia had adapted well to state collapse, and a robust private sector could adapt supply to meet increased demand (Menkhaus 2012: 30). Furthermore, Somalia benefited from a well-developed money transfer system, based on a network of *hawala* companies that directly delivered and insured money transfers, which had been used for previous humanitarian cash transfer programmes and was also responsible for transferring an estimated USD 1.3–2 billion remittances annually from the diaspora into Somalia—including to

[5] For a general discussion of the advantages, risks, and limits of cash transfers as part of humanitarian aid, see 'Cash and voucher assistance' in Chapter 22 on Material Assistance and Direct Service Provision.

remote locations (Ali and Gelsdorf 2012: 59). Nonetheless, it took time and effort for those advocating a cash-based response to persuade key actors in the international community of the feasibility of such an approach as a means to provide recipients with food access.

Three main concerns were cited as reasons for reluctance to fund or implement cash transfers at scale: the risk of diversion—especially to Al-Shabaab, in light of counter-terrorism goals and legislation; difficulties in targeting and monitoring assistance, given insecurity and access issues; and the risk of inflationary effects (Ali and Gelsdorf 2012: 58; Hobbs et al. 2012: 52). The first two concerns were not specific to cash programming. Indeed, prior to the famine, cash transfers were seen by many as a way to *reduce* diversion (Abild 2010: 84). Using the *hawala* system could be expected to reduce risk because money passed through fewer hands than in-kind assistance, was in transit for less time, and handed directly to beneficiaries; hence, advocates argued that cash programming offered less opportunity for beneficiary harassment or for looting or taxation at checkpoints (Ali and Gelsdorf 2012: 59). As for targeting and monitoring, the CBRWG argued that a community-based targeting system had a successful track record on targeting, transparency, and accountability in Somalia, and that while monitoring was difficult without direct access to beneficiaries, post-distribution surveys had been conducted in previous cash programming in Somalia and additionally proposed using peer monitors and/or a contracted third party (Ali and Gelsdorf 2012: 59). With respect to inflation, sophisticated market monitoring provided by FEWS-NET and FSNAU showed that while the prices of local grains had increased enormously in many areas, imported rice—a preferred but normally more expensive staple—was available even at the height of the crisis and its price had only increased slightly (Ali and Gelsdorf 2012: 59). By basing cash transfer amounts on the price of rice rather than sorghum, beneficiaries would be able to afford rice, and the ability to substitute rice for other staples could actually serve as a check on overall food price inflation (Ali and Gelsdorf 2012: 59).

The risks were real, if manageable, and most donors and operational agencies were not prepared to assume such risks until after the declaration of famine, at which point large-scale cash and voucher programming took off almost immediately. Initial small cash-based responses were started in May by a small group of four NGOs, who formed a cash consortium and subsequently a Cash and Voucher Monitoring Group that included six INGOs and eight Somali NGO partners, but broad support for a cash response was still lacking (Ali and Gelsdorf 2012: 59). Eventually the FAO, the United Nations Children's Fund (UNICEF), WFP, and a number of international and national NGOs worked together in a coordinated cash and voucher response, which involved negotiating cash distribution channels at a scale not previously seen by the humanitarian community in south central Somalia (Hobbs et al. 2012: 54).

Within the first eight months after the famine declaration, more than USD 86 m was provided as unconditional grants, vouchers, and cash for work, USD 81 m of which went to more than 1.7m people in south central Somalia (Ali and Gelsdorf 2012: 59). Through an extensive network of cooperating partners, the FAO implemented one of its largest ever cash-for-work programmes to the value of more than USD 25.7 million (Hobbs et al. 2012: 54). UNICEF and the Cash Consortium implemented one of the world's largest ever cash transfer programmes, through which over USD 36.6 million was channelled to between 30,000 and 40,000 households in a series of regular payments (Hobbs et al. 2012: 54). This programme continued beyond the official end of the famine, supported by UNICEF and other donors, and by the end of 2012, USD 92 million had been distributed by 17 NGOs in the form of unconditional cash transfers and vouchers to meet the food and non-food needs of 1.5 million beneficiaries (Hedlund et al. n.d.: 1). Unconditional transfers, cash for work, and related voucher schemes served not only immediate food security goals by supporting purchasing power but also contributed to productive livelihoods, increasing recipients' access to livelihoods inputs, such as tools, fertilizers, and vaccinations (Lautze et al. 2012: 46, 47). An independent evaluation of the UNICEF programme found that it did not result in food price inflation and that corruption and diversion were probably less serious than in comparable in-kind distributions but could have been countered through better risk analysis and preparedness (Hedlund et al. n.d.: 3). Coverage and targeting were more problematic in that the programme did not access the most affected areas sufficiently, and the distribution of aid to the most vulnerable 'minority' populations was inadequate (Hedlund et al. n.d.: 4).

Diversion, manipulation, and instrumentalization of aid

Assistance in Somalia had long been politicized and instrumentalized by all sides in the conflict, including the international backers of the TFG—among them, international military forces and several donor countries. This continued during the famine response.

The TFG and its political backers in the UN sought to instrumentalize aid as a tool with which to bolster the legitimacy and capacity of the government. They insisted that in TFG-controlled areas all humanitarian assistance should be channelled through the government, despite the fact it had been characterized by epic corruption since its creation in 2004 and was too weak even to control its own limited territory (Menkhaus 2012: 33). While Al-Shabaab controlled the overwhelming majority of territory in which famine victims were located, the 5% that was beyond Al-Shabaab's direct control included most of Mogadishu, which was under the nominal control of the TFG and in which IDP camps housed tens of thousands of famine victims (Menkhaus 2012: 33). Humanitarian agencies were generally able to resist pressures to subordinate their aid to stabilization efforts, but they were less successful at resisting the diversion of aid by force. The TFG—and autonomous clan-based militias that were TFG in name only—diverted aid at all phases of delivery, with 'taxes' charged by port authorities to release aid from the port, by the many militia checkpoints in TFG-controlled Mogadishu to allow it to be transported from the port to the IDP camps, and by 'camp managers' taking a proportion from IDPs themselves after rations had been distributed (Menkhaus 2012: 33; see also Bradbury and Maletta 2012: 119).

Al-Shabaab leaders also diverted and instrumentalized aid. As explained above, most of the major international aid agencies paid a levy to operate in Al-Shabaab-controlled areas. Seeking to bolster its own legitimacy, the organization additionally manipulated aid provided by international agencies by, for example, stamping the word 'Al-Shabaab' on vouchers provided to famine victims, then banning the vouchers altogether and insisting that all food aid was handed over to Al-Shabaab's National Drought Committee so that they could control its distribution (Menkhaus 2012: 30–31). While this clearly violated the principle of neutrality in that Al-Shabaab and its leaders were using control of aid to strengthen themselves, it was not an instance of wholesale diversion, and most aid workers believed Al-Shabaab's relief committees to be relatively honest and committed to delivering aid to those in need (Menkhaus 2012: 31).

Impartiality

In fact, there were many other obstacles to providing aid to those most in need. Famine does not affect everyone equally, and minority groups—in particular, the historically marginalized Reewin/Rahanweyn and Bantu/Jarer—were disproportionately affected by the famine in Somalia. According to the humanitarian principle of impartiality, assistance should be provided in accordance with need, meaning that those suffering worst from the famine should be prioritized in the humanitarian response. However, the same marginalization that made minority groups more vulnerable to famine also, in many cases, made them less likely to receive assistance through either informal or formal channels—indeed, the lack of assistance itself contributed to their vulnerability.

Individual and household outcomes were determined in large part by their social capital or connectedness and the resources available within their networks (Maxwell et al. 2016: 63). Given that the international response was not scaled up until the famine was well advanced, support from communities, business groups, diasporas, neighbours, and kin was extremely important (Maxwell et al. 2016: 64). However, not all social groups had equal access to this kind of support. Those most dependent on the rural economy and with fewer relatives and clan members in urban areas or overseas were the most vulnerable (Maxwell et al. 2016: 68). For some Somalis, remittances served as an important source of income that enabled access to food during the famine, but for many of those worst-affected, clan-based support from overseas was more limited or non-existent (Maxwell et al. 2016: 69). Furthermore, dynamics relating to clan identity meant that the displaced—who were predominantly Reewin/Rahanweyn and Bantu/Jarer—were sometimes preyed upon by the more powerful groups who dominated the areas into which they were displaced (Hedlund et al. n.d.: 9; Maxwell et al. 2016: 64).

The most marginalized were also often neglected in the official aid distribution systems, even if this was not the intention of the aid agencies themselves. In part, this reflected operating conditions—aid distribution was, to a significant degree, determined by access constraints rather than need. Food assistance, whether in-kind food aid, cash, or vouchers, was disproportionately focused on Mogadishu and border areas, while the greatest needs were among agro-pastoralists and agricultural households in Bay, Bakool, and the Shabelle regions (Hedlund et al. n.d.: 7). In as far as those with greatest needs were also least accessible, allocating aid on the basis of access is, while perhaps unavoidable, in direct tension with the principle of impartiality. Even where they were operational, aid agencies were not always able to target their assistance to the most needy. This was partly due to inadequate knowledge about the distribution of unmet needs, exacerbated by limited coordination, which means that targeting decisions were made without a clear sense of what other actors were doing (Hobbs et al. 2012: 54). In addition, international aid workers are almost never in a position to monitor who low-status groups are and what aid they have received, and while many Somali national staff sought to target aid to members of marginalized, low-caste groups, their efforts were often resisted or thwarted by powerful local actors with different priorities (Maxwell et al. 2016: 69–70; Menkhaus 2012: 34).

Conclusions

Long-running armed conflict in Somalia, as well as the concerns and interventions of domestic and international political actors, shaped the famine, the international response, and the humanitarian outcomes. The famine was triggered by drought, but long-running armed conflict and an underlying livelihoods crisis were major contributing factors. Counter-terror legislation and restrictions imposed by Al-Shabaab had significantly limited the provision of food assistance in the year preceding the declaration of famine, removing an important safety net. Even when famine was imminent, these same factors contributed to a widespread perception among donors and operational agencies that the risks of responding outweighed the risks of not responding. Only with the declaration of famine did most relevant actors reverse this perception, and funding increased immediately. By definition, this was too late—once conditions reach the threshold of famine, lives have been lost and irreversible damage has been done. Furthermore, even when funding was significantly increased, the conflict and political context limited the scale of humanitarian activities possible, slowed their implementation, and made the impartial targeting of assistance difficult or impossible. Most operational agencies considered that rushing to restart operations in a highly insecure environment was too risky after a long absence during which networks and trust with local communities has been dismantled or eroded. The scaling up of cash transfer programming allowed agencies to overcome some of these obstacles, distributing funding through existing *hawala* networks, and had positive impacts, though targeting of the most marginalized and vulnerable remained inadequate.

11
Ebola in West Africa, 2014–2015

In 2014 and 2015, West Africa suffered an outbreak of Ebola virus disease (hereafter, Ebola) which became a crisis as a result of political factors. The hardest-hit countries were Guinea, Liberia, and Sierra Leone, whose health systems lacked the capacity to mount an effective and timely response to contain the outbreak in its early stages. Prior to the Ebola crisis, their health systems were among the weakest in the world, characterized by poor infrastructure and shortages of medical supplies and personnel (DuBois et al. 2015: 9). Their weakness, in turn, was the outcome of recent armed conflicts in Liberia and Sierra Leone and national and international policies that had impeded the development of robust and resilient health systems in all three countries (Anderson and Beresford 2016; Piot et al. 2014).

Most national and international actors failed to recognize quickly enough the seriousness of the outbreak. The governments of Guinea, Liberia, and Sierra Leone sought to downplay the severity of the crisis because they did not want to be seen as weak or to deter foreign investment. The World Health Organization (WHO), for its part, sought to avoid conflict with these three states and took several months to acknowledge the scale of the crisis. United Nations (UN) agencies and many non-governmental organizations (NGOs) deferred to WHO as the authority on global health. Key Western policymakers largely ignored the outbreak until public fears that Ebola would spread back home combined with electoral dynamics to prompt a significant response, including the deployment of military personnel (DuBois et al. 2015: vi).

In Guinea, Liberia, and Sierra Leone, the Ebola outbreak constituted a humanitarian crisis as well as a health crisis, but the international policy discourse and operational response focused narrowly on the latter. Furthermore, the response—especially the initial response—emphasized biomedical measures without properly taking into account the political and social contexts in which those measures were being implemented. This narrow technical focus not only meant that many important needs went unmet but also limited the effectiveness of those measures and was, in some ways, counterproductive in terms of bringing the outbreak under control.

The Politics and Everyday Practice of International Humanitarianism. Miriam Bradley, Oxford University Press.
 DOI: 10.1093/oso/9780198840695.003.0011

Timeline and trajectory of the outbreak

On 28 December 2013, a two-year-old Guinean boy died of an unidentified haemorrhagic fever. Guinea had no experience of Ebola, health workers could not identify the disease, and the country lacked laboratory capacity for the necessary diagnostic tests (DuBois et al. 2015: 5).

By the time the disease was confirmed as Ebola nearly three months later, it was spreading rapidly in south east Guinea and had crossed into Liberia and Sierra Leone. The Guinean Ministry of Health raised the initial alert of an 'unidentified' disease on 14 March 2014, and a week later laboratory tests in France confirmed it was Ebola. On 22 March 2014, the Ministry of Health alerted WHO to a 'rapidly evolving outbreak' in south-east Guinea, with forty-nine reported cases, of whom twenty-nine had died (DuBois et al. 2015: 5; Grépin 2015: 1). The WHO officially declared the outbreak the next day, despatched a Global Outbreak Alert and Response Network team to Guinea, and alerted authorities in neighbouring Liberia and Sierra Leone to commence surveillance (Kamradt-Scott 2016: 404). That same day, Médecins Sans Frontières (MSF) announced that it had launched an emergency response with the Guinean Ministry of Health (Davies and Rushton 2016: 419). On 27 March, both Liberia and Sierra Leone announced a small number of suspected cases, and by the end of the month, there were 112 confirmed and suspected cases in Guinea (DuBois et al. 2015: 5; Kamradt-Scott 2016: 404).

MSF was sounding the alarm internationally, but WHO was more circumspect. On 31 March, MSF called the spread of the epidemic 'unprecedented' (DuBois et al. 2015: 5). At a press briefing in Geneva on 8 April, WHO officials noted that it was one of the most challenging outbreaks they had ever faced, but the WHO secretariat did not understand or acknowledge the scale and potential severity of the outbreak until June (DuBois et al. 2015: v; Kamradt-Scott 2016: 404; Wenham 2016: 442). Throughout April, WHO mobilized resources and deployed technical experts to assist the health authorities in Guinea, Liberia, and Sierra Leone, though neither the resources nor the expertise were always adequate to the task (DuBois et al. 2015: 18; Kamradt-Scott 2016: 404; Walsh and Johnson 2018: 40).

The failure to contain the Ebola outbreak early on in Guinea, Liberia, and Sierra Leone contrasts with successful containment in countries including Mali, Nigeria, and Senegal—and this failure can be attributed in large part to weaknesses in their health systems (DuBois et al. 2015: 9; MSF 2015: 12;

Wenham 2016: 440). Poor infrastructure in all three countries, with limited access to electricity and running water even in many of the larger hospitals, made good infection control extremely difficult (DuBois et al. 2015: 12; Wenham 2016: 441). Shortages in medical personnel and basic supplies, corruption in the health systems, and frequent non-payment of government health worker salaries all exacerbated the difficulty (Anderson and Beresford 2016; DuBois et al. 2015: 11; Heller Pérache 2015: 5; Wilkinson and Leach 2015: 142). Significant time and resources were required to develop systems of data collection, sharing, and analysis; laboratories had insufficient capacity to meet the demand for case testing, delaying diagnoses and hence increasing the risk of transmission; and hospitals and health centres lacked sufficient space for the proper isolation of Ebola patients, further increasing the risk of transmission (DuBois et al. 2015: 14).

Weaknesses in the Guinean, Liberian, and Sierra Leonean health systems were the consequence of very low per capita health expenditure (a large proportion of which was private, not government, spending), corruption, and the legacy of conflict and state failure (Anderson and Beresford 2016: 472, 474–475; DuBois et al. 2015: 10; Piot et al. 2014: 1034). Several years of international development interventions in all three countries had not served to strengthen their health systems. Conditional loans from the International Monetary Fund (IMF) required reductions in government spending, prioritizing debt service and bolstering foreign exchange reserves, such that targets for increasing health and social spending were routinely missed, especially in Guinea and Sierra Leone (Kentikelenis et al. 2015: e69; Wilkinson and Leach 2015: 142). Heavy dependence on international aid made it difficult for governments to set the agenda independently of constantly shifting donor priorities (Anderson and Beresford 2016: 473). International aid allocated for health had contributed to progress on very specific dimensions of health care, notably those relating to the Millennium Development Goals (MDGs), with donors tending to support programmes which focused on particular diseases (e.g. HIV, malaria, and tuberculosis) or groups of patients (e.g. maternal and child health services), rather than supporting the overall development of the wider health system (DuBois et al. 2015: 12–13; Walsh and Johnson 2018: 28–29; Wilkinson and Leach 2015: 140).

By mid-2014, the scale and severity of the epidemic would have posed a challenge even to well-developed health systems in wealthy countries, and there were concerns that the Guinean, Liberian, and Sierra Leonean health systems would not cope without significant international help. On 21 June,

MSF called on the WHO to deploy massive resources, warning that the epidemic was 'out of control' and that MSF had reached its limits (MSF 2014a).

A failure of global health governance

The institutions of global health governance proved inadequate to the task. After the emergence of the SARS-related coronavirus in 2003, WHO member states had adopted updated international health regulations (IHR) in 2005, which obligated member states to develop national disease surveillance and response capacities, and tasked WHO with providing technical support to countries struggling to do so (Kamradt-Scott 2016: 403). However, the IHR were poorly implemented in many countries, and in Guinea, Liberia, and Sierra Leone, health information and disease surveillance systems were at best weak and at worst non-existent (DuBois et al. 2015: 12). Furthermore, through budget cuts and increasingly earmarked funding, the states that fund WHO had undermined its capacity to mount an epidemic response (DuBois et al. 2015: 15; Wilkinson and Leach 2015: 140). At its creation in 1948, WHO was imbued with considerable authority and autonomy for the containment and eradication of infectious disease (Kamradt-Scott 2016: 402). However, the years prior to the Ebola outbreak saw sharp reductions in WHO's disease outbreak expertise and its Africa crisis team, and its transformation from global health provider to a service organization largely limited to developing guidelines and providing technical advice (DuBois et al. 2015: 15).

The longstanding tension between global health security and respect for state sovereignty within WHO also contributed to its muted response (DuBois et al. 2015: 15; Kamradt-Scott 2016: 403). The 2005 IHR conferred new powers on the WHO secretariat to use non-government sources of information in detecting disease outbreaks, and to 'name and shame' those countries that refused assistance or tried to conceal public health risks (Kamradt-Scott 2016: 403). However, member states insisted on new measures which limited WHO's autonomy, including the explicit requirement for the director-general to convene an emergency committee (selected from a roster of experts nominated by member states) for expert advice before declaring a public health emergency of international concern (PHEIC) or making recommendations such as travel restrictions (Kamradt-Scott 2016: 403). Concern not to offend affected countries who wanted to downplay the Ebola outbreak is widely seen to have contributed to WHO's low-key response in the early months of the outbreak (DuBois et al. 2015: 15; MSF 2015: 8; Walsh and Johnson 2018: 102). Meanwhile, other relevant international actors, including UN country teams and international non-governmental organizations (INGOs) working in the three worst-affected countries, largely deferred to WHO's authority and therefore underestimated the scale of the crisis (DuBois et al. 2015: 15; Walsh and Johnson 2018: 72–74).

A public health emergency of international concern

Things began to change by early August, as people elsewhere in the world became increasingly aware of the potential for Ebola to spread beyond the three most affected countries (McInnes 2016: 387). A US aid worker, Kent Brantly, tested positive in late July—the first white Westerner to do so. Around the same time, Nigeria reported its first cases. Sierra Leone declared a state of emergency in late July, and Guinea and Liberia did the same in August (DuBois et al. 2015: 5). After a meeting of the IHR Emergency Committee on 8 August 2014, WHO declared the Ebola outbreak a PHEIC (Grépin 2015: 1). MSF welcomed the declaration and called on states with the necessary capacities to dispatch infectious disease experts and disaster relief assets to West Africa, highlighting the need for the radical scaling up of 'medical care, training of health staff, infection control, contact tracing, epidemiological surveillance, alert and referral systems, community mobilisation and education' (MSF 2014b). A month later, however, there was still no coordinated international response.

MSF continued to sound the alarm, demanding a much stronger international response to contain the outbreak. In a special briefing on 2 September, MSF called for the immediate deployment of civilian and military assets with expertise in biohazard containment—calling for military intervention for the first time in its forty-three-year history (MSF 2015: 13; Walsh and Johnson 2018: 170). Despite concerns to avoid militarizing the response, MSF made this unprecedented appeal because the biohazard response capacities of states are typically a military capability, and MSF estimated that the scaling up of the INGO- and donor-led response would take around three months, by which time the situation might be catastrophic (Heller Pérache 2015: 5–6; MSF 2015: 12, 14).

In September and October, several countries announced the deployment of civilian and military personnel. Uganda, a country with a history of containing Ebola outbreaks, had sent twenty health experts to Sierra Leone and

Liberia in the first half of August (Paulat 2014). However, a significant global response was launched only in mid-September, when President Obama announced the deployment of nearly 3,000 troops to Liberia under 'Operation United Assistance', and the United Kingdom announced the deployment of 750 troops to Sierra Leone (DuBois et al. 2015: 5, 7; Walsh and Johnson 2018: 198). Russian and French teams went to Guinea (Richards 2016: 24). Cuba sent 165 health-care workers to Sierra Leone at the beginning of October and began preparing another 296 for Guinea and Liberia (Taylor 2014; Walsh and Johnson 2018: 242). At the end of October, China announced a plan to deploy 480 military health personnel to West Africa (DuBois et al. 2015: 7).

Within months, the outbreak was under control, but it is not clear how significant the international response was to this achievement. New cases in Liberia peaked as early as late September 2014, before the influx of international assistance, and in Sierra Leone and Guinea in late November and December, before the international response peaked (DuBois et al. 2015: 3, 22; see also Richards 2016: 125). Nonetheless, the fact that the outbreak was not more catastrophic suggests that the combination of international, national, and community efforts was successful, albeit belatedly. Liberia was declared Ebola-free in mid-2015, Guinea and Sierra Leone on 7 November and 29 December, respectively, albeit with some localized and quickly contained subsequent outbreaks in all three countries (Richards 2016: 2). Projections by the United States Centers for Disease Control and Prevention (CDC) in September 2014 warned that, without intervention or changes in community behaviour, there would be more than 550,000 cases in Liberia and Sierra Leone alone by late January 2015 (DuBois et al. 2015: 17). In the event, there were around 27,000 cases across the three countries and around 11,000 deaths attributed to Ebola (Richards 2016: 20).

At the same time, the Ebola outbreak and the responses to it had a number of indirect or secondary impacts, including effects on education, employment, health and care systems, and food security. For example, there were secondary health effects as some medical staff went to work for international agencies for a better salary and others deserted clinics due to the high risks of contagion and lack of sufficient personal protective equipment (PPE) (Walsh and Johnson 2018: 301). Fear of contagion meant that some patients also stayed away from health clinics, and so non-Ebola illnesses went untreated. In Sierra Leone, it is estimated that 2,819 more people died of AIDS, malaria, and tuberculosis than would have been the case without the Ebola crisis (Walsh and Johnson 2018: 315–316). MSF estimated that the collapse in the health services meant deaths from such causes as untreated malaria, complicated births, and traffic accidents would have multiplied the number of direct Ebola deaths many times over (MSF 2015: 3).

International funding

Part of the explanation for the slowness in scaling up the international response relates to funding. Ultimately, the Ebola response was relatively well funded, but both pledging and disbursement of funds were slow. Had the funds arrived sooner, the outbreak would likely have been brought under control more quickly and at lower cost.

Funding in the crucial six months from April to October 2014 was 'too little and too late to match the scale and speed of Ebola' (DuBois et al. 2015: 19; see also Grépin 2015). In early April, WHO appealed for USD 4.8 million, a target which was exceeded with pledges totalling USD 7 million, but even this was a drop in the ocean compared with what was needed (Grépin 2015: 1). Rapid and significant upward revisions of estimated funding requirements provide an indicator of just how much the scale and consequences of the outbreak itself were, for a long time, underestimated. On 1 August, WHO and the presidents of Guinea, Liberia, and Sierra Leone presented their response plan with a request for international funding of USD 71 million (Grépin 2015: 1). Then, in late August, WHO published the Ebola response roadmap, with an estimated funding requirement of USD 490 million, increased to USD 600 million just a week later (Grépin 2015: 1). On 16 September 2014, the Office for the Coordination of Humanitarian Affairs (OCHA) released a report estimating that USD 1 billion would be needed for humanitarian assistance (Grépin 2015: 1).

From late September 2014, huge pledges from donors exceeded the funding requested by operational agencies, but many donors did not come good on their pledges, and actual contributions consistently fell short of what had been requested (DuBois et al. 2015: 19). By the end of 2014, for example, only USD 1.09 billion had been disbursed out of total pledges of USD 2.89 billion (Grépin 2015: 2). Around 60% of donations came from bilateral donors and 11.5% from multilateral institutions; private sources accounted for a relatively large share,[1] with 10% from private individuals and organizations, 8.3% from foundations, and 6.6% from companies (Grépin 2015: 2). In

[1] On the contributions of public and private donors to humanitarian assistance globally, see 'Donors' in Chapter 18 on Donors and the Funding of Humanitarian Action.

keeping with a general pattern in international funding for humanitarian emergencies, UN agencies received the largest share, in this case 42%, followed by NGOs with around 19%, while the governments of the affected countries received only 11.5%, implying that much of the response was led by international agencies rather than national actors (Grépin 2015: 2, 5).

Discourses and panic

Dominant discourses, including messaging from elites, information campaigns, and media coverage, not only contributed to the prominence of the Ebola outbreak on international and national agendas but also framed how the outbreak was understood and what kind of response was understood to be required.[2]

Stereotyping affected populations

In the global North, Ebola was framed as an exotic and racialized phenomenon. This framing was manifested in stories about bushmeat, for example, which linked Ebola 'to backward African practices that were deemed exotic and disgusting' (Nunes 2016: 551). The use of the term 'bushmeat' by Western media outlets serves to other the person eating it—when an African eats wild animals, it is called bushmeat and depicted as primitive and disgusting, whereas when a European or North American does the same, it is called game and may be considered sophisticated (McGovern 2014). Moreover, stories that emphasized the role of bushmeat in the transmission of Ebola not only overstated the frequency with which most people in Guinea, Liberia, and Sierra Leone ate wild animals but also were completely inaccurate in terms of the mechanisms of transmission—while the first case may have been transmitted from fruit bats, subsequent transmission was almost certainly entirely human-to-human (Wilkinson and Leach 2015: 138).

Western media coverage stereotyped people in the three main affected countries, particularly rural populations, as primitive, irrational, fearful, and violent, both drawing on and reinforcing the trope of Africans as victims lacking in agency and unable to help themselves (DuBois et al. 2015: vi; Monson 2017: 8).[3] Despite the fact that most of the effort on the ground

[2] For a more general discussion of how the media shapes the humanitarian agenda, see 'Setting the agenda' and 'Framing the agenda' in Chapter 17 on Media and Celebrities.

[3] On representations of affected populations in other humanitarian emergencies, see especially 'Media coverage' in Chapter 3 on Drought and Famine in Ethiopia, 1983–1985, 'Media coverage' in Chapter 7 on the 2004 Indian Ocean Tsunami, and 'Western discourses, media coverage, and celebrity engagement' in Chapter 9 on the Haiti Earthquake, 2010.

was local, the media focused on the international response, bolstering a discourse and perception of humanitarian action 'saving' helpless victims (DuBois et al. 2015: vi). Such representations are intrinsically harmful since they serve to reinforce the racist stereotypes on which they are based, and they were also damaging insofar as they shaped international and national efforts to stem the spread of Ebola. At least to begin with, many international actors treated affected communities as a problem to be overcome rather than as a resource—and, as explained below, this impeded measures to reduce transmission (DuBois et al. 2015: vi; Walsh and Johnson 2018: 62).

A global crisis?

With time, the dominant discourses in Western countries came to construct the outbreak as a global crisis or emergency, despite the fact that the overwhelming majority of cases were in West Africa and only a very small number of transmissions occurred elsewhere (McInnes 2016: 386). There was initially low public interest in the West, as slow-onset disasters do not capture public attention and sympathy in the same way that rapid-onset disasters do (DuBois et al. 2015: 20).[4] However, a sense of global crisis was established following cases in Nigeria, the infections of Western aid workers, and especially after transmissions to medical staff in Spain and the United States (McInnes 2016: 387; Wilkinson and Leach 2015: 138). At this point, a discourse of panic emerged, stoked by misrepresentation of transmission and risk in Western media, and generating widespread fear of a pandemic (Monson 2017: 8–10). This 'outbreak narrative', and the media and political spectacles that accompanied and constituted it, served to distort the nature and severity of the risks posed and to obscure the identity and needs of the most vulnerable (Nunes 2016: 543; Wilkinson and Leach 2015: 139).

Linked to this discourse of panic, the Ebola outbreak was securitized internationally, with important implications for the nature and effectiveness of the response. On 18 September, the UN Security Council Resolution 2177 declared that the Ebola outbreak constituted a threat to international peace and security. This was only the third time the Security Council had convened to discuss public health, and the Resolution presented the outbreak as posing a risk of state failure in the three main affected countries and a risk to regional and global security, despite very limited material support for such claims (McInnes 2016: 389–390; Walsh and Johnson 2018: 199). Worldwide,

[4] By contrast, the 2004 Indian Ocean tsunami and the 2010 Haiti earthquake both immediately attracted massive public concern in the global North. See Chapters 7 and 9, respectively.

a conflict or military lexicon was employed to describe the outbreak—'"fighting" the outbreak, "hunting" the virus, healthcare workers on the "frontlines" and so on' (Heller Pérache 2015: 6).

Militarization and securitization

Securitization and the language of warfare can serve to increase attention and funds—and, as in the case of the Ebola outbreak in West Africa, to justify a military response—but also risk turning those infected by the disease and their caregivers into objects of fear and stigma (de Waal 2014). When MSF called for international militaries to respond, it explicitly called on them to deploy their biohazard containment capabilities and 'insisted that any military assets and personnel deployed should not be used for quarantine, containment or crowd control measures, because forced quarantines have been shown to breed fear and unrest, rather than stem the spread of Ebola' (MSF 2015: 14). In the event, however, biohazard teams were not deployed, international military forces were sometimes engaged in such coercive—and often counterproductive—measures as the enforcement of quarantines (IRC 2016: 12), and their wider efforts served only limited goals.

The international military response focused primarily on providing support, coordination, and logistics for local authorities and international aid agencies (MSF 2015: 14). Most commercial airlines had stopped flying, and military flights provided an important air bridge (de Waal 2014). In Liberia and Sierra Leone, the US and UK militaries provided large-scale logistical support for the construction and management of Ebola treatment units (ETUs), supported the management of Ebola response centres, and helped improve command-and-control systems (IRC 2016: 12; Walsh and Johnson 2018: 198). They also built treatment facilities for health-care workers, ensuring a high standard of care, which reassured international and national health-care workers and enabled international agencies to offer stronger assurances about fulfilling their duty of care when deploying staff (DuBois et al. 2015: 21; Heller Pérache 2015: 7; MSF 2015: 14).

At the same time, however, both the UK and US militaries were highly risk averse, so their logistical and medical capacities were not used to full effect. For example, the United States would not use its helicopters to transport laboratory samples for testing or healthy personnel back from working in treatment centres, and military vehicles were not used in any significant way for patient referral (Heller Pérache 2015: 7). The medical facilities they built for the treatment of local and foreign health-care workers were provided so that others could treat patients, and military medics did not offer direct care to the wider community (MSF 2015: 14). British soldiers were explicitly barred from direct contact with Ebola patients, with instructions to ensure a 'zero casualty rate' (Walsh and Johnson 2018: 217–221). Likewise, a UN force deployed in Liberia since 2003 sought to minimize risk to its personnel rather than support the Ebola response. While the United Nations Mission in Liberia (UNMIL) had previously been providing medical services direct to the population and using its logistical capacity to transport national medical staff to the outer provinces, its first reaction when the outbreak began was to return battalions to barracks and cease medical outreach activities (Davies and Rushton 2016: 426).

Beyond militarization, international actors securitized the Ebola outbreak more generally, with negative impacts on the response. For example, hysteria over the risk of bringing Ebola into countries in the global North led some countries to impose travel bans and many airlines halted flights (Richards 2016: 2; Wilkinson and Leach 2015: 139). Cancelling flights was driven by domestic politics in Western countries, but such measures have only a miniscule protective effect and much more significant consequences for the response in the worst-affected countries, making it difficult to get health workers and essential goods in or securely-packaged patient samples out for testing (de Waal 2014; McInnes 2016: 390; Richards 2016: 2, 23; Wilkinson and Leach 2015: 139).

The discourse of panic was not limited to the global North, and the crisis narrative in the three most-affected countries and beyond was accompanied by scaremongering—so-called Ebola panic disease (Richards 2016: 23)—and a securitized response. Securitization was not only damaging for the feared and stigmatized individuals, communities, and countries but also counterproductive for public health. For example, the initial response in Sierra Leone was securitized, with compulsory quarantines, roadblocks, and restrictions on movement enforced by the army and police and overall an emphasis on rules and punishment (Anderson and Beresford 2016: 469; Walsh and Johnson 2018: 127–128). In Liberia, the government declared a state of emergency, postponed elections, and troops fired on demonstrators (de Waal 2014; Richards 2016: 40).

Quarantines may have played an important role in reducing transmission, but they also exacerbated fear and had negative impacts on the lives and livelihoods of those forced to quarantine, and where these different consequences were not well mitigated, people sometimes went into hiding instead of seeking treatment. The quarantining of the 120,000 residents of West Point, an urban slum in the Liberian capital, Monrovia, was enforced by armed police in riot gear and proved catastrophically counter-productive (Walsh and Johnson 2018: 149). It led to violence and increased suffering due to lack of access to food, services, and livelihoods, resulting in a loss of trust in the government such that it may have served to increase rather than reduce transmission (Heller Pérache 2015: 6). The planned twenty-one-day quarantine was cancelled after ten days, and the Liberian President began to promote a more community-based approach to defeating the disease (Walsh and Johnson 2018: 150). In Sierra Leone, however, twenty-one-day quarantines of affected households continued throughout the epidemic, with huge implications for people's livelihoods and sometimes deterring sick people from seeking treatment and incentivizing people to hide bodies rather than cooperate with the official response (Richards 2016: 27, 104; Walsh and Johnson 2018: 151–152).

Just a health crisis?

The Ebola outbreak in West Africa was depicted overwhelmingly as a health crisis and not a humanitarian crisis, despite its many indirect effects, a number of which implied humanitarian rather than strictly health-related needs. Framing the crisis in this way appears to reflect a deliberate strategy on the part of senior UN officials to protect WHO's interests and relationships with the worst-affected states (DuBois et al. 2015: 25). For the governments of those states, labelling the situation as a health crisis was more palatable than labelling it a humanitarian crisis, which could be taken to imply weak governance and greater culpability. However, such a framing obscured many of the most urgent needs of those affected by Ebola and had implications for the kind of international response deemed necessary and appropriate.

International humanitarian response

By mid-2014, it was clear that local capacities plus the combined capacity of MSF and WHO were inadequate, yet other international actors remained hesitant. The Guinean, Liberian, and Sierra Leonean ministries of health were quickly overwhelmed by the size and complexity of the outbreak, and the extremely low numbers of doctors and nurses in each country was made worse by the fact that health-care workers were contracting Ebola (and dying) at alarming rates (DuBois et al. 2015: 21; Heller Pérache 2015: 5). On 8 August, MSF announced that it had mobilized all its Ebola experts and was at the limits of its capacity, with 66 international and 610 national staff working on the Ebola response across Guinea, Liberia, and Sierra Leone

(MSF 2014b). At the same time, however, MSF was publicly and privately advocating *against* other, less experienced international actors seeking to respond, on the basis that non-MSF health facilities were inadequate and unsafe to deal with Ebola patients (Walsh and Johnson 2018: 157–158).

As the epidemic progressed, many aid agencies—including some without a specific health mandate—felt themselves pulled in opposing directions, with growing pressure to act coming from staff, supporters, and donor governments at the same time as a duty of care to staff and fears of infection increased pressure for agencies to leave affected areas (DuBois et al. 2015: 15–16, 20–21; Walsh and Johnson 2018: 173–174). MSF itself struggled with this tension between patient care and staff safety, all the more so because the agency felt pressure to show that it was possible to treat Ebola safely (MSF 2015: 10–11, 17).

Labelling the situation as a health crisis and not also a humanitarian crisis, especially in the early stages, meant that the wider implications of Ebola, including education, livelihoods, and protection, were downplayed or ignored (DuBois et al. 2015: v; Kahn 2015: 11). This, in turn, contributed to uncertainty among many humanitarian agencies as to what role they could play. Only after several months of collective feet-dragging did meaningful international support began to arrive in West Africa in September 2014, and there was a 'haphazard escalation' of involvement by aid actors in the last months of the year, but even then, 'humanitarian agencies struggled to translate energy and effort into relevant programme activities' (DuBois et al. 2015: 25, v). As an agency specializing in water and sanitation, for example, Oxfam struggled to find a constructive role in what was considered a medical emergency—despite the obvious importance of water and sanitation for infection control (Meredith 2015: 15).

Treating Ebola predominantly as a health crisis also had consequences for how the response was coordinated. Senior UN figures allowed or nominated the WHO to lead the international response to the Ebola outbreak, and typical international humanitarian coordination structures were not activated (DuBois et al. 2015: 25, 27). Notably, the cluster system[5] was only activated in Liberia, despite the fact that not only the health cluster, within which WHO is supposed to provide leadership and coordination among the main humanitarian health agencies, but also the logistics and water, sanitation, and hygiene (WASH) clusters, had obvious relevance for the Ebola response (DuBois et al. 2015: 27; Fink-Hooijer 2015: 3). Coordination problems seriously hindered

[5] For an explanation of the cluster system, and an outline of some of its shortcomings in the response to the 2010 earthquake and subsequent cholera outbreak in Haiti, see 'Coordination and ownership' in Chapter 9 on the Haiti Earthquake, 2010.

the response and might have been alleviated by the wider activation of the cluster approach and deployment of OCHA (Walsh and Johnson 2018: 92). The scale of the crisis was too large for WHO to manage, and when this became apparent, the UN Secretary-General appointed his own special envoy and mobilized the UN's first ever health mission—the UN Mission for Ebola Emergency Response (UNMEER) was established on 19 September 2014 and operated until 31 July 2015 (DuBois et al. 2015: 7, 25).

The global health narrative facilitated a biomedical approach aimed at reducing transmission to zero—an approach which was flawed both because it neglected other important priorities and also because ignoring contextual factors reduced the effectiveness of the biomedical approach in reducing transmission. The UNMEER strategy sought to control the outbreak in areas of intensive transmission using a five-part approach named for its acronym, STEPP: (1) Stop the outbreak; (2) Treat the infected; (3) Ensure essential services; (4) Preserve stability; and (5) Prevent further outbreaks. This strategy not only failed to address the wider humanitarian consequences of the outbreak, which was showing many of the indicators of a full-blown humanitarian emergency by the time UNMEER was established, but also failed to put the dignity and humanity of affected communities at the centre of the response and largely ignored the social and cultural context of the crisis (DuBois et al. 2015: 25–26; Kahn 2015: 11–12). More attention to the dignity and humanity of the affected community and the social and cultural context of the crisis would have been valuable in and of itself and could also have improved the efficacy of control measures.

Community outreach and messaging

Community outreach, including awareness raising and surveillance, is an important part of outbreak control and one to which international responders initially dedicated insufficient resources. At their best, social mobilization initiatives could serve both to build capacity among affected communities to prevent and manage Ebola themselves and to increase trust in official outbreak control mechanisms (Meredith 2015: 15). In practice, however, this element of the response—for which the United Nations Children's Fund (UNICEF) has responsibility at the global level—was flawed (IRC 2016: 12). MSF staff were stretched thin treating Ebola patients and lacked capacity to manage essential outreach activities (MSF 2015: 6). Oxfam came to this kind of work somewhat belatedly (Meredith 2015: 16). While governments and other actors had begun large-scale efforts to educate the public as early as April 2014, with government messages about recognizing and avoiding Ebola, seeking treatment in hospital, and safe burial practices, these efforts tended to fail, in part because many communities did not trust the messengers (Anderson and Beresford 2016: 478; IRC 2016: 3).

In the early phases, mutual distrust between international and national responders, on the one hand, and local communities, on the other, served to limit the effectiveness of the response. Trust in government varied within and across the three countries but was in many cases low due to histories of civil war and inter-communal tensions (Anderson and Beresford 2016: 478; Piot et al. 2014: 1034; Richards 2016: 33). At the same time, national and international responders lacked trust in communities, treating people as obstacles rather than resources in addressing the Ebola outbreak, with little confidence in the capacity of local populations to learn quickly about biosafety risks and develop appropriate responses for themselves (Caremel et al. 2017: 74–75; Richards 2016: 122). The assumption that public health experts and scientists, but not local populations, possessed the knowledge to control the outbreak fed into top-down responses that insisted on protocols and procedures that denied valuable input from affected communities and so were often ill-adapted to their priorities, practices, and resources and served to damage already low levels of trust in national governments and international responders (DuBois et al. 2015: v; Meredith 2015: 16; Wilkinson and Leach 2015: 146).

Rumours and conspiracy theories circulated in all three countries, reflecting and reinforcing mistrust of authorities and prompting lack of cooperation with prevention measures and medical teams. Some rumours suggested that Ebola did not exist, while others suggested that it was a virus being spread deliberately by governments and aid agencies (Anderson and Beresford 2016: 478). In Liberia and Sierra Leone, for example, accusations of government corruption fuelled rumours that Ebola was a ruse through which those in positions of power could make money (Anderson and Beresford 2016: 478; Wilkinson and Leach 2015: 144). In Sierra Leone, there were rumours that Ebola was a ruse or strategy by the government to depopulate opposition strongholds and reduce opposition votes (Anderson and Beresford 2016: 478; Richards 2016: 127; Wilkinson and Leach 2015: 144). Likewise, in Guinea, some believed that Ebola was introduced to the country by the government as a means to win elections and legitimize the use of armed forces, and suspicions regarding the political motives of the central government prolonged some infection chains in the forest zone (Caremel et al. 2017: 68; Richards 2016: 40).

Inaccurate messaging, especially early in the outbreak, further reduced trust and sometimes had counter-productive effects. For example, health promotion messaging in all three countries initially focused on not eating bushmeat, and in Guinea, a consumption ban was imposed (Richards 2016: 34–35; Wilkinson and Leach 2015: 137–138). Messaging on the potential for infection through eating bushmeat contrasted with people's lived experiences, in which bushmeat was eaten without adverse effect and served not only to deny people vital sources of protein but also to reduce trust in governments and others spreading the misguided message (Wilkinson and Leach 2015: 138, 145). It also generated a false sense of security among those who did not live near a forest or never ate bushmeat (Richards 2016: 7). Scaremongering—particularly regarding the death rate, with well-intentioned slogans such as 'Ebola kills'—generated fear, which sometimes deterred sick people from seeking help and led people to avoid or flee quarantine and to attack international responders (Richards 2016: 21, 23, 147). In fact, with around 11,000 deaths from around 27,000 cases across the three countries, approximately 60% of those infected with Ebola survived.

Linked to the lack of community engagement, national and international responses were largely designed based on unquestioned, and misplaced, assumptions about social and medical conditions (Richards 2016: 72). Within the three principally affected countries, messaging depended on a biomedical understanding of disease and of the required response, and this was sometimes at odds with the beliefs of affected populations and the resources available to them (DuBois et al. 2015: 33–34; Walsh and Johnson 2018: 43–44). The perfect arguably became the enemy of the good[6] with international responders insisting on the use of chlorine instead of soap, for example, because chlorine kills the virus more effectively, but while everyone had soap, chlorine was distributed and some people stopped using anything at all when they ran out of chlorine (DuBois et al. 2015: 33).

Treatment of Ebola patients

The first to respond to the epidemic were affected communities, and they began implementing protective and isolation initiatives well before significant international measures were deployed (Caremel et al. 2017: 67; Richards

[6] This kind of counterproductive insistence on technical standards echoes the response in Haiti after the 2010 earthquake, when insistence on high-legal and technical standards severely limited the housing solutions international agencies could offer. See 'Recovery and rehabilitation' in Chapter 9 on the Haiti Earthquake, 2010.

2016: 134–135). Yet, to begin with at least, these initiatives were at best ignored and at worst undermined by an international response based on biomedical science, with little attention to context (Caremel et al. 2017: 74). Measures such as barrier nursing, the use of PPE, disinfection, and isolation, failed to contain the outbreak for several months, in part because they did not take into account local capacity, conditions, customs, and politics (McInnes 2016: 387–388).

International agencies generally maintained high standards of protection for expatriate health-care workers, which was important in terms of the duty of care agencies have for their staff and also in terms of enhancing trust and mobilizing others to intervene, but insisting on the highest levels of biosecurity limited the care these actors were able to provide for their patients (MSF 2015: 10–11, 17). Health-care workers from Guinea, Liberia, and Sierra Leone were not so well looked after (Walsh and Johnson 2018: 68–69, 123–125). The national governments were unable to supply adequate protective gear, contributing to the demoralization and death of health workers and further weakening public trust in the health-care systems and their staff (IRC 2016: 14). Outside the specially built Ebola treatment centres, even international agencies were unable to provide adequate and effective protection for local health-care workers, who were sometimes reduced to using plastic bags (Pallister-Wilkins 2016: 514). International staff not only had much better access to PPE and a correspondingly much lower risk of infection, but also could be evacuated to state-of-the-art infectious disease units and receive treatments not available in West Africa, if they were infected (Pallister-Wilkins 2016: 516). By contrast, nearly 500 local health-care workers lost their lives in Guinea, Liberia, and Sierra Leone in the first year of the outbreak (MSF 2015: 2).

Running through the international response was a tension between fulfilling a duty of care to staff and minimizing the risk of transmission, on the one hand, and providing the desired quality and quantity of care to west African Ebola patients on the other. However, high-quality care increased survival chances and made people more likely to seek treatment such that it also served to reduce transmission. Especially early on, care for Ebola patients was often poor quality, not always or only due to inadequate resources, and insofar as this disinclined people to seek treatment, it likely contributed to increased transmission (Walsh and Johnson 2018: 346). Survival rates increased as detection and access to early palliative care improved, and survivors were an important asset; they could provide care to Ebola patients without risk of infection and, as living proof that the disease could be overcome, they were used to spread positive messages about Ebola treatment, but they also suffered stigma and were sometimes shunned by their communities (Richards 2016: 110–113). While treatment and care for Ebola patients improved over time, the international response largely failed to provide non-Ebola health care (Walsh and Johnson 2018: 301). This was problematic not only because people with typhoid, meningitis, or anything else needed treatment, but also because the failure to provide care for those who tested negative for Ebola became a factor in preventing people getting tested, and damaged (already weak) trust in the response (Walsh and Johnson 2018: 346).

Initial resistance to developing a home care protocol or facilitating community-based treatment was another example of top-down decision-making and messaging that failed to take into account social context and material realities. At the outset, low numbers of health workers meant that many Ebola patients had to be turned away from hospitals and cared for by families at home, exposing their contacts to the risk of infection (DuBois et al. 2015: 13). For months after the international response began, there was still insufficient capacity to provide professional care to all those infected with Ebola in hospitals or treatments centres (Wilkinson and Leach 2015: 145–146). For example, MSF, the main international treatment provider, was turning Ebola patients away from its overwhelmed facility in Monrovia (MSF 2015: 4, 10–11). As the outbreak in Sierra Leone reached its peak in November 2014, capacity in ETUs met only 60% of requirements (Michaels-Strasser et al. 2015: e361). Even when professional care was available, many would-be patients distrusted government health systems and feared that going to hospital meant dying alone (Caremel et al. 2017: 68; Wilkinson and Leach 2015: 145–146).

For international responders, however, it was initially unthinkable to develop a protocol for home care (Richards 2016: 9). Where the sick either could not or would not access professional health care, their families were left with the choice of caring for them without any official support or information on best practices or not caring for them at all. In September 2014, WHO published a set of messages for social mobilization and community engagement that was both contradictory, combining 'do not care for a sick person at home' with 'if you provide care', and in many ways highly impractical as they failed to take into account resource constraints, advising the use of plastic bags and raincoats as improvised PPE, without considering who would supply them and insisting that soiled clothes and bedding must be burnt without saying who would replace them (Richards 2016: 124). That same month, with insufficient capacity in treatment centres, MSF began distributing 600,000 home disinfection kits in Monrovia—considered an imperfect, stopgap solution but one that recognized and accepted the inevitability that people were being treated at home (MSF 2015: 18). While some international responders saw advocating home care as unethical due to the risk of transmission, some participants in focus groups in Monrovia—especially the women—felt that it would be unethical *not* to provide care themselves if family members were sick (Richards 2016: 136).

Community care centres (CCCs) offered a middle way between the centralized ETUs and home care. They were able to implement a number of essential safety measures, even as they were not working at the same level of biosecurity as the ETUs, and offered a good but imperfect solution to the dual problem of insufficient ETU capacity and the disinclination of many people to travel far from their homes and communities for treatment (DuBois et al. 2015: 34; Michaels-Strasser et al. 2015: e361). However, in another example of the perfect being the enemy of the good, official resistance to exploring any kind of community-based care was rectified only belatedly and partially (DuBois et al. 2015: 34; Wilkinson and Leach 2015: 145–146). Eventually, CCCs became an important part of the response, first in Liberia and, from November 2014, in Sierra Leone, with UNICEF taking a leadership role and international humanitarian NGOs, such as the International Rescue Committee (IRC) and Oxfam, constructing and running some centres (IRC 2016: 12; Meredith 2015: 16; Michaels-Strasser et al. 2015: e361). Many people were more willing to be referred to a CCC, where patients could remain in the community and be visited by family members, who could see them from a safe distance (Meredith 2015: 16).

Burial practices

Burial practices offer another example of a top-down response based on biomedical science abstracted from context backfiring because it failed to take into account local customs and capacities. Washing the body of someone who died from Ebola to prepare the corpse for burial carried a high risk of infection, and some large-scale funerals led to dozens of infections (Richards 2016: 26; Walsh and Johnson 2018: 59–60). Funerals in much of Guinea, Liberia, and Sierra Leone are not conducted by professional undertakers (Richards 2016: 7–8). Burial rituals demonstrate love, honour, and respect, and many believe that they must be performed the right way and by the right people to ensure that the deceased can find rest and peace in the afterlife—something which was sometimes a greater concern than contracting and dying of Ebola (Walsh and Johnson 2018: 59). To put it another way, 'epidemiologically safe burial [was] unsafe from a social and spiritual perspective' (Richards 2016: 52). Adapting the practices by which bodies were prepared for burial was obviously key to reducing transmission, but heavy-handed policies imposed from above, without sufficient consideration of the resources available or the needs of affected communities, frequently backfired. As such, the 'early instructions on so-called safe burial—rigid and unworkable—were, *in that context*, a textbook manual for unsafe burial' (DuBois et al. 2015: 31).

New policies for 'safe burials' were imposed by the affected governments with the threat of legal sanction and not developed through consultation or with local consent (Richards 2016: 148). In Sierra Leone, for example, the washing of corpses was made a criminal offence, and Liberia ordered mandatory cremations (Richards 2016: 52; Walsh and Johnson 2018: 149). International responders and some of the more elite or urban national responders struggled to understand why people were so reluctant to undertake 'safe burials' when the risks of traditional burials were so high (Walsh and Johnson 2018: 86). In Sierra Leone, teams of mainly younger people from urban areas were equipped with transport and trained to bury corpses in a sanitized way, in a central location, wearing PPE and using chlorine sprays and body bags (Richards 2016: 101; Walsh and Johnson 2018: 85–86). However, burial teams were overstretched and lacked sufficient vehicles to deal with the bodies of suspected Ebola victims in a timely way, and delays of several days in arriving to more remote settlements meant that villagers often dealt with corpses as they saw fit, with unauthorized burials triggering further transmission (Richards 2016: 41, 101).

Apart from insufficient capacity to carry them out in a timely way, 'safe burials' deprived communities of the rituals they needed to perform and, as such, led to talk of 'hidden bodies' and 'secret burials' as well as fostering mistrust of health workers (DuBois et al. 2015: 31; Richards 2016: 52, 95). The plastic of the body bags was seen as particularly disrespectful as it was reminiscent of garbage bags (Walsh and Johnson 2018: 86). In Guinea, some interpreted the suspension of traditional funeral rituals and the requirement to use body bags as a government strategy to prevent the deceased's spirit from leaving the body, exposing family members to vengeance from beyond the grave (Caremel et al. 2017: 68). The mass cremation of corpses in Liberia was feared to have led families to hide their sick (Wilkinson and Leach 2015: 146). There were alternatives to heavy-handed (and ultimately counterproductive) burial policies, with creative adaptions showing that mutually acceptable solutions could be reached (Wilkinson and Leach 2015: 146). For example, training local teams to carry out burials had advantages in that delays were avoided and there was greater trust that local teams could carry out burials with respect because they were socially known to the deceased (Richards 2016: 129, 130, 148). International agencies, however, were slow to support local burial (Richards 2016: 131).

Conclusions

Politics shaped the timing and nature of the international response to the Ebola outbreak in Guinea, Liberia, and Sierra Leone in 2014–2015. While MSF was stretching its capacity to respond from early in the outbreak, most major international actors were slow to respond and risk averse when they did, limiting the scope and impact of their work. Risk aversion concerned not only the risk of staff contracting Ebola but also the risk of blame for recommending imperfect measures—such as washing with soap instead of chlorine, setting up CCCs instead of ETUs, or adapting burial rituals to local concerns—if they led to transmissions, as they inevitably would (albeit probably fewer transmissions than would arise from recommendations of perfect measures being ignored). This risk aversion contributed to top-down messaging promoting isolation, treatment, and burial practices in keeping with the highest levels of biosafety but out of step with the material and social context. International and national responders lacked trust in communities, treating people as part of the problem rather than part of the solution, and as such, they imposed policies from above instead of through consultation and collaboration. The policies were not well adapted to the needs and realities of the people on whom they were imposed, and the lack of community engagement exacerbated already low levels of trust in officials such that many people in all three countries rejected those policies. A lack of trust and cooperation between communities, national governments, and international responders, particularly in the early phases of the response, thus reduced the effectiveness of efforts to prevent transmission and to care for the sick. The eventual successes against the outbreak were, in large part, due to an adaptation of these policies, 'thanks to actors who took into account political realities and cultural sensitivity, particularly around activities like community engagement and burials' (IRC 2016: 3).

12
European 'Migrant Crisis', 2015–2016

The arrival of more than a million migrants to Europe in the space of nine months exposed a crisis of European institutions, policy, and practice. With some important exceptions, the response from European authorities focused mainly on tightening border controls and making life difficult for those migrants who managed to reach Europe's shores. As a consequence, various points of humanitarian crisis emerged, most notably in the Mediterranean itself, where tens of thousands of migrants continue to make dangerous crossings every year, and in camps and settlements in Greece and northern France, where sometimes thousands of migrants are living in dire conditions.

The humanitarian response to this crisis has been undertaken in large part by volunteers working with grassroots organizations, many of which were set up to respond to this particular crisis and lacked prior experience. Newly established non-governmental organizations (NGOs) have taken the lead in search-and-rescue (SAR) missions in the Mediterranean and in providing assistance to migrants in camps in Calais and the Greek island of Lesvos. With the exception of Médecins Sans Frontières (MSF), most of the largest international humanitarian agencies that are the main focus of this book have been absent or have provided a small-scale, low-profile response.

This chapter focuses primarily on 2015 and 2016, but since the crisis is in many ways ongoing, developments up until 2021 are also discussed. While in some ways unique (above all in rescuing people at sea and working in wealthy countries in the continent from which many of the largest humanitarian agencies originate), many of the challenges and dilemmas posed by working in this context are familiar from elsewhere. Above all, these stem from the fact that European policies are the cause of much of the suffering that humanitarian agencies are seeking to alleviate, and thus it is difficult—or perhaps impossible—to work to alleviate that suffering without either becoming complicit with those policies or taking an explicit political position against them.

The Politics and Everyday Practice of International Humanitarianism. Miriam Bradley, Oxford University Press.
 DOI: 10.1093/oso/9780198840695.003.0012

Background

In just nine months from July 2015 to March 2016, more than a million refugees and other migrants arrived in Europe by sea, mainly in Greece but also Italy and—to a much lesser extent—Spain (Borton and Collinson 2017: 1). There are three main routes: the Eastern Mediterranean route, through which migrants cross to Greece, mostly from Turkey and in much smaller numbers from Egypt and Lebanon; the Central Mediterranean route to Italy, mostly from Libya, with smaller numbers from Egypt and Tunisia; and the Western Mediterranean route to Spain, mostly from Algeria and Morocco (Borton and Collinson 2017: 3). The number of migrants arriving to Europe via the Mediterranean Sea decreased significantly from 2016 onwards, as shown in Table 12.1, although the proportion dying en route was much higher in 2016 compared with 2015 and has remained relatively high (more than 1%).

In 2015, 48% of migrants arriving to Greece, Italy, and Spain were from Syria, 20% from Afghanistan and 9% from Iraq (UNHCR n.d.-b: 2). A significant proportion of these likely met the criteria for refugee status, but many others were escaping poverty and lack of opportunity and did not necessarily qualify as refugees.[1] The flows along the Mediterranean routes were thus 'mixed' as they 'were made up of people moving though some combination of force, choice and agency' (Borton and Collinson 2017: 4; see also, del Valle

Table 12.1 Migrants crossing the Mediterranean to Europe, 2014–2020

	Arrivals*	Dead and missing
2014	225,455	3,538
2015	1,032,408	3,771
2016	373,652	5,096
2017	185,139	3,139
2018	141,472	2,270
2019	123,663	1,335
2020	95,031	1,277

Note: *Includes sea arrivals to Italy, Cyprus, and Malta and both sea and land arrivals to Greece and Spain (including the Canary Islands). Numbers are as of 31 December 2020 for all countries except Cyprus, for which they are as of 31 August 2020.
Source: United Nations Refugee Agency (UNHCR) Mediterranean Data Portal.

[1] On the criteria for refugee status, see 'International displacement and migration law' in Chapter 14 on Law and International Humanitarianism.

2016: 28). They were also 'irregular' in that the migrants making these journeys did so without the required authorizations, visas, or other immigration documents and often also without passports or identity papers (Borton and Collinson 2017: 4).

While the number of arrivals in 2015 may have been unprecedented in Europe, it would not have been considered exceptionally high in some other regions of the world—either in absolute terms or relative to the size of host populations. Contrary to the impression created by dominant European discourses, most migration is regional and not orientated to Europe (Andersson and Keen 2019: ii). At the end of 2020, for example, 86% of the world's refugees were hosted in developing countries, and 73% in countries neighbouring their countries of origin (UNHCR 2021c). Other migrants more often travel to higher income countries, but a large proportion nonetheless remain in regions of origin. Around 75% of migration in sub-Saharan Africa, for example, is regional (Andersson and Keen 2019: ii). And although around half a million Syrians crossed the Mediterranean in 2015, the majority of Syrian refugees were hosted in neighbouring countries, with 628,200 in Jordan, 1.1 million in Lebanon, and 2.5 million in Turkey at the end of that year (UNHCR 2016: 16). Given that the numbers of migrants arriving to Europe was not high relative to the numbers arriving elsewhere in the world, and still less so relative to the population of Europe, the so-called European migrant crisis is better understood as a crisis of European policy and institutions rather than of the scale or nature of migrant arrivals. As Ban Ki-moon put it in August 2015, it was 'a crisis of solidarity, not a crisis of numbers' (UN Secretary-General 2015).

European legal and policy framework

Whereas international refugee law (IRL) and the Common European Asylum System (CEAS) provide clear frameworks for the international protection of refugees *qua* refugees, international norms regarding the rights and treatment of other migrants mostly do not relate to their status as migrants.[2] Instead, they are protected as human beings under international human rights law (IHRL) and, depending on their circumstances, may also be entitled to certain rights and protections corresponding to specific categories of migrant such as migrant workers or victims of trafficking. In the context of

[2] See 'International displacement and migration law' in Chapter 14 on Law and International Humanitarianism.

Mediterranean arrivals, a range of international laws that regulate rescue at sea is also relevant.

The principle of *non-refoulement*—which is central to IRL and also features in IHRL more broadly—prohibits the transfer of a person to a country in which he or she will be in danger. Rejection at the border violates the principle of *non-refoulement* if it involves sending people back to danger, but preventing people from reaching the border in the first place is a grey area which European states have exploited through such measures as interdiction at sea, extraterritorial processing of asylum claims, the introduction of mandatory visas, sanctions imposed on carriers (e.g. airlines) who are fined if they carry migrants travelling 'irregularly' without the 'proper' documentation and visas, and the use of force at borders (Borton and Collinson 2017: 4; Goodwin-Gill and McAdam 2007: 270–277). Visa requirements and carrier sanctions mean that many migrants cannot simply buy a plane ticket and board a plane, a cheap and safe means of travel, and instead turn to people smugglers, who charge exorbitant prices to facilitate far more dangerous journeys.

Legal categories are important in determining the rights of different migrants and European policy responses towards them. Those who are granted asylum gain refugee status and with it a number of entitlements including the right to reside and work. However, the outputs of refugee status determination procedures are very different in different European Union (EU) states, and not all individuals protected by the principle of *non-refoulement* are entitled to refugee status (Goodwin-Gill and McAdam 2007: 285–286; Niemann and Zaun 2018: 12). Some may be entitled to another kind of protection, such as 'subsidiary protection', a legal status in the EU since the 2004 European Qualification Directive, to be granted to those fleeing torture and those fleeing indiscriminate violence in internal conflict (Goodwin-Gill and McAdam 2007: 325–328). The rights associated with subsidiary protection are more limited and temporary than those provided to Convention refugees (Goodwin-Gill and McAdam 2007: 332–333). Others are protected by the principle of *non-refoulement* but are not entitled to any protected status and thus end up in a legal limbo whereby they cannot be deported but neither do they have the right to work. Finally, some migrants are not protected by the principle of *non-refoulement* because they do not face danger should they be expelled, and they may be awaiting deportation or living under the radar. All migrants, whether trying to reach Europe or already there, are entitled to dignified treatment in accordance with basic human rights standards.

The fact that states only have a clear obligation to those refugees who reach their territories contributes to a very unequal and unfair distribution of refugees, both globally and within Europe. Under the CEAS, the Dublin III Regulation sets out responsibility for processing asylum claims—usually the first EU state whose territory a refugee enters. This means that for those arriving by land and sea, some countries—notably Italy and Greece—bear the overwhelming share of this responsibility. It also serves to dissuade large numbers of asylum seekers from approaching national institutions that could potentially grant them protected status, because once they are registered in one EU country they are not allowed to apply for asylum in another (Hilhorst et al. 2021: 128). With their asylum systems overwhelmed in 2015, both Greece and Italy waved through asylum seekers, who continued their journeys towards northern Europe (Niemann and Zaun 2018: 4). Countries further north responded in different ways. In August 2015, Germany suspended application of the Dublin Regulation for Syrian asylum seekers and admitted over one million migrants but subsequently backtracked on this open doors policy, implementing border controls, sending people at the border back into Austria, and tightening asylum laws (Niemann and Zaun 2018: 4). Hungary and several other European countries rushed to erect fences, close official crossings, and deploy military and police to prevent the arrival and onward movement of refugees, eventually leading to a closure of the 'Balkan route', effectively trapping migrants in Greece (del Valle 2016: 38; Niemann and Zaun 2018: 4).

The EU responded by deploying Migration Management Support Teams to so-called hotspots in Greece and Italy and introducing a resettlement scheme and a temporary emergency relocation scheme through which newly arriving refugees would be distributed across the EU, although only a small proportion of refugees were actually relocated this way (Niemann and Zaun 2018: 5–8). The schemes did not take much account of the preferences or circumstances of asylum seekers, nor of the fact that protection and welfare standards for refugees vary significantly across EU countries, all factors that contribute to refugees trying to avoid detection and fingerprinting on arrival in Europe so that they can travel onward and claim asylum in a country of their choice. Thus, for example, migrants seek to cross the Channel from France to the United Kingdom, and joint British–French efforts to prevent this have led to the creation of informal migrant camps around Calais, including the so-called Jungle, and to the deaths of at least 194 migrants trying to smuggle themselves across between 2014 and mid-December 2021.[3]

[3] Data collated by International Organization for Migration (IOM) missing migrants project, based on recorded fatalities (as at 13 December 2021). The real number may be significantly higher.

The EU and its member states have also sought to prevent migrants arriving to Europe at all through maritime policies characterized by 'organized hypocrisy', with SAR increasingly securitized and subordinated to border control (Cusumano 2018a; Ghezelbash et al. 2018). Italy's Mare Nostrum operation rescued more than 130,000 people between October 2013 and October 2014, when it was discontinued, in part because it was perceived as a 'magnet' for boat arrivals (del Valle 2016: 32; Ghezelbash et al. 2018: 326). It was replaced by a Frontex (the EU border and coast guard agency) operation, Triton, which focused primarily on border control and initially only operated within thirty miles of Italian waters (Cusumano 2018b: 388; Ghezelbash et al. 2018: 326). Subsequently, in response to the large increase in numbers of migrant arrivals, an EU Naval Force Operation (EUNAVFOR MED) was launched in June 2015 to operate in parallel to Frontex-led activities, with combating smuggling and reinforcing border controls as its primary objectives, and bolstering SAR only a secondary consideration (del Valle 2016: 37–38; Ghezelbash et al. 2018: 333). The duty to provide assistance to those distressed at sea is a well-established norm[4] but one that has been widely and frequently flouted with respect to migrant boats in the Mediterranean, with collective indifference fostered by a system of sanctions on the rescuers of migrants (Basaran 2015).

As part of its efforts to reduce the numbers of crossings and arrivals, the EU also entered into agreements with third countries. In March 2016, the EU concluded a deal with Turkey whereby the latter assumed responsibility to take back and process all migrants and refugees crossing into Greece irregularly in exchange for a commitment from Europe to receive an equal number referred for asylum, and €6 billion to help Turkey secure its borders and host asylum seekers. Human rights organizations were highly critical of the plans, questioning their legality and warning that individuals could be prevented from claiming asylum under the scheme, which more or less sealed the Eastern Mediterranean route (Ghezelbash et al. 2018: 319; Hilhorst et al. 2021: 129). Other routes then became more important, first the Central Mediterranean route between Libya and Italy, with over 180,000 arrivals in 2017, then the Western Mediterranean route between Morocco and Spain, with close to 60,000 arrivals in 2018 (Garcés Mascareñas 2020: 1). Beginning in October 2016, cooperation with Libya, through which SAR was combined with anti-smuggling operations, reduced crossings in the Central Mediterranean. Italy began providing training and resources to the militia-led Libyan Coast Guard

[4] See 'International displacement and migration law' in Chapter 14 on Law and International Humanitarianism.

to keep migrant boats out of European waters, instead forcibly returning them to Libya, and the EU made a deal with the UN-backed Libyan government which favours border control over migrant safety and ultimately seeks to prevent migrants from leaving Libya and, hence, from reaching Europe (Cuttitta 2018: 648, 649; Esperti 2020: 446, 448; Ghezelbash et al. 2018: 334–335; Hilhorst et al. 2021: 129).

European states have also implemented a variety of measures aimed at making life for migrants extremely difficult. Many detain those awaiting deportation (which may include failed asylum seekers, other migrants who are in the country illegally, and migrants who were in the country legally but have committed certain types of crime), and sometimes also asylum seekers awaiting the determination of their status. Detention facilities for migrants are given names that make them sound humanitarian (reception centres, guesthouses, welcome centres, etc.) but they function like criminal detention facilities—often with much less oversight and lower standards than is legally required for criminal detention in the same countries (Kotsioni 2016: 42). Even where migrants are not detained, they may face restrictions on working and on where to live, and lack of access to welfare support or employment leaves many destitute. In one of the more extreme examples, the United Kingdom has worked with France not only to prevent Channel crossings but also to forcibly return some who succeed in crossing and to ensure squalid conditions in the make-shift camps around Calais (Keen 2021: 14–15). Migrants could only access more formal accommodation in France if they claimed asylum, but by doing so they risked being sent back to the first EU country they passed through under the Dublin III Regulation, so many chose to circumvent this process (Hilhorst et al. 2021: 128).

Practices by European states including strict border controls and keeping migrants in harsh conditions, as in the Calais 'Jungle' or Moria refugee camp on the Greek island of Lesvos, have political pay-offs, in that they may serve to deter some arrivals, and they play to those elements of a domestic audience wanting politicians to be tough on immigration (Gordon and Larsen 2021: 428–431; Keen 2021: 13–17). Such practices are also self-legitimating and self-reinforcing: leaving migrants to die at sea through widespread dereliction of rescue duties feeds into an understanding that some lives are worth less than others, and the squalid, undignified conditions that migrants live in as the consequence of strict border controls serves to dehumanize the migrants and hence to legitimize the squalid conditions and justify the strict controls (Basaran 2015; Gordon and Larsen 2021: 432–433; Keen 2021: 18). In several countries, the outsourcing of many migration control operations, including those related to detention, deportation, and the security of reception and processing centres, additionally makes the imposition of strict controls in these areas a lucrative business (Davitti 2019: 34; Kotsioni 2016: 44). Perhaps unsurprisingly, then, private military and security companies have been a driving force behind securitizing migration in the EU, increasing the demand for militarized responses—precisely the kinds of responses they profit from providing (Davitti 2019: 37–41).

A humanitarian emergency?

The first chapter of this book discussed the main types of event that cause and constitute humanitarian emergencies (armed conflict, famine, earthquakes, etc.), none of which apply in contemporary Europe. Yet, the policies of European countries have created a situation in which tens or hundreds of thousands of migrants are in need of humanitarian aid, notably in the Mediterranean, the Greek island of Lesvos, and Calais.

European policies have fostered an environment in which migrants in distress in the Mediterranean are frequently not rescued, either by official SAR missions or by passing commercial vessels (Basaran 2015; Ghezelbash et al. 2018). Many of the tens of thousands who have died in their attempts to cross the Mediterranean over the past ten years could have been saved if nearby boats had stopped to rescue them (Basaran 2015: 206). The large numbers of deaths in the Mediterranean led some to define it as 'the theatre of a complex humanitarian emergency' (Cusumano 2018b: 387). At the same time, some at MSF feared that the discourse of humanitarian emergency could be co-opted and used by states to justify (further) extraordinary measures, including military deployments to intercept boats and deter smugglers, deals with third states, offshore processing of asylum claims, and detention (del Valle 2016: 32). Whether or not a humanitarian emergency narrative is helpful or counter-productive to achieving safety and protection for migrants crossing the Mediterranean, non-governmental SAR missions were widely recognized as necessary in the short term if such safety and protection were to be achieved.

The Greek island of Lesvos, 10 km from the Turkish coast and with a population of 85,000, has served as a landing point for migrant arrivals to Europe for decades, with numbers peaking at 6,000 per day in 2015—half a million (two-thirds of all arrivals to Greece) were registered there that year (Hilhorst et al. 2021: 134; Rozakou 2017: 102). Overcrowding and inadequate assistance made living conditions in some of the camps across the island unhygienic and insecure, and the situation on Lesvos was widely considered a humanitarian emergency (Hilhorst et al. 2021: 135; Rozakou 2017: 102). After the Turkey deal was signed and the Greek border with Macedonia closed in March 2016, migrants were stranded in Greece with no alternative but to seek asylum there, and impromptu settlements were set up across the country (Hilhorst et al. 2021: 134; Rozakou 2017: 103). The designation of Moria, the main reception centre in Lesvos, as a hotspot turned it into what many considered a detention centre (Gordon and Larsen 2021: 431–432; Hilhorst et al. 2021: 134). Migrants were only supposed to stay briefly in Moria, while they were processed and either registered for asylum on the Greek mainland or returned to Turkey, but in practice, residency often lasted up to two years (Hilhorst et al. 2021: 135). What was intended to be a transit centre effectively became a long-term camp but without the kinds of provisions, services, and support a long-term camp requires (Gordon and Larsen 2021: 423–424). Variously described as 'the worst refugee camp on earth', a 'living hell', 'hell on earth', or 'worse than hell', Moria became massively overcrowded and conditions were deplorable (Gordon and Larsen 2021: 423, 424–425).

As the main crossing point from continental Europe to the United Kingdom, Calais has also long hosted migrants and migrant camps or shelters of varying levels of formality—which have frequently been subject to forced closures, harassment, and violence from authorities. The population of the main 'Calais Jungle'—a makeshift camp 5 km from Calais city centre on land (on a chemical dumping ground contaminated with asbestos) which the French government designated as a tolerated space of informal encampment in early 2015—swelled rapidly in 2015–2016, reaching an estimated 10,000 before the French authorities destroyed it in October 2016 (Keen 2021: 13). In the words of a volunteer there, the 'Jungle was not an ordinary refugee camp: regular camping tents used as shelters, piles of rubbish, stench, people wading through thick mud, other people begging for food. Suffering was everywhere' (Sandri 2018: 65). Inhuman treatment of migrants included police violence (teargas, rubber bullets, dogs, beatings), as part of a strategy to dismantle the various camps and 'jungles' that have existed in and around Calais since the 1990s, and police inaction in the face of racist attacks from members of the public (Keen 2021: 12; Sandri 2018: 69). After the 'Jungle' camp was demolished, the French government prohibited new camps at the border, with aggressive policing operations to detect and destroy informal encampments (Amnesty International 2019: 11; Hilhorst et al. 2021: 130). Since many opted out of the state registration and accommodation system due to legitimate fears that, under the Dublin system, they would be returned to their first country of entry to the EU, migrants were criminalized and humanitarian action undermined (Hilhorst et al. 2021: 131). Widespread police abuse has been documented, including excessive use of force against migrants and harassment of humanitarian actors (Amnesty International 2019: 13–15).

Humanitarian response

Across the sites of humanitarian emergency identified in the previous section, the humanitarian response shared three key characteristics: (i) the importance of volunteers and grassroots organizations set up to address this particular crisis and the relative absence of long established international humanitarian agencies; (ii) the centrality of advocacy, especially—but not exclusively—in the response of grassroots solidarity organizations; and (iii) the hostility of state authorities to most kinds of humanitarian response in this context, most apparent in the criminalization and disciplining of those who have sought to rescue and support migrants.

Search and rescue

In the absence of large-scale SAR operations by European states after Italy's Mare Nostrum was discontinued in October 2014, a number of NGOs have conducted migrant rescue missions in the Mediterranean, mainly offshore Libya (Cusumano 2018b: 387; Esperti 2020: 441). Broadly speaking, they operate on one of two models: fully fledged SAR missions, which include rescuing migrants in distress and transporting them to a European port (usually in Italy); or smaller boats, which provide lifejackets to migrants and host them aboard until they can be transferred to a larger vessel to shuttle them to port (Cusumano 2018b: 389; Cuttitta 2018: 642–643). While MSF and Save the Children both operated migrant rescue boats at different times, most of the major international humanitarian agencies that are the primary focus of this book did not engage directly in rescue missions.[5] Instead, many of the rescue vessels were operated by smaller, single-issue organizations set up specifically to respond to the plight of migrants crossing the Mediterranean, often staffed entirely by volunteers (Cusumano 2018b: 389, 391; Esperti 2020: 442). The MSF operation alone rescued and provided medical assistance to more than 15,700 people between May and August 2015, and it is estimated that

[5] Distinct from SAR, but related, the International Committee of the Red Cross (ICRC) has worked to train and assist authorities in Greece and Italy with what it calls humanitarian forensics work, recovering, handling, documenting, and identifying the bodies of migrants who have died en route (ICRC 2017: 2).

from the outset of the crisis in 2015 to 2018, NGOs were responsible for up to half of all rescues (del Valle 2016: 36; Ghezelbash et al. 2018: 347).

These missions have responded to real and immediate needs to rescue migrants in danger and also provided a platform from which to advocate for changes in European policies. Even without an explicit advocacy component, the mere operation of such missions may be deemed to imply a criticism of the practices of European states and a demand for safety for those risking travel across the Mediterranean (see, e.g. del Valle 2016: 31; Esperti 2020: 444). For most of the NGOs undertaking SAR operations, however, public communications were seen as an integral part of their missions, and for MSF, bearing witness was part of the rationale for engaging in SAR in the first place (Cusumano 2018b: 393; Cuttitta 2018: 640–641; del Valle 2016: 27, 37). As a minimum, advocacy by SAR NGOs aimed to pressurize EU states to step up their SAR efforts, and some additionally had more ambitious policy goals, advocating for EU states to create more safe and legal options for migrants trying to reach Europe. MSF, for example, had two clear advocacy messages: first, that EU states had the capacity and responsibility for SAR in the Mediterranean and that MSF efforts could not replace those that were lacking at the EU level; and second, that given the push and pull factors driving people to migrate, EU states should create safe and legal routes for migration to Europe (del Valle 2016: 31, 37). MSF's advocacy thus went beyond its usual approach and included commenting on European policies, which MSF saw as the cause of the crisis. Advocacy on these issues was not limited to those NGOs operating SAR missions, with human rights organizations, citizen activist groups, and, to some degree, international humanitarian agencies also criticizing European policies (Hilhorst et al. 2021: 127). The International Rescue Committee (IRC) and UNHCR, for example, were among those calling for EU states to open up more legal pathways for migrants to seek asylum in Europe (Hilhorst et al. 2021: 138). UNHCR was also highly critical of the criminalization of SAR (Hilhorst et al. 2021: 141). In addition, the boats operated by several of the SAR NGOs had journalists on board since the NGOs needed to publicize their work to attract donations, and some also sought to advocate against the border regime (Cuttitta 2018: 645).

Official policies towards non-governmental SAR missions varied within and across countries as well as across time. The actions of Italy and Libya have the most direct impact on SAR NGOs because they have the responsibility for coordinating SAR operations in the area in which most of these NGOs are operating, and lack of capacity in Libya means that Italy has had to assume responsibilities in what is theoretically the Libyan SAR area

(Cuttitta 2018: 642). However, Italian (and Libyan) actions are shaped by the EU policy environment (and, conversely, policies across the EU are shaped by their actions). After it became clear that Triton had neither the resources nor the mandate to replace Mare Nostrum, Italian authorities had welcomed SAR NGOs as important additional capacity, but political and media attitudes shifted as the numbers of arrivals from Libya increased in 2016 and in the absence of any mechanism or serious political will for the relocation to other EU countries of migrants disembarking in Italy (Cusumano 2019: 107–108).

The EU trained, equipped, and pressured the Libyan Coast Guard to intercept and return migrants to Libya before they were rescued by other vessels, and as the SAR NGOs posed an obstacle to interception, they were attacked in violent incidents involving the Libyan Coast Guard and in a media, political, and judicial campaign involving European authorities (Cusumano 2018b: 392; Cuttitta 2018: 648). A series of hostile acts by Libyan authorities began when Libyan Coast Guard officials fired several bullets and then boarded and searched Sea-Watch 2 in April 2016, and threats from Libyan forces led Save the Children, Sea Eye, and MSF to suspend operations in August 2017 (Cuttitta 2018: 647–648; Maritime Executive 2017). For their part, various European authorities accused SAR NGOs of acting as a pull factor, facilitating the work of smugglers or even collaborating with them, and being responsible for their use of increasingly unseaworthy vessels (Cuttitta 2018: 648; Ghezelbash et al. 2018: 347). After its calls for greater EU solidarity in the reception of migrants again fell on deaf ears at the June 2017 European Council meeting, Italy circulated a code of conduct for SAR NGOs, which imposed several limitations and could be expected to weaken their rescuing capabilities, compromise their commitment to humanitarian principles, and pose a risk to the safety of their personnel (Cusumano 2019). With the threat that failure to sign may result in refusal to disembark in Italian ports, some SAR NGOs—including Save the Children—signed the code of conduct immediately, others later and reluctantly, while others—including MSF—refused to sign (Cusumano 2019: 111).

Port closures and prosecutions have further contributed to a hostile environment for SAR operations, reducing their capacity and sometimes preventing them altogether. In June 2018, Italy's new government immediately stated its intention to close Italian ports to NGO boats and foreign-flagged merchant vessels carrying migrants rescued off the shore of Libya, and in the following months, disembarkation was denied or delayed for a number of vessels (Cusumano and Gombeer 2020: 245–248). The decision to close the Italian ports to migrant rescue vessels may not be illegal (Cusumano and

Gombeer 2020: 247–250; cf. Moreno-Lax et al. 2019: 726–732), but it has significant humanitarian implications if SAR vessels are forced to travel further (e.g. to Spanish ports, which have been more ready to allow them to disembark), as this means spending lengthy periods outside the main rescue areas (Cusumano and Gombeer 2020: 250–251). While denying disembarkation altogether has been rare, delays are common. In 2020 alone, there were 22 incidents involving 3,597 rescued people (including at least 954 children) in which a vessel was made to wait more than 24 hours before eventually being allowed to dock (FRA 2021). SAR has also been increasingly criminalized,[6] with Germany, Greece, Italy, Malta, the Netherlands, and Spain initiating fifty-eight criminal or administrative proceedings against SAR NGO vessels or personnel between 2016 and mid-2021, with criminal charges including 'attempted human trafficking' and 'aiding and abetting illegal immigration' (FRA 2021; Ghezelbash et al. 2018: 347). Such measures reduce SAR capacity when they deter SAR NGOs and staff from operating and when rescue boats are blocked in ports pending legal proceedings (as at June 2021, nine vessels were thus out of action; see FRA 2021).

Often linked to these challenges from European and Libyan authorities, SAR missions in the Mediterranean posed dilemmas for the NGOs undertaking them. Most of these dilemmas were familiar from humanitarian responses around the world, but some were exacerbated by the specific nature of SAR. As with humanitarian action in general, SAR missions raised a number of concerns about unintended consequences. The argument that SAR serves as a pull factor for migrants is a maritime variation on the general argument that humanitarian aid serves as a pull factor, and experienced agencies such as MSF tend to reject the idea that aid is ever a primary trigger or driver of displacement (del Valle 2016: 32). In terms of SAR in the Mediterranean, the evidence suggests that SAR missions neither increase the number of crossings nor result in more dangerous crossings, which undermines the 'pull factor' argument and suggests that SAR does not prompt smugglers to use more unseaworthy vessels (Cusumano and Villa 2021; Steinhilper and Gruijters 2017).

Other concerns about unintended consequences were not so easy to dismiss. On the one hand, there remained a risk that the SAR missions ended up serving to benefit smugglers and traffickers. MSF explicitly acknowledged and accepted this risk, likening it to the familiar dilemma over the potential for humanitarian aid in conflict settings to benefit the warring parties, but

[6] For a wider and more detailed discussion of this issue, see 'Criminalization of support for migrants' in Chapter 14 on Law and International Humanitarianism.

sought to minimize the risk by avoiding contact, negotiation, or exchange of information with smuggling networks (del Valle 2016: 33). On the other hand, non-governmental SAR missions risked benefiting European authorities, and the nature of SAR missions required contact and even cooperation with those authorities, not least regarding docking and disembarking. In the Central Mediterranean route, the Italian authorities had responsibility for coordinating SAR activities, and all of the NGOs were constantly in touch with the Maritime Rescue Coordination Centre in Rome (Cuttitta 2018: 646). While this was inevitable, and humanitarian agencies almost always have to coordinate to some degree with the governments of states[7] affected by humanitarian agencies (as a minimum, for consent to operate, visas for international staff, etc.), several NGOs were concerned about being embedded in the very system of European border control policies that they were advocating against (Cusumano 2018b: 391–392; del Valle 2016: 38–39).

Interactions with law enforcement raised particular concerns. Disembarking rescued migrants requires some kind of interaction with police authorities, who have the right to board the vessels in port and start identification procedures there (Cuttitta 2018: 643). Beyond this, however, the question of sharing information or images with the authorities divides the SAR NGOs. Pictures taken by drones operated by the Migrant Offshore Aid Station to increase search capacity and shared with the Italian authorities led to the arrest of a migrant identified as the driver of a migrant boat in May 2015 (Cuttitta 2018: 644). As a matter of policy, MSF does not share sensitive images and has sought to explain to border authorities that cooperation with law enforcement is beyond its mandate and could jeopardize the safety of its teams at sea (del Valle 2016: 39). Nonetheless, SAR NGOs could be legally bound to hand over material in the event of a criminal investigation, and for this reason, MSF has avoided taking pictures and videos (Cusumano 2018b: 392; Cuttitta 2018: 644).

There is also a significant risk that, in rescuing migrants in distress at sea, SAR NGOs facilitate what the NGOs see as dereliction of duty by EU authorities—precisely the opposite of one of their central goals, which is to pressure the EU and member states to take responsibility for SAR (Cusumano 2018b: 393). Furthermore, to the extent that they are relieving governmental actors from their responsibilities, their assistance may be freeing up official resources for other purposes, including purposes the SAR NGOs actively disagree with—a dilemma which was particularly acute when SAR NGOs were

[7] See 'State sovereignty, responsibility, and practice' in Chapter 19 on Government and Civil Society in Affected States.

requested by the Italian authorities to transfer to port migrants rescued by EUNAVFOR MED vessels, freeing the latter to continue with their military border security activities (Cuttitta 2018: 643, 650). More generally, by reducing mortality in the Mediterranean, there is a risk that they contribute to maintaining restrictive asylum and migration policies that might otherwise be relaxed in the face of public backlash against lives lost at sea (del Valle 2016: 30).

Lesvos

The initial response in Lesvos was provided mainly by the island population, who were soon joined by other Greek and international volunteers (Hilhorst et al. 2021, 134; Papataxiarchis 2016: 8; Rozakou 2017). Only a few of the large, established international humanitarian agencies were operational in Lesvos in 2015, among them MSF, Médecins du Monde (MDM), IRC and UNHCR, while thousands of volunteers and an estimated 250 smaller organizations were contributing to the humanitarian effort (Hilhorst et al. 2021: 134; Rozakou 2017: 101). After the Moria reception centre was designated a hotspot in March 2016, some of the agencies that had previously been operating there, including UNHCR and MSF, refused to work in what they considered to be a detention-like context (Hilhorst et al. 2021: 134). When the policy was changed several months later, giving Moria residents the freedom to leave the camp and move around Lesvos, but not to leave the island, UNHCR considered the compromise satisfactory and resumed work there, while MSF provided assistance to migrants from outside the camp (Hilhorst et al. 2021: 134). As with the SAR NGOs, then, those humanitarian actors working in Lesvos faced the dilemma that providing assistance to migrants could constitute complicity with the policies and practices that cause suffering (Gordon and Larsen 2021: 437).

Other camps functioned on very different models. Olive Grove, an unofficial settlement emerged outside Moria, accommodating those fleeing overcrowding or violence in the main camp (Gordon and Larsen 2021: 424; Hilhorst et al. 2021: 135). Several agencies provided services there, and paid rent for the land, but conditions were still poor, with extremely limited access to water, sanitation, washing facilities, or electricity (Gordon and Larsen 2021: 424; Hilhorst et al. 2021: 135). Kara Tepe, which was run by the municipal authorities with many humanitarian agencies operating inside the camp, received refugees based on referral, limiting numbers to match the resources available and working in the tradition of Greek hospitality, was referred to as a 'model camp' but was forced to turn people away in order to maintain standards (Hilhorst et al. 2021: 135). In every way the opposite of Moria, Pikpa was a small, open facility initiated by a local organization, Lesvos Solidarity, grounded in an ethic of solidarity, which—like Kara Tepe—had to compromise its commitment to unconditional solidarity and only accepted people in need of special care, usually referred by UNHCR (Hilhorst et al. 2021: 135).

The situation in Lesvos was bad in 2015–2016, and subsequently deteriorated further. In mid-2017, the EU decided to channel its response solely through government actors, many NGOs closed their programmes, and the flow of international volunteers also slowed (Hilhorst et al. 2021: 134). In the absence of proper accommodation and support for migrants on the island, solidarity from the local population dwindled and tensions with the permanent local population increased, culminating in violent protests by residents and violent attacks by far right groups on migrants, humanitarian actors, doctors, and journalists in 2020 (Gordon and Larsen 2021: 429; Hilhorst et al. 2021: 136). As a consequence of the violence and perceived lack of protection from the local authorities and police, many humanitarian actors left Lesvos, reducing the overall provision of food and medical assistance, and leaving migrants, especially the most vulnerable of them, exposed to further harm and harsher conditions (Gordon and Larsen 2021: 429).

Some humanitarian actors working on the island reported that police not only failed to protect them, turning a blind eye to violent attacks against them, but also targeted and harassed them in what they understood was intended to serve as a deterrent to humanitarian action (Gordon and Larsen 2021: 429). While such harassment falls short of criminalizing (and prosecuting) humanitarian actors,[8] it fits a broader pattern across Greece and Europe of policing and disciplining those who provide assistance to migrants (Carrera et al. 2019; Gordon and Larsen 2021: 429–43013). At the height of arrivals in 2015, for example, the Greek government made it illegal to transport migrants in a move that was subsequently overturned by the courts but nonetheless contributed to antagonistic relationships between the state and humanitarian actors (Hilhorst et al. 2021: 135–136). Police objected to photos or video footage being taken of Moria camp, and vocal critics of the treatment of migrants by Greek authorities were sometimes arrested in what many saw as an attempt to silence criticism (Gordon and Larsen 2021: 432).

[8] For more discussion of the criminalization of solidarity action within and beyond Europe, see 'Criminalization of humanitarian action' in Chapter 14 on Law and International Humanitarianism.

Against this background, advocacy was an important component of the humanitarian response, its focus and goals changing over time, from scrutinizing EU policies and seeking to influence the implementation of the EU–Turkey deal in 2016 to a focus on local policies and seeking to improve living conditions for migrants on the island by 2018—a change that seems to have been driven by perceptions of the likelihood of their advocacy yielding results (Hilhorst et al. 2021: 136). Differences between the grassroots organizations and the large international humanitarian agencies were particularly apparent when it came to advocacy. UNHCR worked in a somewhat invisible way, engaging in private dialogue with the authorities, and many other actors criticized UNHCR for not speaking out publicly to help defend humanitarian space, condemn violent attacks on migrants, or openly side with Pikpa camp when threatened with legal action (Hilhorst et al. 2021: 136–137).

Grassroots solidarity initiatives in Greece were sometimes overtly antagonistic to established international humanitarian agencies, scrutinizing their professionalism and uncertain as to whether or not they should cooperate with them and receive funding from them (Rozakou 2017: 103). Yet, the differences and antagonisms have arguably been overstated. Different types of humanitarian actors sometimes worked together, and very often individuals moved between categories, as when former 'solidarians' or 'volunteers' took jobs with the large international humanitarian agencies, for example, or agency staff volunteered in activist organizations (Hilhorst et al. 2021: 136; Papataxiarchis 2016: 8).

Calais

The camps around Calais also attracted large numbers of volunteers, especially as the 'Jungle' grew enormously in 2015–2016, while the established agencies mostly remained absent. Organizations such as Secours Catholique (Caritas France), Calais Migrant Solidarity, and Auberge des Migrants had provided basic assistance—mainly food and clothes—to migrants in Calais for several years but lacked the resources to provide other kinds of support (Sandri 2018: 68–69). The 'Jungle' camp was unofficial, lacking approval from the local authorities or the French government, and most major international humanitarian agencies will not operate without official permission such that the French agencies MDM, MSF, and Secours Catholique were the only large, long-established organizations operating there, while the smaller grassroots organizations that did so were subject to high levels of scrutiny by governments (Keen 2021: 15; Sandri 2018: 65–66, 72). While repeatedly expressing concern over conditions in the 'Jungle', UNHCR took the view that, as France had the capacity to provide accommodation, it was the French government's responsibility to do so and not UNHCR's (Hilhorst et al. 2021: 132; Sandri 2018: 69). Instead, grassroots organizations run by mostly inexperienced volunteers from France, the United Kingdom, and elsewhere took responsibility for supporting the migrants in the 'Jungle', responding to immediate needs and improvising solutions (Hilhorst et al. 2021: 131–132; Sandri 2018: 66, 74). While the work they did was enormously important, their informality meant that volunteers were not accountable, and their lack of experience limited their ability to deal with more complex situations that arose, such as trafficking, exploitation, and violence (Sandri 2018: 74).

Much like the SAR NGOs, those providing assistance to migrants in northern France were routinely accused of creating a 'pull factor' and sometimes of complicity in violence and criminality, with such accusations used to justify restrictions on their work (Amnesty International 2019: 18). Authorities sought to discipline humanitarian actors through tactics including 'smearing attacks, orders and instructions imposing arbitrary restrictions as to where and when aid can be provided and by whom; frequent ID checks; numerous parking fines; abusive language; threats of arrest; assault; and in some cases, detention and prosecution on a variety of grounds, including defamation, contempt, and assault' (Amnesty International 2019: 17). As in Greece, French courts offered some protection from efforts by the authorities to prevent or criminalize the provision of basic assistance to migrants, with two March 2017 municipal orders forbidding food distribution in Calais annulled by the courts, for example, but such efforts nonetheless contributed to a hostile environment for humanitarian response (Amnesty International 2019: 12, 16–17; Keen 2021: 24).

Advocacy was, once again, an important part of the response and, as in Greece, exposed some of the differences between grassroots organizations and established agencies, especially UNHCR. Some of the volunteer organizations in Calais were explicitly non-political, but many emphasized border contestation and solidarity activism (Sandri 2018: 74–75). Even those that viewed themselves purely as providers of humanitarian aid ended up becoming 'vocal in contesting restrictive policies of migration and refused to enact mechanisms of control on the refugees' (Sandri 2018: 75). Set up to respond to the Calais crisis specifically, grassroots organizations did not fear the repercussions of speaking out as much as longer-established organizations with international mandates and were able to take a confrontational approach, bringing legal action against the state, and being outspoken about police violence and the failure of the authorities to provide water, sanitation, and other basic services (Hilhorst et al. 2021: 132–133). Many of the grassroots organizations operating in Calais comprised large numbers of British volunteers, who took their stories and advocacy to the United Kingdom, where they organized demonstrations, ran awareness campaigns, and challenged violent border practices and the asylum regime (Sandri 2018: 75, 76).

The longer-established agencies varied in their approach to advocacy. As in Greece, UNHCR prioritized behind-the-scenes, case-specific advocacy with the French authorities, through which it sought to resolve particularly serious cases involving extremely vulnerable individuals rather than pushing for structural shifts or publicly denouncing abuses on the part of state authorities (Hilhorst et al. 2021: 133, 134). Furthermore, and again similar to criticisms levelled against UNHCR in Lesvos, UNHCR was criticized for failing to speak out on behalf of other humanitarian actors, including volunteers experiencing police brutality and agencies denied permission to provide shower facilities to migrants, and more generally for failing to defend humanitarian space (Hilhorst et al. 2021: 141). The French Red Cross stuck to service delivery rather than advocacy, and other established French agencies sought to encourage the French national human rights ombudsman and international human rights organizations to visit the border zone and report on what they saw (Hilhorst et al. 2021: 133).

Advocacy seems to have had some effect in Calais, where daily food distribution, subsidized by the state and provided by an NGO, began in March 2018 following sustained pressure by civil society organizations involving legal challenges and advocacy, both at the local and international levels (Amnesty International 2019: 17). This advocacy, however, came at a dual cost—diverting time and resources from the direct provision of assistance and seemingly exacerbating harassment and intimidation by the authorities, much of which occurred when individuals witnessed police abuses of migrants (Amnesty International 2019: 17).

Conclusions

Many of the largest international humanitarian agencies were founded in Europe, have their foundations and donor bases there, and felt their identities shaken by the border security measures implemented by 'their' governments (Hilhorst et al. 2021: 126–127). With suffering and humanitarian need largely understood to be the consequence of European policies and practices, humanitarian actors responding to that suffering and need risked becoming complicit in those policies and practices, above all if they did not accompany their practical response with advocacy. MSF embraced this, launching SAR missions and working in both Lesvos and Calais with advocacy as an important and integral component in those responses. By contrast, UNHCR—the agency with a specific mandate for refugee protection—largely avoided anything which could be deemed confrontational. In working with narrow definitions of vulnerability, and without mounting a robust challenge to EU policies, UNHCR was arguably complicit in those policies and contributing to a politics of abandonment (Hilhorst et al. 2021: 141). Many other international agencies opted not to respond or to offer only a very limited response, raising questions as to their effectiveness in their 'home' region rather than in Africa, Asia, or elsewhere (Borton and Collinson 2017: 1). Grassroots organizations stepped in to fill the gap and tended to adopt a much more solidarist humanitarian approach, in which advocacy was central—not only in the sites this chapter has focused on but also in places hosting migrants across Europe.

II
CONCEPTS AND TRENDS

Introduction to Part II

The Boundaries of Humanitarianism

The chapters in this section turn to the legal and normative frameworks for humanitarian action and explore the boundaries, ethics, and ethos of international humanitarianism. The boundaries of humanitarianism are contested along three main dimensions. First, what constitutes a humanitarian emergency and hence demands or legitimates a humanitarian response? What contexts, in other words, should be humanitarianized? Second, should humanitarian action be limited to saving lives and reducing suffering today or should a humanitarian response also seek to address the causes of suffering? More generally, what goals can properly be considered *humanitarian* goals? Third, what is the appropriate way to pursue those goals? For example, how, and to what extent, should humanitarian actors engage with political and military actors? Under what conditions should they work without state consent? How, and to what extent, should they speak out about violations of international humanitarian law (IHL) and abuses of human rights? The chapters in this part of the book show how these boundaries have been contested and how they have shifted over time.

Another way of thinking about these boundary questions is to ask whether, and where, lines are drawn between humanitarianism, on the one hand, and development, human rights, or peace-building work on the other (Barnett 2018). A useful starting point here is to distinguish the ideal type humanitarian ethos from the human rights ethos and the development ethos. Humanitarianism is most associated with compassion and the charitable imperative, whereas development work may emphasize utilitarian social policy, and human rights initiatives largely focus on principles of justice and rights (African Rights 1994: 9). Correspondingly, humanitarian action concerns itself with alleviating suffering and addressing its proximate causes, while development and human rights initiatives often seek to improve the human condition through structural and institutional change. In this way, humanitarianism focuses on symptoms rather than root causes and has a less transformative ambition than human rights or development work.

The case study chapters in Part I of this book contain several examples of these lines being blurred and of partial shifts in the boundaries of humanitarianism. For example, Chapter 12 on the 2015–2016 European 'Migrant Crisis' shows some international humanitarian agencies working to support migrants in the absence of conflict, famine, or disaster, implying a stretching of the concept of the 'humanitarian emergency'. In other emergencies, international humanitarian agencies sought to do more than address the symptoms of crisis and pursued goals more associated with peace-building (see Chapter 6 on Afghanistan, 2001–2014) or development, as in recovery and rehabilitation efforts after sudden-onset disasters (see Chapter 7 on the 2004 Indian Ocean Tsunami and Chapter 9 on the Haiti Earthquake, 2010). They have often also used means and methods beyond traditional humanitarian modus operandi. This includes working closely with political and military actors in Afghanistan (see Chapter 6) and placing advocacy centre stage in the European 'migrant crisis' (see Chapter 12).

The chapters in this second part of the book explore in more detail the positioning and contestation of humanitarianism's boundaries, arguing that humanitarian action very often does seek to address causes and may concern itself with justice, rights, and equality, albeit usually to a more limited degree than human rights or development actors. The boundaries between human rights and humanitarianism, or between development and humanitarianism, are not so much clear dividing lines but large, contested grey areas.

Chapter 13 on Politics, Principles, and Humanitarian Action maps out the lines of contention regarding the goals and means of humanitarianism—that is, the second and third dimensions of humanitarianism's boundaries. On the second dimension, this chapter introduces the main debates between the so-called new humanitarians, who believe that humanitarian actors can, and should, leverage the assistance they provide in pursuit of broader, transformational goals, and classical humanitarians, who believe that it should not be used that way. On the third dimension, this chapter discusses the perspectives of different agencies with respect to speaking out about violations of IHL and human rights abuses and working without the consent of the state.

Chapter 14 on Law and International Humanitarianism introduces the bodies of law that have most bearing on humanitarian action and outlines the ways in which they both facilitate and limit humanitarian response and outcomes. Different bodies of law apply depending on the nature of the humanitarian emergency. For example, IHL regulates the conduct of armed conflict, while international disaster response law (IDRL) applies in disaster contexts and international refugee law (IRL) sets our the rights of refugees and the corresponding obligations of states. To the extent that international law provides a legal basis for humanitarian response, it also provides—either explicitly or implicitly—some answers to the questions of what goals and activities can properly be considered humanitarian.

Chapter 15 on the Shifting Parameters of Humanitarian Emergencies shows how conceptions of what constitutes a humanitarian emergency—of what requires a humanitarian response—have expanded, with several of the major international humanitarian agencies increasingly responding to protracted and urban crises, working in Europe, and seeking to address the needs of people on the move. These shifts in the first dimension of humanitarianism's boundaries are also linked to shifts in the second and third dimensions—put simply, humanitarian agencies very often undertake activities akin to development programming in protracted and urban crises, and adopt some of the goals and methods of human rights agencies when working in Europe and/or to support migrants.

Chapter 16 on the Nexus Concept charts the main ways in which international actors have sought to coordinate or integrate humanitarian, development, and peace-building responses to crises over the past forty years. In doing so, it focuses less on the erosion of boundaries between these different fields of practice and more on coordination and collaboration across the boundaries. At the same time, however, it reveals a general trend towards expanding definitions of what constitute humanitarian goals and practices.

13
Politics, Principles, and Humanitarian Action

International humanitarian agencies are principled actors, but different agencies emphasize different principles, and none is driven exclusively by the principles to which it subscribes. The focus of this chapter is on the four main principles of humanitarian action: humanity, impartiality, neutrality, and independence. According to some understandings of what it means to be humanitarian, these four principles are constitutive of humanitarianism itself, and those actors and actions that do not subscribe to them should not be labelled 'humanitarian'. Other understandings or definitions of humanitarianism allow for greater flexibility, particularly with regards to the principles of neutrality and independence. Among the proponents of these more flexible definitions are those who argue not only that it is possible to be humanitarian without being neutral and independent but also that truly neutral and independent action is either impossible or undesirable, or both.

This chapter does not seek to take one side or another in these arguments but rather to explain what each principle requires of its adherents and what the main lines of contention are. Through an analysis of an apparently diverse selection of international aid agencies, it additionally seeks to show that, while there are some differences between their respective positions towards these principles, these differences have often been overstated. Most of the large international humanitarian agencies continue to profess their commitment to all four principles. The biggest differences arise not in their rhetorical commitment to the principles but in the ways in which they interpret and implement the principles in practice. Accordingly, this chapter explains some of the ways in which the interpretation and implementation of each principle vary across different humanitarian agencies and different operational contexts.

The Politics and Everyday Practice of International Humanitarianism. Miriam Bradley, Oxford University Press.
 DOI: 10.1093/oso/9780198840695.003.0013

Principles of humanity, impartiality, neutrality, and independence

The Red Cross and Red Crescent (RC) Movement, of which the International Committee of the Red Cross (ICRC) is a part, has seven 'Fundamental Principles', officially proclaimed in 1965. Here, we are concerned with the first four of those principles, namely, those of humanity, impartiality, neutrality, and independence.[1] Contrary to what is often assumed, these principles 'were not part of humanitarianism's original DNA; rather, they [fell] into place over decades of action and debate' (Barnett 2011: 5). That said, these principles are now central to the identity of the ICRC, and many other agencies have also adopted them. In 1991, the United Nations (UN) General Assembly adopted these principles as the operational basis for humanitarian action, and in 1994, the wider humanitarian system adopted them in the *Code of Conduct for the International Red Cross and Red Crescent Movement and NGOs in Disaster Relief* (Bennett et al. 2016: 48). Even alternative perspectives on the proper role of principles in humanitarian action often state their position in contradistinction to the ICRC perspective. As such, the ICRC position is a good starting point for understanding what each of these principles means and the ethics and logics behind them.[2]

In his 1979 commentary on the Fundamental Principles, Jean Pictet classifies the principle of humanity as 'the essential principle' from which the others flow, and it is best understood as an aspirational rather than an operational principle, deemed to be of moral or intrinsic rather than instrumental value. The principle of humanity is stated as follows:

> The Red Cross, born of a desire to bring assistance without discrimination to the wounded on the battlefield, endeavours—in its international and national capacity—to prevent and alleviate human suffering wherever it may be found. Its purpose is to protect life and health and to ensure respect for the human being. It promotes mutual understanding, friendship, co-operation and lasting peace amongst all peoples.
>
> **(Pictet 1979: 12)**

[1] The other three are voluntarism, unity, and universality. Pictet (1979: 9) explains that these 'are standards for application, relating to the structure and operation of the institution, coming into play primarily in connection with specific tasks. They are less far-reaching than the previous principles.' They are also specific to the Red Cross and do not apply to international humanitarian action more broadly.

[2] For a full explanation of the Red Cross doctrine with respect to these principles, see Jean Pictet's 1979 Commentary. In his excellent book on humanitarian ethics, Hugo Slim provides a more up-to-date and sophisticated commentary on the principles of humanity, impartiality, neutrality, and independence in humanitarian action in general, that is, not specific to the Red Cross (Slim 2015: chs 2 and 3).

The objective is to 'prevent and alleviate suffering *wherever it may be found*' (emphasis added), yet the mention of Red Cross origins on the battlefield reminds us that humanitarian action is focused on humanitarian emergencies.[3] Pictet was clear about the limits and focus of the Red Cross, saying that as 'the world confronts new needs, it is natural to attempt to meet them. Not all suffering however can be alleviated by the Red Cross [. . .] The essential mission of the Red Cross remains that of protecting human beings in the event of conflict and of relieving their suffering' (Pictet 1979: 11, 20). While the ICRC has expanded its mandate beyond the battlefield to include some other situations of violence and the International Federation of Red Cross and Red Crescent Societies (IFRC) responds to disasters, the scope of their action retains limits. Thus, two senior ICRC staff explain that the ultimate and sole aim of humanitarian action 'is to prevent and alleviate suffering, protect life and ensure respect for the dignity of people *in desperate situations as a result of conflict or disaster*' (Labbé and Daudin 2015: 186, emphasis added).[4]

As formulated by the Red Cross, the principle of humanity goes beyond concern for bare life to encompass dignity and respect for the human being. Pictet breaks the principle down into three elements: to prevent and alleviate suffering, to protect life and health, and to assure respect for the individual (Pictet 1979: 14–18). Hugo Slim discusses this notion in some depth, emphasizing that 'human life is to be understood biographically as personhood and individuality, as well as biologically as flesh and blood' (Slim 2015: 48). As seen in Chapter 1 of this book, however, the construction of humanitarian emergencies tends to focus on bare life, with consequences for the kinds of humanitarian response that ensue.

For the Red Cross, both the principles of humanity and impartiality are substantive principles, defining the objectives of humanitarian action (Pictet 1979: 8). Hugo Slim similarly characterizes these two principles as collectively constituting the humanitarian goal (Slim 2015). However, the principle of impartiality can additionally be understood as a more practical principle in that it further specifies and operationalizes the principle of humanity, stipulating that aid must be provided in accordance with need and need alone. For the Red Cross, impartial humanitarian action is understood to be that which 'makes no discrimination as to nationality, race, religious beliefs, class

[3] On what constitutes a humanitarian emergency, see Chapter 1. On how the parameters of humanitarian emergencies have shifted to encompass situations beyond conflicts and disasters, see Chapter 15.

[4] National Red Cross and Red Crescent Societies very often do provide a range of services, in particular health-related services, in their own countries outside of emergencies (see, e.g. O'Callaghan and Leach 2012). However, *international* humanitarian action both within and beyond the Red Cross Movement is largely limited to so-called humanitarian emergencies.

or political opinions [. . .] It endeavours to relieve the suffering of individuals in proportion to the degree of their suffering and to give priority according to the degree of urgency' (Pictet 1979: 4, 27). Thus impartiality consists of two main components: non-discrimination and proportionality.

Whereas the principles of humanity and impartiality are seen as being of intrinsic value, the principles of neutrality and independence are usually only considered to be of instrumental importance. They are characterized as operational rather than moral principles, indicating the ways and means rather than the objectives of humanitarian action, and their proponents expect them to facilitate the acceptance of humanitarian actors by armed parties to conflict (Pictet 1979: 9, 31). As such, they are seen as less important than the 'moral' principles of humanity and impartiality (Pictet 1979: 8).

The principle of neutrality prohibits taking 'sides in hostilities or [engaging] at any time in controversies of a political, racial, religious or ideological nature' (Pictet 1979: 4). A moral argument for this kind of neutrality can be made that outsiders have no right to impose their preferences and that, by taking sides in a conflict or in any contentious political matter, they would be doing so (Fox 2001: 281). However, as indicated above, arguments in favour of neutral humanitarian action mainly revolve around access and acceptance in conflict contexts (see also Slim 2015: 66–67). The logic is that the parties to armed conflict will be more likely to accept humanitarian action if it does not interfere with that conflict and that such acceptance will facilitate the access which is necessary if aid is to be provided impartially.

The principle of independence entails operating independent of influence from other actors with a vested interest, including states and international organizations. The primary logic here is that if action is to be impartial, based on criteria of need alone, it cannot also be driven by other motives and, hence, must not be influenced by actors with ulterior motives. The secondary logic is that if action is to be neutral, it should not be influenced by anyone with an interest in the conflict or other relevant contentious political issue. In short, in order to act impartially and neutrally, humanitarian organizations need to be able to operate as they see fit rather than to further the aims of others.

Political or apolitical humanitarian action

Collectively, the four principles discussed above define a form of humanitarianism that is often said to be apolitical and is intended to be non-threatening to authorities and armed actors in the context of armed conflict. As Nicholas Leader has put it, 'they are rules for supping with the devil without getting eaten, or corrupted; for humanitarian agencies they are the "long spoon" of the proverb' (Leader 1998: 290). Whether even this kind of principled humanitarian action is deemed sufficiently non-threatening to facilitate acceptance by combatants is an empirical question, and we lack the necessary evidence to answer it conclusively. On the one hand, it is widely acknowledged that, in many countries, the ICRC has better access than other international agencies. However, the ICRC is not by any means immune to rejection and threats by parties to conflict, as targeted attacks on ICRC staff and facilities in Baghdad and Afghanistan in 2003, to take just two examples, have painfully demonstrated. Moreover, to the extent that the ICRC does have better access than other international humanitarian organizations, this may be due to a variety of factors other than its strong commitment to impartial, neutral, and independent action. For example, the ICRC devotes significant time and energy to building relationships with all parties to conflict in the contexts in which it is working and tends to have more contacts and longer-standing relationships with armed actors than most other aid agencies (Bradley 2016a: 63, 139, 141–148).[5]

Whether or not they facilitate access, the idea that action conforming to these four principles is truly apolitical is a fiction. First, even the kind of strictly principled humanitarian action espoused by the ICRC represents a liberal project and can only be understood as apolitical from *within* a liberal framework (Barry 1990; Hopgood 2019). The humanitarian ethic of valuing human life above other goals and of valuing all lives equally is a fundamentally liberal ethic, with the principle of humanity characterized by 'radical equality' (Bradley 2019a; Slim 2015: 56). Second, the ICRC seeks to influence public policy and international law to the extent that it relates to the organization's mandate to protect civilians in conflict contexts and other situations of violence, in particular when it relates to the development or implementation of international humanitarian law (IHL) (Bradley 2016a: 57). Such efforts to influence policy and law can only be understood as not engaging in controversies of a political nature if the value and mandate being promoted are accepted as given—acceptance which cannot be assumed outside of a liberal outlook. Thus, while the ICRC and other organizations committed to the same interpretation of these four principles seek to avoid involvement in partisan politics and international realpolitik, defining their approach as apolitical depends on prior acceptance of basic liberal ideals (Barry 1990).

[5] On the importance of trust and relationships with authorities and communities, see 'Relationship-building interactions' in Chapter 23 on Dialogue, Negotiation, and Advocacy.

Furthermore, while seeking to avoid involvement in partisan politics and international realpolitik, even the most strictly principled humanitarian action has unintended political consequences in many—if not all—contexts. These unintended political consequences provide the starting point for many critiques of 'classical' or strictly principled approaches to humanitarian action. According to several commentators, the ICRC perspective was widely shared until the 1990s, at which point a division arose between 'classical humanitarians' like the ICRC and those who advocate a more 'political humanitarianism', sometimes referred to as 'new humanitarianism' (Barnett 2005; Fox 2001; Weiss 1999). The so-called new humanitarianism is characterized by 'the integration of human rights and peace building into the humanitarian orbit; the ending of the distinction between development and humanitarian relief; and the rejection of the principle of neutrality' (Fox 2001: 276).

Between those who defend a classical stance and those who advocate more political engagement and the pursuit of more transformative goals, there are two main, interconnected, lines of dispute. The first is about the extent to which humanitarian action can be disassociated from political action, and the second is about the extent to which any such disassociation is desirable.

Feasibility of apolitical action

Humanitarian action often has unintended political consequences, as several of the case study chapters in this book have illustrated. In the Nigerian Civil War, for example, Biafran secessionist leaders manipulated and exaggerated the effects of famine caused by the Nigerian government's blockade. The famine kept them in the news and made them seem a worthy political cause, and negotiating with relief agencies gave these leaders legitimacy. The material benefits of humanitarian aid to the Biafran people enabled the insurgency to continue much longer than it would have been able to in the absence of such support.[6] In the context of famine in Ethiopia in 1983–1985, humanitarian assistance was used to facilitate forced relocations away from rebel-held areas, and to feed government forces.[7] After the 1994 Rwandan genocide, refugee camps in neighbouring countries were militarized as the perpetrators of genocide hid among the civilian refugees, gained control of the aid distribution systems within the camps, taxed and diverted aid, and

[6] See 'Political impacts: Supporting the secessionists and prolonging the war' in Chapter 2 on the Nigerian Civil War, 1967–1970.

[7] See 'Use of food aid for non-humanitarian ends' in Chapter 3 on Drought and Famine in Ethiopia, 1983–1985.

used the camps to regroup, restrengthen, and re-arm.[8] In the aftermath of the Sri Lankan Civil War in 2009, international aid was used to support military-controlled camps in which internally displaced persons (IDPs) were interned by the government with complete restriction on movement.[9]

In short, action that is neutral in intent may be far from neutral in outcome, unintentionally assisting one side or another in a conflict and, in some cases, increasing violence and prolonging conflict. Indeed, it is probably impossible to operate in a conflict context without having some impact on conflict dynamics, and this is widely recognized by proponents of classical and more political humanitarianisms alike. However, it is unclear just what the magnitude of such impacts are. In a few high-profile cases, the effects of aid on conflict dynamics may have been significant, and in the worst cases, aid may have done more harm than good for the people it was intended to support, but in most cases, aid is a relatively small component in the wider political economy of conflict (Keen 1998b; Slim 2015: 185).

Even those who seek to adhere closely to the principles of neutrality and independence are aware that the aid they provide is likely to have political consequences and may be instrumentalized by other actors in pursuit of political ends. Two senior ICRC officials demonstrated such awareness in 2015: 'Humanitarian action has never taken place in a political vacuum—it has always been politicized and instrumentalized' (Labbé and Daudin 2015: 184). There may be some disagreement as to the *extent* to which aid shapes political outcomes, but the bigger difference is between those who believe that humanitarian actors can, and should, leverage aid in pursuit of broader, more transformational aims and those who believe that it cannot, or should not, be so used. In other words, even if it is not feasible to be apolitical in the sense of neutral outcomes, is it desirable for aid agencies to maintain neutral intentions and minimize the political impacts of their work as far as possible, or should they eschew neutrality and independence in an effort to do greater good in the longer term?

Desirability of apolitical action

Greater good in this sense means moving from a focus on addressing the symptoms of humanitarian crises to addressing their causes. As Fiona Fox puts it, it 'is easy to see traditional humanitarian aid as a vicious circle in which agencies continue to apply sticking plasters without healing the

[8] See 'Diversion, manipulation, and unintended consequences of aid' in Chapter 5 on Genocide in Rwanda and Its Aftermath, 1994–1996.

[9] See 'International humanitarian response' in Chapter 8 on Sri Lanka, 2008-2009.

wound' (Fox 2001: 284). Certainly, there is a moral logic to seeking 'to prevent suffering rather than simply repair it' (Slim and Bradley 2013: 4). Preventing the suffering that arises from humanitarian emergencies usually means ending or reducing conflict and/or fostering development. International humanitarian agencies can seek to contribute to these goals by using their aid and influence to affect the course of conflict or development. The provision of aid becomes conditional on its contribution to these broader goals, and a decision is made to 'trade lives and suffering today for political gains tomorrow' (Stockton 1998: 356).

The pursuit of these more transformative goals raises three main objections from those defending a more apolitical humanitarianism. First, a normative argument questions whether unelected and often unaccountable outsiders should be deciding what economic, political, and social goals should be pursued (Barnett and Weiss 2011: 107; Fox 2001: 281). Second, even if one accepts that certain goals reflect sufficiently universal values as to be decided on by outsiders, it is argued that international humanitarian agencies are not capable of predicting the long-term impacts of implementing conditionality policies (Stockton 1998: 356). Indeed, some believe that efforts to use aid to manipulate political actors actually leave humanitarian actors more, rather than less, exposed to political manipulation themselves. Third, and in light of the first two objections, classical humanitarians argue that suffering and death in the short term are an unacceptable sacrifice for what is only a potential reduction in suffering in the longer term. They point out that pursuing a transformative agenda risks creating 'deserving' and 'undeserving' victims, with the latter group denied aid because they are not expected to contribute to peace and development (de Torrenté 2004, 4; Fox 2001: 282; Stockton 1998: 354–355). In short, using aid to further political goals implies accepting some avoidable suffering in the immediate term in the hope of having a bigger impact on reducing suffering in the longer term. Proponents and opponents of such an approach differ in the extent to which they believe it right to define those political goals, possible to achieve them, and acceptable to sacrifice lives and suffering in order to do so.

Even without advocating the pursuit of transformative goals, there are two further objections to 'apolitical' humanitarianism in as far as being apolitical is understood to require silence on the subject of abuses witnessed in the course of humanitarian work and respect for borders and state sovereignty. Much of the discussion of politics and principles in humanitarianism conflates different ways of being 'political', but pursuing transformative goals is distinct from speaking out about human rights abuses and violations of IHL, and both are also distinct from compromising state sovereignty

(Bradley 2022). Humanitarian agencies can adopt none, or one, or any combination of these 'political' ways of working.

In many humanitarian operations, especially in conflict contexts, aid workers witness abuses of human rights and violations of IHL. It is sometimes argued that on witnessing abuses, there is a moral imperative to speak out, with failure to do so seen to imply complicity with the abuses. There is also a consequentialist argument based on the idea that public criticism of one or more actors may contribute to a reduction in abuses.[10] Public criticism is a very different way of being 'political' from pursuing transformative goals, and the relationship between neutrality and speaking out is contested. According to some analysts, neutrality imposes silence, and advocacy is central to the move to incorporate transformative goals (Bridges 2010: 1251; Fox 2001). However, advocacy can be neutral in intent and does not necessarily imply the pursuit of transformative goals. While public condemnation *can* have transformative goals, where a humanitarian agency supports one side or criticizes the other with the objective of affecting the balance of power between them, it can equally well have the limited goal of saving lives in the short term, bringing pressure to bear on those actors contravening IHL with the aim of changing the way they conduct conflict but without seeking to affect the conflict outcome. Whether neutral in intent or part of a broader strategy for peace or development, public criticism of authorities or armed groups may have repercussions that threaten access to the victims who aid agencies seek to support.[11]

Where states deny humanitarian agencies access to some or all of their territory, humanitarian agencies may consider undertaking operations without state consent.[12] For example, Joint Church Aid operated relief flights into Biafra from São Tomé during the 1967–70 Nigerian Civil War, without the consent of the government.[13] During the conflict and famine in Ethiopia in the 1980s, working under the auspices of the Emergency Relief Desk (ERD) in close collaboration with the relief arms of the relevant rebel groups, several non-governmental organizations (NGOs) engaged in cross-border operations from Sudan into rebel-held areas.[14] More recently, during the Syrian Civil War, some aid agencies have undertaken cross-border operations,

[10] See 'Threat reduction' in Chapter 23 on Dialogue, Negotiation, and Advocacy.

[11] For more detailed discussion of the relationship between advocacy and access, see 'Neutrality, access, and humanitarian space' in Chapter 23 on Dialogue, Negotiation, and Advocacy.

[12] For a discussion of the requirement for state consent and cross-border operations, see 'State sovereignty, responsibility, and practice' in Chapter 19 on Government and Civil Society in Affected States.

[13] See 'International humanitarian response' in Chapter 2 on the Nigerian Civil War, 1967–1970.

[14] See 'Non-cooperation and access restrictions imposed by the government' in Chapter 3 on Drought and Famine in Ethiopia, 1983–1985.

mainly from Turkey, into rebel-held areas of Syria. While operating on the territory of a state without the consent of that state can be understood as 'political' in the sense that such action contravenes state sovereignty, it need not be any more 'political' than classical humanitarianism in terms of pursuing transformative goals and seeking to influence the outcome of conflict or speaking out about violations of IHL and human rights. While many of the agencies involved in Joint Church Aid (JCA) explicitly and vocally supported the Biafran cause, many of those involved in the ERD were committed both to limited goals and to discretion (Duffield and Prendergast 1994: 49–52; Hendrie 1989; Pérouse de Montclos 2009). Indeed, in a discussion of cross-border operations in the Syrian context, Hugo Slim and Emanuela-Chiara Gillard argued that the ethics and legality of such operations depended on them being for limited humanitarian purposes, not part of the pursuit of broader political goals, and strictly neutral and impartial (Slim and Gillard 2013: 8).

Perspectives of the largest international humanitarian agencies

At the rhetorical level, then, there exists significant debate on the value of principled humanitarian action—especially the principle of neutrality. Notably, much of this debate comes from external commentators—journalists, academics, policy analysts—rather than from people who work in operational aid agencies. Indeed, for all the claims that a new and politicized humanitarianism has come to dominate the sector, the major international humanitarian agencies mostly maintain a stance much closer to that of the ICRC.

Stated commitments to principles

The stated positions of the two largest (by financial expenditure) UN humanitarian agencies—the United Nations Refugee Agency (UNHCR) and the World Food Programme (WFP)—and the 'big five' humanitarian international non-governmental organizations (INGOs)—the International Rescue Committee (IRC), Médecins Sans Frontières (MSF), Oxfam, Save the Children, and World Vision—suggest that there is broad consensus on the appropriateness of apolitical humanitarian action.

According to the UNHCR emergency handbook, 'underlining all humanitarian action are the principles of humanity, impartiality, neutrality and independence' (UNHCR n.d.-a). The handbook goes on to confirm that UNHCR 'is guided by humanitarian principles in its response to all humanitarian crises, whether caused by conflict, violence or natural disaster'. In 2004, the Secretariat of WFP outlined ten principles to guide the agency's work. These include the 'core principles' of humanity, impartiality, and neutrality and an additional seven 'foundations of effective humanitarian action' (WFP 2004: 9). Although the principle of independence is not included, the document in which these principles are set out includes the following statement: 'We shall endeavour not to act as instruments of government foreign policy' (WFP 2004: 8).

While MSF is sometimes characterized as the polar opposite of the ICRC in terms of institutional attitude towards politics and principle in humanitarian action (see, e.g. Fox 2001; Weiss 1999), this is not by any means how MSF and its staff view the institutional position. Instead, they affirm their commitment to neutral, independent, and impartial humanitarian action and emphasize the similarities rather than the differences between the ICRC and MSF positions on the merits of apolitical humanitarianism (Brauman 2012; MSF n.d.; Tanguy and Terry 1999).

Oxfam asserts its commitment to the principles of humanity, impartiality, and independence but rejects the principle of neutrality, claiming that, 'Impartial advocacy does not mean saying that every party to violence is always equally to blame. Nor is Oxfam neutral in the sense of avoiding anything that could be construed as a policy controversy' (Oxfam 2013: 2, 3). One of the many national identities that collectively comprise the World Vision partnership, World Vision Australia produced a document in 2017 setting out its policy on the principles of humanitarian action and affirming that it 'carries out all its humanitarian work in accordance with the humanitarian principles: humanity, neutrality, impartiality and independence' (World Vision Australia 2017).

Public statements of IRC and Save the Children policies with regard to the principles of humanitarian action are somewhat harder to find. A commitment to principles does not seem to feature as such a central part of their identities as it does for the agencies discussed above. Nonetheless, both the IRC and Save the Children—and, interestingly, considering its rejection of neutrality, Oxfam—were signatories to a 2017 press release signed by several NGOs in Myanmar which specified a commitment 'to the principles of humanity, independence, neutrality, and impartiality' (NGOs 2017). A 2016 speech by the IRC president, David Miliband, likewise confirmed this commitment: 'The humanitarian community has always embraced fundamental principles of action. The principles of independence, neutrality, impartiality and humanity are important every day in our work—protecting our staff and supporting our beneficiaries' (Miliband 2016).

In summary, all but two of the eight international humanitarian agencies with the largest budgets proclaim their commitment to the principles of humanity, impartiality, neutrality, and independence. Oxfam is the only real objector, and even Oxfam only rejects one of these four principles, namely, neutrality. WFP does not commit itself absolutely to independence but will 'endeavour' to avoid instrumentalization by governments. In light of such commitments, any claims that principled humanitarianism has been replaced with a new, more political variant should be treated with caution. Yet, what any actor—aid agencies included—says and does may be two different things. In practice, alongside this widespread rhetorical commitment, different agencies understand the main principles of humanitarian action in different ways (Bradley 2016a: 40–68; Slim 1997).

Interpretation and implementation of these principles in practice

Examining the ways in which international humanitarian agencies interpret and implement these principles reveals that there is more nuance than their rhetorical commitments to these principles might suggest.

First, while the ICRC and MSF reject the pursuit of more transformative goals altogether, the extent to which the other large agencies pursue a transformative agenda depends very much on the operational context. In the aftermath of the US-led intervention in Afghanistan, for example, many agencies eschewed neutrality and independence in favour of an approach that integrated humanitarian action with the political goals of the UN and the United States.[15] Elsewhere, however, many maintain a much more apolitical stance.

Second, with the exception of the ICRC and MSF, the large international humanitarian agencies do not just have humanitarian mandates. They are what are commonly referred to as multi-mandate agencies in the sense that they have a humanitarian strand of work but also development or human rights strands of work (Schenkenberg van Mierop 2015; Slim and Bradley 2013). Thus, while they often seek to be neutral in their humanitarian

[15] See Chapter 6 on Afghanistan, 2001–2014.

work, they may pursue a transformative agenda in the other strands of their work (Slim and Bradley 2013: 4). While this may be unproblematic in theory, it can be difficult to distinguish different strands in practice (and there may also be some tension between them). For example, an evaluation of WFP humanitarian principles and access policy found that neither the institutional policy nor the majority of staff members interviewed for the evaluation distinguished between emergency and development operations, generating confusion as to the application of humanitarian principles (Steets et al. 2018: paras 41–42). In Afghanistan, some funders maintained distinct humanitarian and development funding strands, and some agencies accepted ostensibly development funding to meet what were essentially humanitarian needs, with conditions that contravened the principles of impartiality and neutrality, highlighting the complexities of delineating development from humanitarian assistance (Featherstone 2012: 14).

Third, agencies often adopt other ways of being 'political' without embracing transformative goals (Bradley 2022). Speaking out against human rights abuses and IHL violations is now considered by most international agencies to be necessary and important, with Hugo Slim including advocacy as one of the main strands of humanitarian action alongside assistance and protection (Slim 1997, 2015).[16] While Oxfam cites its commitment to advocacy as a reason for rejecting the principle of neutrality, MSF makes explicit that the institutional commitment to principles does not imply silence in the face of serious human rights abuses:

> The principles of impartiality and neutrality are not synonymous with silence. When MSF witnesses extreme acts of violence against individuals or groups, the organisation may speak out publicly. We may seek to bring attention to extreme need and unacceptable suffering when access to lifesaving medical care is hindered, when medical facilities come under threat, when crises are neglected, or when the provision of aid is inadequate or abused.
>
> (MSF n.d.)

Even the ICRC, which has a reputation for silence, does, in some cases, publicly criticize the authorities of the countries in which it works, as it did, for example, in Myanmar in 2007 (ICRC 2007a, b). And while MSF is often seen as having *témoignage*, or speaking out, at the heart of its mandate, the role of public statements has been contested among the senior management, and more often than not the organization undertakes its medical work without

[16] See 'A failure of protection' in Chapter 8 on Sri Lanka, 2008–2009, and 'Public advocacy and criticism of authorities' in Chapter 23 on Dialogue, Negotiation, and Advocacy.

making public statements about abuses in its zones of operations (Weissman 2011a). Of course, speaking out about the causes of crises, and as part of a rights-based approach to poverty, suffering, and injustice, does imply the pursuit of more transformative goals (Bennett et al. 2016: 52). However, it is the goals and content of the advocacy, and not the fact of advocating in itself, that make such action non-neutral in intent.

A willingness to undertake operations without state consent is less widespread, but in practice most agencies will do so under some circumstances. As an intergovernmental organization, respect for state sovereignty is a fundamental organizing principle of the UN. If an international organization carries out an unauthorized relief operation, it violates the affected state's sovereignty and territorial integrity (Gillard 2013: 370). Unsurprisingly, then, UN humanitarian agencies do not, for the most part, undertake activities on the territory of a state without the consent of that state, even where such activities would be practicable (e.g. where they could access opposition-controlled territory from a third state without the need to transit through territory controlled by the affected state whose consent has not been granted). However, they may do so with the authorization of a political organ of the UN. Thus, in Syria, for example, UNHCR and WFP began cross-border operations only after a series of UN Security Council resolutions in 2014 demanded that the Syrian authorities allow access and eventually authorized cross-border operations by UN agencies without Syrian government consent.

NGOs are often more flexible, and many were already operating in opposition-controlled territory without the consent of the Syrian authorities prior to 2014. The 'sans frontières' of MSF's name refers precisely to this idea, and for MSF, if it is deemed useful to set up medical services in opposition-controlled territory without state consent, 'then the only considerations that are taken into account are practical ones' (Brauman 2012: 10). ICRC practice demonstrates a strong preference for working with state consent, but some flexibility, with occasional operations without state consent, including flying relief into Biafra without permission from Lagos and participating in the cross-border relief operations from Sudan into rebel-held areas in Ethiopia from 1976 to 1987 (Forsythe 1997: 243–245). In Syria, the ICRC resisted operating without state consent, prioritizing its presence in government-controlled territory and seeking to reach opposition-held areas by negotiating to cross frontlines (Krähenbühl 2013). The ICRC preference for state consent is based on expectations about the consequences of overriding sovereignty rather than respect for the principle per se (Sandoz 1992: 224).

Fourth, beyond these big themes, different individual agencies understand the principles of neutrality and independence as having different implications for their practice. These principles are interpreted to mean a range of quite different things for humanitarian engagement with national governments and with armed actors.[17] For example, while the ICRC and MSF seek to avoid the use of armed escorts wherever possible, other agencies take a much more flexible approach (see, e.g. Labbé and Daudin 2015; Thompson 2008). On the other hand, the ICRC especially, but also MSF, often engage more with armed actors in order to negotiate access and, in the case of the ICRC, to try to increase compliance with IHL (Bradley 2016a). The principle of independence is also understood quite differently in terms of its implications for funding and coordination. MSF attaches importance to both financial and political independence from states, while the ICRC is financially dependent on a relatively small number of donor governments and emphasizes its formal independence from the UN and from humanitarian coordination mechanisms. UNHCR seeks to maintain independent decision-making despite both political dependence on the UN General Assembly and financial dependence on a few donor governments.

While most of the debate focuses on the 'political' principles of neutrality and independence, the interpretation and implementation of the 'moral' principles of humanity and impartiality merit serious attention. Hugo Slim has highlighted two 'heresies' in the way that the principle of humanity has been interpreted—or, in his words, distorted—by several international humanitarian actors. The first of these 'commodifies humanitarianism and relates it solely to material help' (Slim 1997: 345). This is problematic because the reduction of humanitarian action to the provision of material assistance largely serves to exclude protection, which is not a physical item and is not susceptible to a commodity-based approach (Bradley 2016a: 41; Darcy and Hofmann 2003: 6). Humanitarian agencies often prioritize the delivery of material assistance over protection, as demonstrated very clearly in the case studies of war in Bosnia (Chapter 4) and Sri Lanka (Chapter 8). Moreover, such a focus on material concerns tends to prioritize bare life over biographical life, and an exclusive focus on physical welfare ignores important aspects of people's lives and dignity (Slim 1997: 346). The second distortion 'tends towards making humanitarianism non-negotiable in war', and here Slim argues that the notion of a non-negotiable 'humanitarian imperative' may 'undermine the very serious negotiation which needs to take place between

[17] Engagement with these kinds of actors is discussed in detail in Chapter 19 on Government and Civil Society in Affected States and in Chapter 20 on Armed Actors.

warring leaders and humanitarians to ensure that humanitarian action is fair' (Slim 1997: 345–346).

While the principle of impartiality offers greater specificity than the principle of humanity, there is sufficient ambiguity in the way the principle is formulated and understood for it to be consistent with multiple different distributional commitments. Non-discrimination will often be a necessary condition for prioritizing those with greatest need or responding in proportion to need, but it is not a sufficient condition. Moreover, these last two goals can conflict with one another, and aid agencies may incorporate other considerations—such as efficiency—under the rubric of impartiality (Parfit 1997; Rubenstein 2008). In practice, aid agencies do not respond to suffering wherever it may be found but limit themselves to suffering in humanitarian emergencies.[18] Moreover, they interpret impartiality in different ways according to their mandates and expertise. The ICRC mandate, for example, limits it largely to contexts of conflict and violence, while UNHCR is mandated to respond to people who have been displaced, and Save the Children focuses on children. Furthermore, they are limited in their efforts to implement their own interpretations of the principle, given the limited availability and comparability of information on needs and suffering,[19] difficulties accessing populations in need,[20] and insufficient or restricted funding[21] (Bradley 2016a: 45–49). At the global level, given most international funding comes from a small set of donor governments, aid is not always distributed in proportion to need.

Conclusions

Even the most principled humanitarian action is political in that it is grounded in a liberal ethic of valuing all lives equally and seeks to inject that ethic into the very contexts in which it is furthest from being realized. Furthermore, all kinds of humanitarian action can, and do, have unintended political consequences and can impact conflict dynamics or debates on contentious issues. According to the 'new humanitarians', humanitarian actors should embrace the potential for political impact, eschew neutrality,

[18] See Chapter 1 on Humanitarian Emergencies, and Chapter 15 on the Shifting Parameters of Humanitarian Emergencies.

[19] See 'Assessing needs, contexts, and capacities' in Chapter 21 on Needs Assessment, Evaluation, and Response Decisions.

[20] See, e.g. 'International humanitarian response' in Chapter 8 on Sri Lanka, 2008–2009 and 'Background' and 'International humanitarian response' in Chapter 10 on Famine in Somalia, 2011–2012.

[21] See Chapter 18 on Donors and the Funding of Humanitarian Action.

and leverage the assistance they provide in pursuit of broader, transformational goals. By contrast, 'classical humanitarians' believe that humanitarian assistance should not be used that way, among other reasons because subordinating aid to other goals undermines impartiality. Although there has been significant debate, in particular regarding the value of neutrality, most of the largest international humanitarian agencies continue to profess their commitment to all four principles, even as they differ in terms of how they interpret and implement them.

14
Law and International Humanitarianism

Law interacts with humanitarian action in multiple and sometimes contradictory ways. The operational agencies whose work is the focus of this book sometimes play a role in developing international law on humanitarian issues. Under certain conditions, international law, in turn, provides a legal foundation for the work of these agencies and, in some cases, also helps define their mandates. International law additionally specifies a number of rights and protections for the victims of humanitarian emergencies and corresponding obligations of states and other actors. Relevant laws can be used by humanitarian agencies in their efforts to persuade states and de facto authorities to meet their obligations, notably those obligations relating to permitting access to populations in need, refraining from targeting civilians in conflict, and granting asylum to those fleeing persecution. These laws are also sometimes enforced through harder measures, including the use of force in UN peacekeeping mandates and the administration of justice through international and regional tribunals. However, law does not only serve to facilitate the pursuit of humanitarian goals. It can also conflict with them, and counter-terror and anti-smuggling legislation have served to criminalize humanitarian action in some contexts.

Processes and politics of law-making

International humanitarianism is shaped by both international and domestic laws. This chapter focuses mainly on the former since they have wider reach. It is often said that domestic law has a vertical structure, with laws that imply rights and obligations for the people being made by a 'higher' authority, and that international law has a horizontal structure, with laws implying the rights and obligations of states, who are also the lawmakers. Such a distinction, however, oversimplifies the nature of international law. While states are the primary lawmakers in the realm of international law, other actors, including courts, international organizations, and non-governmental

The Politics and Everyday Practice of International Humanitarianism. Miriam Bradley, Oxford University Press.
 DOI: 10.1093/oso/9780198840695.003.0014

organizations (NGOs), also play a role. Furthermore, contemporary international law implies not only rights and obligations of states but also rights and obligations of individuals and other entities, including corporations, international organizations, and non-state armed groups.

The sources of 'hard' international law include treaties and the customary practice of states. International treaties are formal agreements which states choose to enter into, and they function much like contracts in that states consciously and explicitly consent to them, and the obligations they contain are binding only on those states that are party to them. Treaties are negotiated and agreed by states, but other actors, including intergovernmental organizations and international NGOs, can play a role in drafting laws and campaigning for states to adopt them. For example, the development of the core international humanitarian law (IHL) treaties—the 1949 Geneva Conventions and their 1977 Additional Protocols—was driven in no small part by the International Committee of the Red Cross (ICRC). International NGOs played crucial roles campaigning for more recent treaties prohibiting the use of particular weapons such as landmines and cluster munitions (Petrova 2018). The United Nations Children's Fund (UNICEF) and several international NGOs, including Save the Children, engaged extensively and consequentially in the drafting of the 1989 Convention on the Rights of the Child (Cohen 1990).

Customary international law emerges where there is a widespread and consistent practice that is accepted as law. There are two constituent elements, and evidence for both of them is required to establish the existence of a rule of customary law: general practice, primarily of states and, under certain conditions, of international organizations; and *opinio juris*, which refers to the conscious conviction on the part of those engaged in that practice that they are doing so with a sense of legal right or obligation, as distinct from mere habit. Rules of customary law bind all states (or international organizations), except where a state has persistently objected to a rule during its formation. Local or regional customary law may also exist and be binding on the relevant actors within the region. Actors other than states have been able to draw on their expert and moral authority in particular domains to influence state practice through the interpretation of treaty-based law and to establish the existence of customary rules through the identification of state practice and *opinio juris*. The ICRC and the United Nations Refugee Agency (UNHCR) have been very active in this regard and, as such, have contributed significantly to the development of IHL and international refugee law (IRL), respectively (Cryer 2017; Goodwin-Gill 2020; Nicholls 2006).

Some of the most important rules supporting humanitarian practice and outcomes are binding on all states (and, where applicable, on non-state actors within them). Many of the most relevant treaties have been ratified almost universally. Indeed, the four Geneva Conventions of 1949 have achieved universal ratification. The vast majority of states have also ratified the 1977 Additional Protocols (albeit with some important exceptions, including Israel and the United States): at the time of writing, there were 174 states parties to Additional Protocol I and 169 states parties to Additional Protocol II. The core treaties of IRL—the 1951 Convention on the Status of Refugees and its 1967 Protocol—were not far behind with 146 and 147 states parties, respectively. Furthermore, many of the rules specified within these treaties are also now considered to be customary law such that they are also binding on and in states that are not party to the relevant treaties. Beyond the rules of customary law, there are some peremptory norms which are considered so fundamental that they are absolute obligations that apply to everyone at all times and override all other laws and norms. These are referred to as *jus cogens*, and they are norms of general international law which have been accepted and recognized by a very large majority of states as rules from which no derogation is permitted (UN General Assembly 2019). Several *jus cogens* norms have potential relevance in humanitarian emergencies, including the prohibition of genocide, the prohibition of torture, and the basic rules of IHL.

International standards, including guiding principles and declarations, represent an important and voluminous source of 'soft' international law with a bearing on humanitarian action. These are negotiated texts which have been developed and agreed at the international level, sometimes by states and sometimes by other actors, including operational humanitarian agencies, which set out recommendations and non-binding commitments. They are important because they can have a direct influence on the conduct of states and other actors, and they can also contribute to the development of customary law.

International legal frameworks

Four main bodies of law have particular salience for humanitarian action and outcomes: IHL, international disaster response law (IDRL), IRL (together with other laws relevant for displaced persons and migrants), and other parts of international human rights law (IHRL). The state has primary responsibility for meeting the basic needs of the people on its territory or under its effective control (Akande and Gillard 2016: 11–13). Where a state is unwilling or unable to meet its obligations, whether in a context of armed conflict, disaster, or other humanitarian emergency, international humanitarian agencies may seek to respond. Their work is, to varying degrees, facilitated and governed by international treaties, customary law, jurisprudence, and United Nations (UN) resolutions. As well as providing a legal basis for humanitarian action in certain situations, international law helps to define the mandates of some aid agencies who have corresponding obligations relating to the implementation or supervision of particular treaties or bodies of law (Bradley 2016a: 79). Furthermore, these different bodies of international law articulate a range of rights and protections for the victims of humanitarian emergencies and/or corresponding obligations of states and other authorities.

International humanitarian law

Also called the law of armed conflict or the laws of war, IHL seeks to balance military necessity with concerns to limit the human cost of war and imposes limits on how armed conflicts can be fought. It sets out the duties and obligations of both state and non-state parties to conflict.

IHL provides a legal basis for international humanitarian action in situations of occupation and armed conflict—both international and non-international.[1] In situations of occupation, the occupying power is obliged to provide for the survival of the civilian population of the occupied territory and, where it is unable to meet this obligation, it is obliged to accept relief operations that are humanitarian and impartial (Akande and Gillard 2016: 18; Gillard 2013: 357–358). In both international and non-international armed conflicts, each party to conflict has the primary responsibility for meeting the basic needs of civilians on the territory which it controls (Akande and Gillard 2016: 12–13). Where such a party is unable or unwilling to meet those needs, impartial humanitarian bodies can offer to help do so, through relief actions, to provide such necessities as food, water, medical supplies, clothing, and means of shelter (Akande and Gillard 2016: 14). In general, the consent of the affected state is required before such relief actions can be carried out, but such consent may not be arbitrarily withheld (Akande and Gillard 2016: 16). In the case of cross-border operations, consent is additionally required from—and may not be arbitrarily withheld by—the states from which relief actions are taken and those through whose territory the operations must transit (Gillard 2013: 364).

There is less clarity over what exactly constitutes arbitrary withholding of consent, and even where consent is withheld arbitrarily, unauthorized relief

[1] For a thorough analysis, see the Oxford guidance on the law relating to humanitarian relief operations in situations of armed conflict (Akande and Gillard 2016).

operations are only lawful in limited circumstances (Gillard 2013: 360, 369). Unauthorized operations undertaken by states or international organizations would constitute a violation of the affected state's sovereignty and territorial integrity, whereas private actors—such as NGOs—are not directly bound by the rules of public international law on sovereignty and non-interference (Gillard 2013: 370). However, while representatives of states and international organizations usually have certain privileges and immunities which include immunity from legal processes in domestic courts, the staff of NGOs could face proceedings in any state in which they carry out unauthorized operations (e.g. on grounds of illegal entry or provision of support to the enemy), although they may not be punished for providing medical assistance (Gillard 2013: 371).

IHL also helps define the mandate of the ICRC—the self-described 'promoter and guardian' of IHL—providing a legal foundation for its protection and assistance work in armed conflicts (Giladi and Ratner 2015). That foundation is stronger in international armed conflict than in non-international armed conflict and especially robust when it comes to the protection of those deprived of their liberty in international conflicts: Article 126 of the Third Geneva Convention of 1949 gives the ICRC the right to visit prisoners of war and civilian internees with no requirement for consent from the relevant state. In other situations of violence (e.g. riots, internal disturbances, organized criminal violence) that do not meet the IHL thresholds for non-international armed conflict,[2] the ICRC maintains a right of initiative, which it considers to be customary law, allowing the organization to propose a response (Bradley 2020: 1066). However, where IHL is applicable, the ICRC has a stronger institutional commitment to offer a response, and states have a greater obligation to consent to that response. In other situations of violence, any ICRC response is at the discretion of both the ICRC and the state in question (Bradley 2020: 1067).

Where the ICRC—or any other impartial humanitarian body—offers its services to a country experiencing armed conflict, and the state in question consents to their operations, IHL specifies rights, roles, and responsibilities of states and humanitarian actors with respect to relief actions. For example, once relief operations have been consented to, parties to conflict and other relevant states must allow and facilitate rapid and unimpeded passage, and IHL sets out rules and recommendations regarding the entry of personnel, customs clearance, and the taxation of relief, among other things (Akande

[2] These thresholds relate to the intensity of violence and the level of organization and territorial control of participating non-state armed groups.

and Gillard 2016: 26–29). Humanitarian agencies must provide relief in an impartial manner, and the parties to conflict must allow the free passage of relief items that are essential to the survival of the civilian population throughout the territory under their control (Akande and Gillard 2016: 26). The parties to conflict may prescribe technical arrangements, and they have the right to search relief consignments and convoys (e.g. to check for weapons or military equipment), to require relief convoys to use prescribed routes at specific times, and to insist that an impartial third party supervises the distribution of relief consignments (Akande and Gillard 2016: 28–29). However, these kinds of technical specifications must be applied in good faith, be necessary and proportionate, and not be applied in such a way that they make it impossible to provide relief items that are essential to the survival of the civilian population (Akande and Gillard 2016: 29).

The bulk of IHL focuses not on facilitating the work of humanitarian agencies, however, but on the conduct of hostilities, in particular with respect to protected persons (civilians, sick and wounded combatants, and prisoners of war). As a body of law which defines duties and obligations (above all, of the parties to armed conflict) rather than rights, it does not specify entitlements of individuals, but these are implicit in the duties and obligations of others—duties and obligations which are underpinned by three main principles. The principle of distinction prohibits means and methods of warfare that do not distinguish combatants, who are legitimate targets of attack, from protected persons, who must not be targeted. Under IHL, civilians and civilian property can legitimately be collateral damage, so their presence in and of itself does not make an attack unlawful, but the principle of proportionality prohibits attacks when the expected military advantage does not outweigh the risk to civilians and civilian property. The principle of precaution requires combatants to take into account the presence of civilians or civilian property prior to any attack.

International disaster response law

In disaster contexts that are not subject to IHL (i.e. that do not occur in the context of armed conflict), an emerging body of IDRL provides a foundation for humanitarian response, although most of this is soft law.[3] For the most part, international rules and recommendations on international disaster response do not seek to impose obligations on states to accept international

[3] For a detailed discussion of IDRL, see IFRC (2007), but note that this was published fifteen years ago and so omits subsequent developments.

assistance but rather aim to facilitate international responses (through such measures as expediting visas, work permits, and the export/import of goods and equipment, as well as cutting costs and taxes on all these processes) and to ensure their quality (through setting certain minimum standards for international responders).

At the global level, binding rules related to the provision of humanitarian assistance to the victims of disasters are scattered across a number of conventions. In some cases, these deal with specific categories of disaster, such as nuclear disasters, oil pollution, and industrial accidents (Cubie 2017: 254). In others, they deal with very specific aspects of disaster response. Thus, the Convention on Facilitation of Maritime Traffic requires states to facilitate the arrival and departure of vessels engaged in 'natural disaster' relief work and, to the greatest extent possible, the entry and clearance of the people and cargo arriving in those vessels (Cubie 2017: 254). There are also conventions on telecommunications in disasters and civil defence assistance (Cubie 2017: 273–277).

At the regional level, more progress has been made in developing hard IDRL. In 1991, for example, the Organization of American States (OAS) agreed the Inter-American Convention to Facilitate Disaster Assistance, which came into force in 1996. It does not place any kind of obligation on states either to offer or accept assistance in the event of a disaster but sets out rules for the facilitation and coordination of such assistance should both the stricken and assisting states agree to it. However, to date this Convention has only six states parties. The Association of Southeast Asian Nations (ASEAN) adopted the Agreement on Disaster Management and Emergency Response in 2005 following the Indian Ocean tsunami in December 2004. This is widely considered a progressive and comprehensive example of best practice and encompasses both risk reduction and disaster response (Simm 2018: 117–118, 125). It affirms the primacy of the affected state and allows for international assistance only with the consent of that state, but the ASEAN Secretary-General was able to use the Agreement to help negotiate that consent when the Myanmar government was reluctant to accept international assistance after Cyclone Nargis in 2008 (Simm 2018: 26, 134–135).

Since the mid-1980s, much soft law (in the form of non-binding resolutions, guidelines, and minimum standards) has been developed to facilitate and regulate international disaster response. Among the most important developments are the 2007 Guidelines for the domestic facilitation and regulation of international disaster relief and initial recovery assistance and the 2016 International Law Commission (ILC) draft articles on the protection of persons in the event of disasters. The Guidelines were drafted by the

International Federation of Red Cross and Red Crescent Societies (IFRC) with input from over 140 governments, 140 National RC Societies and 40 international organizations, NGOs, and NGO networks. They were unanimously adopted by the International Conference of the RC Movement, which brings together the ICRC, the national societies, the IFRC, and states party to the 1949 Geneva Conventions, that is, all recognized states (IFRC 2011: 7). Essentially, they are a set of recommendations to states concerning the regulation of international assistance, with the goal of minimizing bureaucratic bottlenecks and ensuring minimum quality standards and effective coordination. The Draft Articles were adopted by the ILC following several years spent examining the protection of persons in disasters and consultations with states and relevant intergovernmental organizations, and with the idea that the United Nations General Assembly (UNGA) should use them as the basis for elaborating a binding treaty in the future (Cubie 2017: 266). Several provisions pertain to humanitarian response and recovery, addressing coordination, cooperation, the role of the affected state, and the facilitation of external assistance (Cubie 2017: 271–272). The ILC Draft Articles go beyond much IDRL by specifying that states have a duty to seek external assistance when the scale of a disaster exceeds national response capacity and that their consent to external assistance is required but must not be arbitrarily withheld.

The Hyogo Framework for Action and its successor, the Sendai Framework for Disaster Risk Reduction (DRR) focus primarily on preventing disasters but also encompass disaster preparedness and thus have some bearing on disaster response.[4] The Hyogo Framework, adopted by UN member states in January 2005, set out goals and priorities for governments and the wider international community over the 2005–2015 period which included the development and strengthening of national frameworks and improving contingency planning for rapid and effective disaster response (IFRC 2007: 15–16). However, there is very little detail on disaster response (UNISDR 2005: 12–13). While it calls on international organizations to support states with the provision of relief assistance, the recommendations to states themselves are concerned only with risk reduction and make no mention of response (UNISDR 2005: 14–16). The Sendai Framework sets out a follow-up DRR plan for 2015–2030 and has similarly little to say about international responses to disasters. What it does have to say about response is focused on national- and local-level disaster preparedness for effective response and on incorporating DRR thinking into any response.

[4] On DRR thinking in general, see 'Disaster risk reduction and "building back better"' in Chapter 16 on the Nexus Concept.

International displacement and migration law

The victims of humanitarian emergencies often seek to flee to safer ground, whether within the borders of their own country as internally displaced persons (IDPs) or by crossing borders to neighbouring countries or further afield. In addition, a broader category of migrant is increasingly seen to be in need of humanitarian assistance in transit and/or at their destinations.[5] International law with implications for protection and assistance for other displaced persons and migrants is dispersed across different legal frameworks, the most central of which is IRL.

IRL is a subset of IHRL, and offers a clear framework for the protection of those who meet the criteria for refugee status.[6] The core IRL treaties—the 1951 Convention on the Status of Refugees and its 1967 Protocol—define a refugee as someone who is outside of his or her country of nationality or former habitual residence due to a 'well-founded fear of being persecuted for reasons of race, religion, nationality, membership of a particular social group or political opinion'. Those who fit this definition are referred to as Convention Refugees. With the 1969 Organisation of African Unity (OAU) Convention Governing the Specific Aspects of Refugee Problems in Africa and the 1984 Cartagena Declaration on Refugees, states in Africa and Latin America developed regional approaches with expanded refugee definitions that better reflected the reality of refugee flows in their respective regions which were not, for the most part, the result of persecution but of insecurity associated with armed conflict and generalized violence.

IRL, together with certain complementary strands of IHRL, impose obligations on states regarding the treatment of refugees, above all through the principles of *non-refoulement* and asylum. The principle of *non-refoulement* prohibits returning or expelling people to places where their lives or freedom would be threatened. It is widely agreed that this principle is customary law, and it has also been argued that the principle has acquired the status of *jus cogens* (Allain 2001). In addition to the principle of *non-refoulement*, those who meet the criteria for refugee status have the right to enjoy asylum. The content of asylum is largely defined by analogy to the rights and benefits accorded to nationals or to other aliens in the country of refuge. Asylum provides for a refugee to reside legally in the country of refuge and not to be discriminated against in terms of economic and social rights, but the exact nature of those economic and social rights can, and does, vary from one

[5] See 'People on the move' in Chapter 15 on the Shifting Parameters of Humanitarian Emergencies.
[6] For a wide-ranging explanation and analysis of IRL, see Goodwin-Gill and McAdam (2007).

country to another. However, not all individuals protected by the principle of *non-refoulement* are entitled to refugee status (Goodwin-Gill and McAdam 2007: 285–286). Some countries thus offer one or more kinds of complementary protection to a broader category of forced migrant (e.g. to people fleeing generalized violence), often with less expansive rights than are attached to refugee status (see, e.g. Goodwin-Gill and McAdam 2007: 325–333).

While IRL provides a clear framework for the legal protection of those who qualify as refugees and a supervisory mandate for UNHCR, neither the 1951 Convention nor the 1967 Protocol provide an explicit legal foundation for the provision of material assistance by UNHCR, nor do they specify any role for other humanitarian agencies. State representatives participating in the UN Conference at which the draft of the 1951 Convention was completed and signed unanimously recommended that governments and inter-governmental bodies 'facilitate, encourage and sustain the efforts of properly qualified organizations'—with special reference to NGOs—to provide welfare services to refugees (UNHCR 2010: 11). However, this is only a recommendation, and it is not clear whether it is intended to apply to *international* NGOs. UNHCR itself was conceived as a small, non-operational agency whose task was to provide or negotiate protection and solutions for refugees but has since grown enormously and become increasingly operational (Bradley 2016a: 25–26). States hosting large numbers of refugees often want or need assistance from UNHCR and other humanitarian agencies, but there is no requirement in international law for them to request or accept it. In seeking to achieve full harmony with the principles of the 1951 Convention, the 1969 OAU Convention arguably missed an opportunity to legislate comprehensively for the provision of quality humanitarian assistance to refugees in Africa (Okoth-Obbo 2001: 99, 101–105). Only the 1989 Convention on the Rights of the Child stipulates the right of (child) refugees to receive appropriate protection and humanitarian assistance (Cubie 2017: 290). Implicit in that right is an obligation of states hosting refugee children to provide or facilitate such protection and assistance.

With respect to migrants who do not qualify as refugees, there is no body of international migration law as such. All migrants, *qua* human beings, are entitled to dignified treatment in accordance with basic human rights standards. Additionally, rights and obligations salient to particular groups or categories of migrants can be found across different international legal frameworks. For example, the International Convention on the Protection of the Rights of All Migrant Workers and Members of Their Families, which came into force in 2003, details an extensive range of civil and political, as well as economic and social, rights that are relevant to migrant workers and their families in countries of transit and employment, but it has not been widely ratified (Desmond 2015). For migrants travelling by sea, the duty to rescue at sea has obvious relevance. In this regard, the main obligations of states are contained in three UN treaties, rules of customary law, and a set of guidelines that are soft law.[7] Each state has the obligation to require all ships flying its flag to rescue and assist persons in distress at sea, regardless of their nationality or status or the circumstances in which they are found, and to treat them 'with humanity' (Ghezelbash et al. 2018: 322). States must cooperate to ensure that, following any rescue, ships are able to continue on their intended voyage 'with minimum further deviation', arguably implying that the state in charge of the most conveniently located port should allow disembarkation. Coastal states must additionally promote the establishment, operation, and maintenance of SAR operations, and the world's oceans have been divided into SAR regions, with responsibility for each region assigned to a nearby coastal state (Ghezelbash et al. 2018: 321). States that are responsible for a SAR region have primary responsibility to ensure that persons who have been rescued are disembarked and delivered to a place of safety (Ghezelbash et al. 2018: 322).

When it comes to those displaced without having crossed an international border, the most important instrument at the international level is the 1998 Guiding Principles on Internal Displacement, which provide, among other things, a framework for humanitarian response. Drawn up by Francis Deng, an individual appointed by the Secretary-General as UN representative on IDPs, the Guiding Principles are soft law, but they specify a number of rights based on hard law. These include, for example, the right not to be arbitrarily displaced and the rights to life, dignity, liberty, and security of person. Several of the principles concern the provision of humanitarian assistance, stipulating, *inter alia*, that national authorities have the primary responsibility for protection and humanitarian assistance' that international humanitarian agencies may offer their services, for which consent should not be arbitrarily withheld; and that humanitarian assistance should not be diverted, in particular for political or military purposes. At the regional level, the African Union (AU) Convention for the Protection and Assistance of IDPs in Africa (the 'Kampala Convention'), which is the first binding convention on IDPs, was adopted by the AU in 2009 and came into force in 2012 (African Union 2009). It incorporates key aspects of the Guiding Principles—but here as hard law. Among other things, it stipulates that states should ensure assistance to IDPs both by meeting their basic needs and by facilitating humanitarian agency access.

[7] For a helpful outline of the international legal regime for search and rescue (SAR) and an explanation of its humanitarian dimensions, see Ghezelbash et al. (2018: 317–323).

International human rights law

IHRL is a large body of law, which encompasses IRL as well as a range of treaties and customary rules concerning issues from the most fundamental rights (e.g. the right to life) to the more aspirational (e.g. the right to work). This brief section focuses on those rights with most salience for humanitarian responses and outcomes and that have not yet been discussed. Unlike IHL and IDRL, which apply only in particular contexts, IHRL is generally considered to be applicable at all times, although states can derogate from some obligations in times of emergency. Whereas IHRL was once largely understood to apply in times of peace, there is now increasing agreement that it applies concurrently with IHL during armed conflict.

Several of the most fundamental human rights, contained in the International Covenant on Economic, Social and Cultural Rights and the International Covenant on Civil and Political Rights, correspond to basic needs and, hence, are closely related to humanitarian needs and response. For example, the rights to life, adequate food and housing, and health imply that where people cannot realize these rights due to a humanitarian crisis, the state must either meet their basic needs or allow others (such as international humanitarian agencies) to do so.

IHRL also provides a legal foundation for much of the advocacy work of international humanitarian agencies. The Declaration on the Right and Responsibility of Individuals, Groups and Organs of Society to Promote and Protect Universally Recognized Human Rights and Fundamental Freedoms, which was adopted by consensus by the UNGA in 1998, recognizes the right to defend human rights as an autonomous and independent right and encourages everyone to defend human rights. This means that humanitarian agencies and workers, among others, have a right to speak out to prevent human rights violations from taking place and to monitor and criticize the activities of authorities and other powerful actors without fear of attack or retaliation (Amnesty International 2019: 25). As a declaration, this is soft law, but its adoption by consensus by the UNGA indicates strong support among states for the rights it contains.

Some human rights instruments establish specific rights for specific categories of person, such as migrants or children (OHCHR 2011: 20). Thus the previous section discussed a range of IHRL applicable to different categories of displaced persons and migrants. Likewise, the 1989 Convention on the Rights of the Child (CRC) applies to children and has relevance for humanitarian response and outcomes where children are among the victims of a humanitarian emergency. UNICEF is mandated to promote and protect child rights, as enshrined in the CRC, by supporting the work of the Committee on the Rights of the Child. The remit of UNICEF and the CRC goes well beyond humanitarian emergencies, but humanitarian response is an important part. For example, the CRC specifies the right of refugee children to protection and humanitarian assistance, and its Optional Protocol on the Involvement of Children in Armed Conflict requires states to provide appropriate assistance for the recovery and reintegration of former child soldiers (Cubie 2017: 160–161).

Finally, it is important to mention two widely used soft law frameworks, which were developed within the humanitarian sector as voluntary professional codes with the aim of increasing quality and accountability. The 1994 Code of Conduct for the International Red Cross and Red Crescent Movement and NGOs in Disaster Relief asserts the right to receive humanitarian assistance and commits its signatories to a number of principles, including—roughly speaking—the principles of humanity, impartiality, neutrality, and independence and others concerning participation, accountability, and sustainability. In 1997, a group of professionals from different humanitarian agencies began drafting the Sphere Humanitarian Charter and Minimum Standards in Disaster Response (which have since been updated several times), which takes a different approach from the Code of Conduct. On the one hand, the Charter sets out relevant legal rights and obligations together with a set of beliefs shared by the drafters (and those who have subsequently committed to it) about the principles that should govern humanitarian response and the roles and responsibilities of different actors (Slim 2015: 257); on the other hand, the minimum standards sets out quite technical standards for practical humanitarian operations that were agreed by the drafters to be the minimum content of a right to life with dignity (Darcy 2004: 113). These two voluntary codes are difficult to classify as they are concerned with disaster response, relevant in armed conflict, and set out rights and standards which are specific to camps for refugees and IDPs.

Implementation and enforcement of the law

It is often said that international law has no 'teeth' in that it lacks the enforcement mechanisms of domestic law (police, courts, prisons), and the lack of teeth has been seen as especially problematic when it comes to issues relating to security and defence, including with respect to armed conflict. However, both hard and soft mechanisms for the enforcement of international law, including the law of armed conflict, have developed enormously in recent decades.

'Soft' enforcement measures

A range of actors engage in soft enforcement measures, sometimes with authority explicitly delegated to them by states (as in the case of UN specialized agencies and the ICRC) and sometimes without a formal mandate from states, as in the case of international non-governmental organizations (INGOs) but with expert and moral authority acquired over time. These measures can include monitoring, investigation, persuasion, public naming and shaming, and litigation (Eilstrup-Sangiovanni and Sharman 2021).

The ICRC has long played a part in implementing IHL, primarily through training for armed parties to conflict and bilateral confidential dialogue with alleged perpetrators of IHL violations, and since the late 1990s, additionally draws on IHRL in this kind of dialogue (Bradley 2013; 2016a, 76, 92–98). UNHCR also favours behind-the-scenes negotiation when seeking to persuade states to comply with their obligations relating to displaced persons. The legal frameworks discussed in this chapter can be used by international humanitarian agencies in this kind of bilateral dialogue and also in private and public advocacy[8]—though the ICRC and UNHCR are the two most centrally concerned with implementing law per se. International human rights organizations and UN special rapporteurs engage in extensive public advocacy, 'naming and shaming' violators of human rights, and sometimes those rights relate to humanitarian needs. In response to appalling living conditions and mistreatment of migrants in northern France, for example, UN Special Rapporteurs and Amnesty International called out violations of the rights to housing, health, food, and physical integrity, among others (Amnesty International 2019: 12). Since the 1990s, international human rights organizations like Amnesty International and Human Rights Watch have increasingly also spoken out about violations of IHL.

Harder enforcement measures

To the extent that enforcement refers to the use of force, there is little enforcement in the international order, although this kind of enforcement can occur when the United Nations Security Council (UNSC) authorizes peacekeeping missions under Chapter VII of the UN Charter. Such action, which primarily seeks to stop or prevent large-scale violations or abuses, is relatively rare. Beyond UN peace operations, the UNSC has also responded to large-scale

[8] For a more general discussion of the uses of different kinds of dialogue and advocacy by international humanitarian agencies, see Chapter 23 on Dialogue, Negotiation, and Advocacy.

human rights abuses and IHL violations through UN sanctions regimes, monitoring and reporting processes, the creation of international tribunals, and the promotion of national and regional accountability mechanisms (UN OCHA 2019: 6). The UNSC can also refer a situation to the International Criminal Court (ICC), which empowers the Court to investigate.

On occasion, the UNSC has played a role in enforcing laws relating to relief operations. It can require states to consent to the implementation of humanitarian relief operations on their territory or to the transit of such operations through their territory (Akande and Gillard 2016: 18; Gillard 2013: 359). In 2014, the Security Council made such a binding demand for the first time, requiring states to allow humanitarian relief operations in Syria (Akande and Gillard 2016: 19; Gillard 2013: 378). On the few occasions that the Security Council had adopted binding measures in relation to relief operations prior to this, the measures had focused on creating security conditions conducive to implementing relief operations rather than requiring states to allow access (Akande and Gillard 2016: 18–19).

It is important to note, however, that UNSC action in response to major humanitarian crises is highly selective (Binder 2009). Given the veto power of the permanent five members of the UNSC, this is not surprising, and it means that while Security Council action may serve as a strong enforcement mechanism in some cases, it is not consistently so.

International justice

International justice offers another mechanism for the enforcement of international law. Since the 1990s, a range of new international and regional institutions have been invested with supervisory and adjudicatory powers such that they contribute to the clarification, development, and enforcement of international criminal law, notably the International Criminal Tribunal for the Former Yugoslavia (ICTY), the International Criminal Tribunal for Rwanda (ICTR), and the ICC (Bothe 2005: 147; OHCHR 2011: 8). In addition, under the principle of universal jurisdiction, domestic courts can, and do, consider many of the most serious claims of IHL violations, regardless of where the alleged crime took place or the nationality or residence of the alleged perpetrators (Mills 2013: 606–607). Through these different tribunals, individuals are prosecuted for, among other crimes, serious violations of IHL. A central aim of international criminal justice is deterrence (Cryer 2015: 190–193). To the extent that the threat of prosecution serves as a deterrent to would-be violators of IHL, then international justice can contribute to protection and assistance for people in armed conflict, although the magnitude of any deterrent effect is unclear (Cryer 2015: 199–205). Courts may also strengthen compliance indirectly as a mechanism of dissemination, increasing knowledge of what the law says and acceptance of the norms contained within it (Cryer 2015: 209–213).

International criminal justice mechanisms are often criticized for their selectivity, which may take different forms. The Nuremberg trials after the Second World War largely ignored those crimes perpetrated equally by the victors as the vanquished and focused instead on the crimes committed only by the vanquished (Best 1997: 184, 195–196). Established by the UN with mandates to prosecute all serious violations of IHL, this problem of 'victor's justice' was supposed to be avoided in the ICTY and ICTR, but in practice, the victor governments were able to influence the prosecutorial agenda by withholding cooperation on certain matters (Peskin 2005). The ICC is widely perceived to be biased, with a focus on Africa both in referrals from the UNSC and in prosecutions by the Court. As a consequence, many African countries that were strong supporters of the creation of the Court have since lost confidence in it (Mbaku 2013). Selectivity highlights the role of political factors shaping the process (and hence outcomes) of international justice. Selective enforcement (or perceptions thereof) can also be expected to limit any positive impact these courts have in terms of both deterrence from breaking the law and understanding and accepting the substance of the law (Cryer 2015: 207, 214).

Whereas international *criminal* justice is about holding individuals to account, other international justice mechanisms exist to hold states to account. The principal judicial organ of the UN, the International Court of Justice (ICJ), adjudicates inter-state disputes and provides advisory opinions on legal matters referred to it by UN organs and specialized agencies. Unlike the ICC, it is widely recognized and respected, although some states nonetheless reject its jurisdiction. Some treaties, including the 1951 Refugee Convention, have a clause which states that disputes over its interpretation or application can be taken to the ICJ. The clause is not repeated in the 1967 Protocol, and for this reason, a small number of states (including the United States) have signed the 1967 Protocol but not the 1951 Convention. This clause is considered customary law anyway, but in practice no state has ever been taken to the ICJ on the basis of the 1951 Convention. With respect to humanitarian action and outcomes more generally, the ICJ has probably done more to clarify the law than to hold states to account. Additionally, there are a number of regional human rights courts and other compulsory mechanisms for the resolution of disputes. For example, regional bodies such as the European Court of Human Rights, the Inter-American Court of Human Rights, and the Inter-American Commission on Human Rights have long been delegated authority by member states to enforce human rights, in many cases going beyond IHRL *stricto sensu*. There are also mechanisms to monitor the compliance of states with, for example, various human rights treaties, which are intended to serve more as a preventative measure, deterring instead of punishing violations.

The relationship between international justice and humanitarian action is not straightforward. Beyond deterrence, the goals of international justice are quite different from the goals of classical humanitarianism and reflect a different underlying ethic. ICRC staff members do not participate or testify in judicial investigations on the basis that any such activity would compromise ICRC neutrality and confidentiality and may endanger people who provided the ICRC with information about violations of IHL—and they have an exemption explicitly written into the procedures of the ICC (Forsythe 2005: 195). While some international humanitarian agencies, such as Oxfam, are much concerned with justice, their ability to provide assistance and meet needs on the ground could be severely compromised if they participate in international justice proceedings, particularly in connection with an ongoing crisis. After the ICC issued an arrest warrant for Sudanese president Omar al-Bashir on charges of war crimes in March 2009, for example, thirteen international NGOs were immediately expelled from Darfur and three national NGOs had their licences revoked by the government of Sudan (IRIN News 2009). Although these agencies had taken care not even to appear to be providing information to the ICC, the government had long suspected that they were providing evidence to prosecutors (Addario and Polgreen 2009).

Criminalization of humanitarian action

While a range of international law facilitates international humanitarian action, and even requires some kinds of response from particular organizations, some legislation—at the national, regional, and international levels—poses *obstacles* to international humanitarian action.

Counter-terror legislation

In the aftermath of the 9/11 attacks, several countries and intergovernmental institutions strengthened their existing counter-terror legislation and developed additional legal measures aimed at preventing attacks and the preparation of attacks (Burniske et al. 2014: 3). This kind of legislation generally entails listing individuals and/or groups as terrorists and implementing a range of sanctions (e.g. asset freezes, travel bans, and arms embargoes) against those included in each list. For example, in 1999, UNSC Resolution 1267 designated specific individuals and groups affiliated with the Taliban or al-Qaeda and imposed sanctions on them (Burniske et al. 2014: 3). Related resolutions subsequently designated additional groups, and all UN member states are obliged to impose sanctions on them. In addition, different countries and regional institutions maintain their own lists of designated individuals and groups against whom they specify particular sanctions. After the 9/11 attacks, UNSC Resolution 1373 (28 September 2001) broadened this counter-terror sanctions regime. Without defining terrorism, it requires UN member states to implement laws and measures to criminalize the financing of terrorism, freeze the funds of individuals involved in acts of terrorism, deny financial support to listed groups, and cooperate with other governments to prevent and prosecute terrorist acts (Burniske et al. 2014: 3; Pantuliano et al. 2011: 2).

Of serious concern to international humanitarian agencies, several countries have enacted legislation that criminalizes acts in support of listed individuals and organizations, even where such support is unintentional. In territory under the de facto control of one of these organizations, any kind of humanitarian operation risks providing support—whether through theft or diversion of aid before or after distribution or because an aid agency is unaware that a particular recipient is connected to a listed organization. A small number of countries, notably New Zealand, have enacted exemptions for humanitarian aid in their counter-terrorism legislation, but these are the exception rather than the rule (Burniske et al. 2014: 9–10). It is perhaps the US legislation that is the greatest source of concern. It makes it an offence, punishable by up to fifteen years' imprisonment, for individuals, irrespective of their nationality, to provide material support—excluding medicine and religious materials but including other kinds of humanitarian assistance—to an entity listed as a foreign terrorist organization, regardless of intention (Burniske et al. 2014: 4; Pantuliano et al. 2011: 4).

This is not merely a hypothetical worry. There may not have been any prosecutions of personnel from major international humanitarian agencies to date, but smaller Islamic organizations have been targeted and found guilty—for example, of supporting Hamas through contributions to West Bank *zakat* committees (Pantuliano et al. 2011: 4). In 2010, the US Supreme Court confirmed that the intention of a provider of support to a designated foreign terrorist organization is irrelevant, so long as the provider of support knows that the organization is thus designated, and specified that material support could include training or expert advice meant to promote peaceable, lawful conduct, on the grounds that any such support frees up resources of the group in question for nefarious activities (Burniske et al. 2014: 4; Pantuliano et al. 2011: 5). The potential for prosecution is clear, and the criminalization of support for listed groups is impacting humanitarian action in practice. There are several mechanisms through which counter-terror legislation has led to a reduction in the quality or quantity of programming in areas under the control of groups listed as terrorists.

First, reduced funding and increased administration have served to limit the scale and speed of humanitarian response in those countries in which one or more listed organizations operate (Pantuliano et al. 2011: 7). Two years after listing Al-Shabaab in 2008, for example, the United States had cut its funding to Somalia by 88%, with significant impacts for the severity of famine in 2011 and for the humanitarian response to that famine.[9] Sometimes, it is not the donors per se but the banks transferring funds from operational agency headquarters to country offices stopping and investigating these transactions (Pantuliano et al. 2011: 7). Islamic organizations, especially local or national NGOs in contexts such as Pakistan and the occupied Palestinian territories, have been most affected as they have come under greater scrutiny than other agencies—but other organizations have also seen funding for some programmes cut and an increased administrative burden (Pantuliano et al. 2011: 7). Among both donors and operational agencies, counter-terror legislation has also reduced transparency and coordination, with minutes deliberately not recorded in cluster meetings in some contexts to avoid official acknowledgement of proscribed organizations, and donors giving policy advice on non-headed paper or preferring not to discuss these issues openly (Mackintosh and Duplat 2013: 104; Pantuliano et al. 2011: 8).

Second, the threat of prosecution under counter-terror legislation, or fear of falling foul of related clauses in donor contracts, sometimes leads operational agencies to self-censor. In this way, the designation of Hamas as a terrorist group had major implications for humanitarian action in Gaza. A number of donors introduced clauses into grant agreements that require operational agencies to ensure that funds do not support terrorism or to vet local implementing partners and suppliers against numerous lists of designated groups (Burniske et al. 2014: 6; Pantuliano et al. 2011: 5). After Hamas won legislative elections in January 2006, major donors made aid grants

[9] See 'Background' in Chapter 10 on Famine in Somalia, 2011–2012.

conditional on assurances that there would be no contact with, or benefit for, Hamas, but with the group being the governing authority, such assurances were impossible, and some NGOs had to suspend or limit their operations (Mackintosh and Duplat 2013: 94; Pantuliano et al. 2011: 10). Aid agencies also stopped training programmes for elected municipal officials, for example, which could potentially have exposed the agencies and their staff to prosecution in the United States (Pantuliano et al. 2011: 10). This kind of self-censorship was seen in Somalia too and, in both contexts, was exacerbated by a lack of clarity as to what counter-terror regulations permitted, proscribed, and required (Mackintosh and Duplat 2013: 84, 99).

Third, in some cases, the listed groups themselves restrict humanitarian access in response to being listed. For example, Al-Shabaab expelled, restricted, or banned the work of several international humanitarian agencies during the 2011 famine.[10] Although this was probably the consequence of a variety of different factors, it is likely that the listing of the group by the United States in 2008 contributed to an environment of mistrust in which Al-Shabaab saw aid agencies as politically aligned and taking sides against the group (Mackintosh and Duplat 2013: 80–82). In Gaza, NGOs found themselves caught between Hamas demanding an audit and the United States Agency for International Development (USAID) announcing that it would suspend funding to any NGO allowing the audit to take place, and when the International Medical Corps (IMC) duly objected to the audit, Hamas closed down the agency's office (Pantuliano et al. 2011: 10).

Fourth, relations between humanitarian agencies and local communities have sometimes been undermined as a result of counter-terror laws and regulations, with negative impacts on the quality of humanitarian response. Donor-imposed requirements to disclose personal information on partner organizations and beneficiaries undermine relations between humanitarian organizations and local communities and, hence, potentially compromise access to populations in need (Pantuliano et al. 2011: 8).

Through one mechanism or another, then, the quality or quantity of aid is reduced in areas where there is a risk that its impact could run counter to the interests of the 'war on terror'. In this way, counter-terror legislation has obvious implications for principled humanitarianism.[11] Action is not independent if those carrying out that action do not have the freedom to decide how and where to operate. It is not neutral if it is being used to support the 'global war on terror', even if support is limited to abstaining from assisting

[10] See 'International humanitarian response' in Chapter 10 on Famine in Somalia, 2011–2012

[11] On the meaning of the different principles, see 'Principles of humanity, impartiality, neutrality, and independence' in Chapter 13 on Politics, Principles, and Humanitarian Action.

people in territory controlled by those designated as enemies by the UNSC and powerful and donor states. Moreover, it is not impartial if assistance is being provided on the basis of where designated terrorist groups are (not) active. Counter-terror legislation also differs from, and conflicts with, IHL in important ways. The designation of a group as terrorist makes one group criminal per se under counter-terror legislation, while IHL is grounded in the doctrine of belligerent equality and regulates the conduct of all parties in the same way (Pantuliano et al. 2011: 6). Furthermore, while IHL seeks to balance military necessity with humanitarian concerns, and in doing so imposes limits on how hostilities can be conducted, counter-terror frameworks threaten to erode those limits, as seen in Afghanistan, where the ends were seen to justify the means in the Bush administration's 'war on terror', and war against the Taliban effectively had no limits.[12]

Criminalization of support for migrants

The criminalization of humanitarian action goes beyond counter-terror legislation and, in recent years, has become a significant issue for efforts to assist migrants in several places, including Australia, Canada, the United States, and a number of European countries.

In several countries, anti-trafficking and -smuggling efforts have extended to the criminalization of not-for-profit assistance to migrants. The 1985 Schengen Agreement and the 2000 UN Smuggling Protocol both explicitly stipulate that facilitating the entry or stay of irregular migrants should only be considered a criminal offence if the facilitation is done for financial gain (Carrera et al. 2019: 2). In 2002, however, the European Union (EU) adopted the so-called Facilitators Package, which removed the requirement for financial or material gain as a requirement for criminalization and made it optional for EU member states to exempt those providing humanitarian assistance to irregular migrants from criminal penalties (Carrera et al. 2019: 2; Ghezelbash et al. 2018: 348). Similarly, in Australia, strict anti-smuggling laws require only that a person organizes or facilitates the entry (or proposed entry) of a migrant who has no right to enter Australia, even if there is no financial or material benefit to that person, and even if Australia may owe the migrant(s) in question protection obligations under IRL or other IHRL instruments (Ghezelbash et al. 2018: 348–349). Where more than five migrants are involved, this constitutes an aggravated offence punishable by

[12] See 'Addressing violations of international humanitarian law' in Chapter 6 on Afghanistan, 2001–2014.

up to twenty years' imprisonment (Ghezelbash et al. 2018: 348). In Canada, the definition of smuggling similarly does not require material benefit or the exchange of money, which are simply additional factors to consider in sentencing (Watson 2015: 45). US legislation explicitly states that the crime of smuggling can be for other purposes (such as family reunification) and does not require financial gain (Watson 2015, 45).

Individuals and civil society groups providing assistance to migrants at sea have been prosecuted several times in Europe. The first prosecution of the personnel of a SAR NGO operating in the Mediterranean took place in 2004, when personnel from the German NGO Cap Anamur were prosecuted for abetting illegal immigration, forcing the organization to suspend its activities (Cusumano 2018b: 388). More prosecutions followed, and other NGOs suspended their activities, although as at 2020, all the NGOs which had been investigated had also been acquitted (Cusumano and Villa 2021). NGOs have not been active in migrant SAR off Australia's coast, so while they could be liable for prosecution under existing laws, it is not clear how active the authorities would be in prosecuting them or whether they would be able to rely on the defence of a 'sudden and extraordinary emergency' (Ghezelbash et al. 2018: 348–349).

The provision of humanitarian assistance on land has also resulted in police action in several countries. In addition to the examples from France and Greece discussed in the chapter on the European migrant crisis,[13] individual volunteers providing basic assistance (food, clothes, and shelter) in countries as diverse as Belgium, Denmark, Switzerland, and Italy have been arrested on charges of human smuggling (Vandevoordt 2019: 249). In both Australia and Canada, courts have affirmed that family members and NGOs are not exempt from prosecution for smuggling, and humanitarian actors have been charged for assisting recognized refugees to access protection in these states (Watson 2015: 45). In the United States, volunteers with grassroots aid organizations providing support to migrants in southern Arizona have been fined for littering when leaving supplies of water along known migrant trails and charged with transporting and conspiring to transport 'in furtherance of an illegal presence in the United States' when driving migrants to a clinic for medical treatment (Cook 2011: 562). Another volunteer in Arizona was arrested in 2019 and charged with harbouring and conspiracy to transport illegal aliens after providing water and food to migrants crossing the desert (Oztaskin 2020). Other policies stop short of formal criminalization or prosecution but create suspicion and social stigmatization, with the effect of intimidating and

[13] See 'Lesvos' and 'Calais' in Chapter 12 on the 2015–2016 European 'Migrant Crisis'.

disciplining rescuers (Carrera et al. 2019; Cusumano and Villa 2021). Such policies include harassment by the police and public statements by officials suggesting that those providing assistance serve to attract migrants or are actively collaborating with smugglers and traffickers.

Conclusions

International law provides a basis—sometimes a relatively strong basis, sometimes weaker—for humanitarian agencies to respond to humanitarian emergencies as well as a number of important rights and protections for the victims of those emergencies. However, some humanitarian emergencies and needs are not well addressed in existing legal frameworks such that some people in need benefit from few protections in international law, and working to support them sometimes lacks a clear legal basis or is even criminalized. Even for those categories of person who are well protected on paper, in practice, the implementation of the law depends in large part on the willingness of domestic actors to apply and enforce it. Although international enforcement mechanisms have expanded significantly in recent decades, it remains the case that the vast majority of victims of humanitarian emergencies have only very limited possibilities to see their rights enforced.

15 Shifting Parameters of Humanitarian Emergencies

Individual humanitarian agencies have tended to expand their mandates over time, incorporating new contexts and issues. In parallel, the sector as a whole has shifted its parameters such that states of affairs that were previously not seen to constitute humanitarian emergencies are now seen to demand a humanitarian response. That is to say, there has been a shift in the first dimension of humanitarianism's boundaries, outlined in the introduction to this part of the book: what constitutes a humanitarian emergency and hence demands or legitimates a humanitarian response. This chapter explores four relatively recent shifts in the kinds of contexts or crises that humanitarian agencies seek to respond to, linked to protracted crises, urbanization, humanitarian needs in Europe, and migration. Taking each of these in turn, the chapter outlines the main debates about whether, and how, humanitarian actors should respond in these contexts and discusses how expanding the parameters of what constitutes a 'humanitarian emergency' is connected to shifts in other dimensions of humanitarianism's boundaries, in particular with regard to goals, principles, and advocacy.

Protracted 'emergencies'

Crises that endure for several years are not a new phenomenon, but they account for an increasing share of humanitarian assistance, stretching the meaning of the label 'emergency', which arguably becomes inappropriate as the time horizon of a crisis situation extends. Thus, the International Committee of the Red Cross (ICRC), for example, has stopped using the word 'emergency' in the title of its annual funding appeals (ICRC 2016: 20). Behind this semantic question lies a more substantive debate about the kinds of response that are appropriate or required, and an expansion of the temporal scope of humanitarian action beyond the emergency present has occurred in parallel with moves to incorporate human rights and development discourses and goals into humanitarian practice. However, it is not clear which change

The Politics and Everyday Practice of International Humanitarianism. Miriam Bradley, Oxford University Press.
© Miriam Bradley (2023). DOI: 10.1093/oso/9780198840695.003.0015

is driving the other or whether these two kinds of expansion are mutually reinforcing.

Protracted crises account for a large proportion of humanitarian funding and a large part of the work of international humanitarian agencies. In the six years to 2021, the number of countries with five or more consecutive years of United Nations (UN)-coordinated appeals doubled to thirty-four (Development Initiatives 2021: 12, 89). Around two-thirds of the ICRC budget is spent on protracted conflicts, and in the countries hosting the ICRC's ten largest operations, the ICRC has been present an average of more than thirty-six years (ICRC 2016: 5). At the end of 2014, it was estimated that one-third of the United Nations Refugee Agency (UNHCR) caseload (or, combining the UNHCR and United Nations Relief and Works Agency for Palestine (UNRWA) caseloads, half of all global refugees) had been in exile for over ten years (Crawford et al. 2015: 13).[1] In almost 90% of the countries monitored by the International Displacement Monitoring Centre in 2014, some internally displaced persons (IDPs) had been displaced for over ten years (IDMC 2015: 63). In some cases, as with many Palestinian and Somali refugees, generations have been born and raised entirely in camps.[2]

As the length of time that people live in crisis increases, their needs and priorities often change. For example, in addition to the more immediate and direct suffering resulting from attacks, deprivation, and displacement, protracted conflict typically results in the cumulative deterioration of infrastructure, services, livelihoods, and living conditions (ICRC 2016: 5, 10; Policinski and Kuzmanovic 2019: 968). Impoverishment can become entrenched, and protracted conflicts such as those in the Democratic Republic of the Congo, Syria, Libya, South Sudan, and Somalia have made large numbers of people 'war poor' (ICRC 2016: 11, 16). The cumulative impact of protracted crises additionally makes those people living in them especially vulnerable to new shocks. For example, twenty-five of the thirty-four countries experiencing protracted crises in 2020 were deemed at high or very high risk of the effects of COVID-19 and, at the same time, had some of the lowest rates of vaccination against COVID-19 worldwide (Development Initiatives 2021: 17).

Arguably, the ethical duties of those international humanitarian actors seeking to help people living in crisis also change or increase as crisis becomes protracted. Where the response to protracted refugee situations is limited to

[1] Note that these averages were skewed downwards by large recent displacements from Syria and South Sudan.

[2] For a discussion and critique of camps as a response to (protracted) displacement, see 'Politics of camps' in Chapter 22 in Material Assistance and Direct Service Provision.

basic needs, for example, refugees are effectively kept in long-term limbo, with significant costs in terms of wasted potential and human suffering (Lindley 2011: 49). Thus, some have argued that it is necessary to think beyond limited goals of saving lives when crises become protracted (Brun 2016: 400). For Hugo Slim, the ethical responsibilities of aid agencies thicken the longer they spend alongside any given community, stretching beyond concern for the present to include things like resilience programming (Slim 2015: 97). According to these perspectives, then, in protracted crises, humanitarian actors have a moral duty to think beyond keeping people alive to incorporate considerations of how well they are able to live, not only now but also in the future. Another perspective focuses on cost effectiveness, exemplified in the commitments made at the 2016 World Humanitarian Summit (WHS). Partly motivated by the costs of providing assistance over years and decades, the 'new way of working' seeks not only to meet people's immediate humanitarian needs in protracted crises, but also to reduce those needs so that assistance is not required in the future (UN Secretary-General 2016: 10).

On the other hand, insofar as humanitarian action is legitimized through emergency claims (as in the humanitarianization process described in Chapter 1) or through reference to international humanitarian law (IHL), it should arguably be minimalist in scope, limited to the most urgent and important tasks (i.e. keeping people alive and well in the immediate term). Among other things, emergency claims may allow for the suspension of normal rules in order to return a state of affairs affected by some sudden or unexpected occurrence to the status quo ante (Rubenstein 2015). In underscoring the urgency and importance of responding to conflict and disasters (e.g. to save lives under imminent threat) such claims allow for exceptions to the general rule that interventions are not legitimate without the consent of the intervened (Barnett 2012: 513). In line with this logic, IHL stipulates that in contexts of armed conflict, states must not arbitrarily withhold consent for relief actions that are humanitarian, impartial, and necessary (Akande and Gillard 2016: 16–20). Where such commodities are lacking, the direct provision of food, water, medical supplies, clothing, and means of shelter fits indisputably into the category of necessary relief actions, but whether or not indirectly contributing to the provision of these goods by reforming systems or structures can also be considered necessary relief actions is less clear (but cf. ICRC 2016: 12).

In practice, aid agencies adapt their work to protraction by employing one or both of two main strategies (instead of, or in addition to, their traditional care-and-maintenance models of assistance). First, through interventions focused on affected individuals, households, and communities (including the distribution of productive assets, vocational training, and microeconomic initiatives, grants, and loans), they seek to support sustainable livelihoods and self-reliance (Crawford et al. 2015: 2; ICRC 2016: 28). Such interventions have been relatively ineffective, in part because they are often short-term with an overly technical focus on providing assets or developing skills, and insufficient consideration of the market viability of the activities such interventions are intended to stimulate, and the financial services that people (the displaced, in particular) are able to access (Barbelet and Wake 2017; Crawford et al. 2015). Second, through interventions focused on the structures that affect those individuals, some agencies seek to build resilient systems. Thus, the ICRC, for example, designs and maintains urban water supply infrastructure and upgrades government hospitals, among other system-level interventions (ICRC 2016: 19, 27).

As these examples illustrate, in responding to protracted crises, the boundaries between humanitarian and development work can become blurred. The ICRC, for example, has explicitly acknowledged that in protracted conflicts it undertakes 'activities that would, in peacetime, be understood, in international policy terms, as development activities' (ICRC 2016: 12). However, it is not clear whether, or how much, structural engagement or development programming can be incorporated into a response without compromising humanitarian principles, and in particular the principle of neutrality. The ICRC seeks to maintain its neutrality by engaging in systems support and long-term activities only with the humanitarian purpose of meeting essential needs and not in pursuit of a particular ideology of political and social change (ICRC 2016: 13). Whatever the intentions of the ICRC, support for state-owned infrastructure and hospitals can be expected to strengthen the state in practice. Such unintended consequences are not unique to structural support projects but may be of greater magnitude than when international actors provide relief items directly to those in need.[3] For many agencies, the protraction of crises has also led to a more rights-based approach and a move from attending only to the symptoms of crisis to addressing the causes (Barnett 2018: 327). Such a move invariably implies a departure from neutral intentions and is likely to be in tension with an impartial needs-based approach.[4]

[3] On unintended consequences of humanitarian action, see 'Feasibility of apolitical action' in Chapter 13 on Politics, Principles, and Humanitarian Action.

[4] On the merits and risks of incorporating transformative goals into humanitarian response, see 'Desirability of apolitical action' in Chapter 13 on Politics, Principles, and Humanitarian Action.

Urbanization

Over the past ten years, international humanitarian agencies have increasingly turned their attention to urban areas. This follows the ever greater presence in cities of the people these agencies have traditionally sought to assist, namely, victims of armed conflicts, disasters, and health crises. This shift to urban response implies challenges related to identifying 'beneficiaries', addressing the specific needs arising from city living, and adapting to the constraints and opportunities of urban environments. The urban turn also reflects increasing international concern with 'urban violence'. High levels of violence in cities which does not meet the IHL thresholds for non-international armed conflict is now seen by many international humanitarian agencies as requiring their response, and that response must be adapted not only to the specifics of working in urban environments but also to addressing widespread violence without the support of the IHL framework.

Meeting the needs of their traditional beneficiary populations increasingly requires humanitarian actors to respond in urban areas. According to the World Bank, 55% of the world's population lives in urban areas, and that proportion is expected to increase to nearly 70% by 2050 (World Bank 2020). As more and more people live in urban areas, so more and more of those who have traditionally been the objects of humanitarian response can be expected to be living in urban areas. Furthermore, two factors suggest that the percentage of traditional recipients of humanitarian assistance living in urban areas may be growing at an even faster pace than that of the world's population as a whole. First, those living in informal settlements in urban areas may be especially vulnerable to the devastating consequences of disasters and health crises, as exemplified in the earthquake in Haiti in January 2010 and the spread of Ebola in Monrovia in 2014 (Fallah et al. 2018; Patrick 2011: 2; see also ALNAP 2012: 6–8). Second, those displaced by humanitarian crises increasingly move to urban areas, with best estimates suggesting that at least 60% of refugees lived in urban areas by the end of 2016 (Bradley 2017: 8). Reliable data on the numbers of internally displaced persons (IDPs) living in urban areas does not exist, but it is widely believed that the proportion is increasing, with available information suggesting that it reached around two-thirds in 2020 (UNHCR 2021a: 27).

Most international humanitarian agencies have far greater experience in rural settings, and working in urban areas poses an initial challenge in terms of targeting assistance. Urban populations tend to be diverse and their needs differentiated such that targeting by geography is relatively ineffective (ALNAP 2012: 9; Cotroneo 2017: 288). The vast majority of refugees and IDPs in urban areas are dispersed in individual accommodation rather than living in organized camps. Humanitarian agencies often struggle to get data on these 'hidden' or 'invisible' populations to identify who to assist and what they need, and many displaced persons living in urban areas receive little or no international assistance (Cotroneo 2017: 303–308; Crisp et al. 2012: S27–S28; Haysom 2013: 6). The interaction of the protracted and the urban nature of displacement in many cases generates specific challenges, and the lack of assistance for those displaced in cities is not just due to difficulties in identifying beneficiaries. It can also be deliberate, with some municipal or central authorities seeing aid as a pull factor, refusing to accept the long-term presence of displaced populations in their cities, and resisting initiatives aimed at assisting, or integrating, the displaced (Crisp et al. 2012: S26; Haysom 2013: 1; Lindley 2011: 38). UNHCR, for its part, long shared the assumption held by many states that refugees belong in camps and should be deterred from settling in urban areas but since the 1990s has adopted a more proactive approach to assisting the displaced in cities (Crisp et al. 2012: S25; Lindley 2011: 38).[5]

Whether in the context of disasters occurring in cities or the arrival in cities of people displaced by emergencies elsewhere, traditional humanitarian response models need adapting to urban environments.[6] For example, governance institutions are concentrated in cities such that affected states more often have the expectation and capacity to lead the response to urban disasters, and international actors may need to work in a support role (ALNAP 2012: 9; Cotroneo 2017: 289). As well as liaising with national governments, an effective response in urban areas requires that humanitarian agencies develop relationships with municipal authorities, service providers, and urban police forces (Crisp et al. 2012: S26). Meeting such basic needs as shelter and water requires not only a different technical skill set (designing and maintaining urban water infrastructure is a very different task from drilling bore holes) but also knowledge of institutional, normative, and policy frameworks regarding complex uses of, *inter alia*, lend tenure, tenancy, and housing rights (Cotroneo 2017: 290).

The urban turn in humanitarian action also goes beyond adapting traditional models to respond to the needs of the victims of conflict, disasters, and epidemics living in, or displaced to, urban areas. Many of those who have been displaced to cities live among impoverished, and marginalized residents, and, while some of their needs and vulnerabilities are specific to displacement, other needs and vulnerabilities are shared with the broader

[5] See 'Politics of camps' in Chapter 22 on Material Assistance and Direct Service Provision.
[6] For a handbook on good practice in urban humanitarian response, see Sanderson (2019).

urban poor population (Bradley 2017: 15–19). They are also often competing for scarce resources and overburdened services with the host population (Cotroneo 2017: 300–303). In recognition of this, urban programming by organizations traditionally mandated to address the needs of the displaced frequently now also seek to support their host communities, who may not be victims of a humanitarian emergency. Extending humanitarian assistance to host communities can be done to ease (potential) tensions between the displaced and host communities but also reflects an interest in notions of fairness and equity. It means that, in at least some of this work, humanitarian agencies are not responding to an emergency in the sense of a rupture from what is normal but rather to a state of affairs that is normal for the host population. Accordingly, the response is not about returning the situation to the status quo ante but raising the standard of 'normal' and, as such, it looks a lot more like longer-term development work than the traditional tasks of providing food and shelter in camps (Crisp et al. 2012: S25).

Aid agencies have also increasingly incorporated 'urban violence' outside armed conflict into their mandates. Violent deaths in conflict now represent only a fraction of violent deaths worldwide, and the high concentration of non-conflict collective violence in cities has made urban violence a particular concern, especially in Latin America, which has 8% of the world's population and 33% of homicides (Moser and McIlwaine 2006; Muggah and Aguirre Tobón 2018: 2). In 2017, the five most violent cities in the world (as measured by homicide rate) were all in Latin America, and each of them had a homicide rate of around fifty times the global average (Seguridad Justicia y Paz 2018). In addition to the more general adaptations required for an effective response in urban areas, working in contexts of violence in which IHL does not apply, humanitarian organizations lack a clear legal basis for that response. In the case of the ICRC, whose core mandate is for working in armed conflicts, where its work is structured by IHL, the move to address urban violence has involved an expansion of its goals to address not only the symptoms but also the causes of violence and a move away from strict neutrality as it seeks to deter people from joining gangs, for example (Bradley 2020).

Europeanization

Europe is primarily a provider rather than a recipient of international humanitarian assistance. Many of the largest international humanitarian agencies were originally created to respond to the fallout of wars in Europe. Several have their headquarters in Europe, and are, in large part, funded by European governments and staffed by European citizens, but for the past several decades, the focus of their operations has been in Africa, Asia, the Middle East, and, to a lesser degree, Latin America. In recent years, however, some international humanitarian agencies have turned their attention to European counties, providing a response—albeit a limited one—to needs arising there. Since the 1980s, for example, Médecins sans Frontiéres (MSF) and Médecins du Monde (MDM) have worked to support vulnerable populations on 'home' territory in France through medical aid, socio-legal support, and public advocacy (Hanrieder and Galesne 2021). Oxfam GB launched a UK Poverty Programme in the mid-1990s (Pickering-Saqqa 2019). Other international humanitarian agencies with headquarters in Europe also have some limited domestic programming. Most recently, with the so-called migrant crisis in Europe from 2015,[7] some have responded to large-scale humanitarian needs in Europe of the kind more commonly associated with armed conflicts and refugee or IDP camps in Africa or Asia (Gordon and Larsen 2021: 435; Keen 2021: 10; Sandri 2018: 66).

Many humanitarian agencies seem unsure of their role in Europe, with decisions to launch operations reached only after extensive internal discussion and debate (del Valle 2016; Hanrieder and Galesne 2021, 1720–1723; Pickering-Saqqa 2019: 480–481). For some humanitarian actors, the fact of it being their 'home' region implies a stronger obligation to respond to humanitarian needs in Europe. When MSF was debating launching search-and-rescue (SAR) missions in the Mediterranean, for example, some staff felt that they had a special responsibility there, given the weight of MSF's reputation and support base in Europe (del Valle 2016: 34). Operating on home-territory can also be seen as a way to counter the 'othering' and neocolonialism that characterize a Western humanitarian project focused on faraway places (del Valle 2016: 34; Pickering-Saqqa 2019: 483–484). For other actors, the fact that European countries have the wealth and capacity to meet everyone's basic needs is reason to leave the response to them. This was apparently UNHCR's rationale for not providing material assistance to migrants living in dire conditions in makeshift camps in northern France (Hilhorst et al. 2021: 132). Some in MSF argued against launching SAR missions precisely because states in Europe had the infrastructure and resources to respond themselves (del Valle 2016: 34). By contrast, for those comfortable taking a confrontational stance, the fact that European countries have the wealth, capacity, and responsibility but are not adequately meeting everyone's

[7] See Chapter 12 on the 2015–2016 European 'Migrant Crisis'.

basic needs is a reason to pressurize the relevant authorities—and a practical response is often the basis for pressurizing them.

For some humanitarian agencies, decisions about whether or not to respond to humanitarian needs in Europe may come down to more prosaic concerns, given that many of their main funders are the same European governments whose policies and practices are the primary cause of humanitarian need there. In some cases, those governments are actively opposed to the provision of assistance to the migrants most in need of it, as in northern France and with SAR operations in the Mediterranean.[8] David Keen suggests that their dependence on UK government funding might be a factor explaining Oxfam and Save the Children's absence from Calais, despite the level of humanitarian need there (Keen 2021: 15). Individual donors provide the vast majority of MSF's funding, which likely made it easier to support migrants in northern France and other European countries and to carry out SAR operations in the Mediterranean in the face of opposition from several European authorities.[9] MSF actually lost some of these individual donors over the SAR operation but gained others who had never previously supported the agency (del Valle 2016: 35). While its heavy dependence on donations from private sources ensured that donor governments could only exert limited pressure on MSF through their funding decisions, in June 2016, MSF went a step further and, in protest at European Union (EU) migration policies, announced that it would no longer accept funding from the EU or member state governments for any of its operations worldwide (MSF 2016).

Compared with their operations elsewhere in the world, advocacy has played a more central role in the work of international humanitarian agencies in Europe. If European countries have the resources and infrastructure to respond to everyone's needs, it is a lack of will, not a lack of ability, that is preventing them from doing so. In this context, humanitarian agencies generally do not want to substitute for the state, which may facilitate its dereliction of duty, but rather seek to hold the state to account and convince it to fulfil its responsibility.[10] For many international agencies working in Europe, therefore, advocacy has been a central part of their response and often a motivating factor.[11] For example, leaders in both MSF and MDM see their organizations'

[8] See 'Humanitarian response' in Chapter 12 on the 2015–2016 European 'Migrant Crisis'.

[9] For a more general discussion of how donors influence the focus and decisions of operational agencies, see 'Independence, earmarking, and conditionality' in Chapter 18 on Donors and the Funding of Humanitarian Action.

[10] On the different ways humanitarian agencies relate to the governments of affected states, see 'International aid agencies and the governments of affected states' in Chapter 19 on Government and Civil Society in Affected States.

[11] See 'Humanitarian response' in Chapter 12 on the 2015–2016 European 'Migrant Crisis'.

efforts to document and bear witness to suffering as especially important in their work in France and have sought to keep assistance short-term in order to minimize substitution for the state (Hanrieder and Galesne 2021: 1716). On its work in Greece, for example, the International Rescue Committee (IRC) also makes explicit reference to the fact that these needs are arising in Europe, where authorities have the capacity to do better (Phillips 2019: 15).

Debates about humanitarian neutrality arguably also shift when European humanitarian actors are working in Europe—on 'home' territory.[12] One argument in favour of neutrality makes the case that unelected and often unaccountable outsiders should not be deciding what economic, political, and social goals are pursued. When Europeans are working in their home countries or region, though, they are not outsiders but citizens. Perhaps unsurprisingly, then, some MSF workers reported feeling a greater sense of responsibility to speak out on issues relating to immigration detention in Europe because this advocacy was targeted towards their 'home' societies (Kotsioni 2016: 55). Linked to its SAR operations in Europe, MSF additionally goes beyond advocating for authorities to comply with the law (e.g. regarding the duty to rescue at sea) to demand policy change, in terms of creating legal migration routes.[13] Another argument in favour of neutrality emphasizes the instrumental role adherence to the principle is expected to play in generating acceptance and access, especially in armed conflict contexts. While there are plenty of examples of humanitarian advocacy in Europe generating a hostile reaction from the authorities, with a negative impact on humanitarian response,[14] the fact that there is no armed conflict and that many aid workers hold European citizenship (and so do not need visas and cannot be expelled) make access less problematic, even when advocating on contentious issues.

People on the move

Again, partly linked to the arrival of significant numbers of migrants in Europe, migration has increasingly come to be seen as a state of affairs requiring an international humanitarian response. Whereas refugees in refugee camps in the global South have long been a central humanitarian concern,

[12] For a discussion of the principle of neutrality, see 'Principles of humanity, impartiality, neutrality, and independence' in Chapter 13 on Politics, Principles, and Humanitarian Action.

[13] See 'Search and rescue' in Chapter 12 on the 2015–16 European 'Migrant Crisis'.

[14] See 'Humanitarian response' in Chapter 12 on the 2015–2016 European 'Migrant Crisis'.

international humanitarian agencies have not historically prioritized refugees arriving to wealthy countries in the global North or migrants who do not qualify for international protection as refugees. Over the past ten to fifteen years, however, that has changed. For example, the ICRC undertook to work more consistently to alleviate the suffering of vulnerable migrants in 2007, and then, in 2014, identified migration as an institutional priority in its 2015–2018 Strategy (Bradley 2023). As explored in more detail in Chapter 12 on the 2015–2016 European 'Migrant Crisis', some international humanitarian agencies have responded by launching SAR missions, working with detained migrants in European countries, and providing assistance to those in camps, as in northern France and Greece. The ICRC, IRC, MSF, and others provide different kinds of support to detained migrants around the world (ICRC 2015: 5; Kotsioni 2016: 47; Phillips 2019: 6).

Protecting and assisting people on the move generates challenges and dilemmas for humanitarian agencies. Much of the migration that requires a humanitarian response is 'irregular'—or 'irregularized' by the border policies of states. The principle of impartiality requires the provision of assistance in proportion to need, but needs do not correspond to bureaucratic labels and legal statuses. Indeed, undocumented migrants are often vulnerable precisely because they lack legal status. While UNHCR has a particular mandate focused on refugees, asylum seekers, and stateless persons, and for which legal statuses are central, agencies such as the ICRC and MSF explicitly take an inclusive approach, providing assistance on the basis of needs and vulnerability (del Valle 2016: 29; ICRC 2015: 1–2). Even when taking this kind of approach, however, their work can still be conditioned by these policy categories. Where migrants have a legal status which affords them particular rights or protections in international or domestic law (e.g. as a refugee or asylum seeker), humanitarian agencies can provide legal assistance to help ensure that they receive the protection to which they are entitled. Insofar as they make use of these legal categories, however, they risk reinforcing them, potentially to the detriment of those migrants who do not benefit from a particular status. For those agencies that see these policy categories as part of the problem, this implies a dilemma.

Humanitarian actors also face a challenge in working to assist migrants who fall into the 'wrong' policy category without endangering those migrants or coming into conflict with the law themselves. The risk of endangering migrants is an issue which has arisen in the context of NGO SAR operations in the Mediterranean, with some agencies taking care to avoid collecting any information that could potentially incriminate migrants, while others have shared images with the Italian authorities, including, in at least one case, that led to the arrest of a migrant identified as the driver of a migrant boat.[15] Even in providing material assistance on land, where interactions with authorities are more easily avoided, there remains a risk that assisting vulnerable migrants serves to make those migrants more visible to the authorities, and in the United States, for example, migrants crossing the Arizona desert have been arrested at posts where water and other supplies are positioned (Oztaskin 2020). Moreover, many states have gone so far as to criminalize those who provide assistance to 'irregular' migrants, and in a number of cases, humanitarian actors providing assistance to migrants have been prosecuted.[16]

In addition, humanitarian agencies face a dilemma when their efforts to mitigate harm and relieve suffering serve to facilitate the very policies that are causing the harm and suffering. This issue arises in all kinds of humanitarian emergencies but is perhaps especially prominent in the case of migration, where there is not always some other crisis event (an earthquake, famine, armed conflict, etc.) driving humanitarian needs. Official policies are a major factor in making migration dangerous—and in generating the needs which humanitarian agencies now seek to address. For example, strict border policies in Europe and the United States have pushed migrants to undertake more dangerous journeys (Cook 2011: 569–570; Pécoud 2020). Neglect and harassment by authorities have led to migrants living in dire conditions, and immigration detention policies are a further cause of suffering. Humanitarian action to relieve that suffering can facilitate the policies that cause it. By making the consequences of those policies less awful and allowing authorities to point to the humanitarian response as evidence of their humanity, for example, it can deflect criticism from authorities. By relieving the authorities of some of their responsibilities, humanitarian actors can also serve to free up capacity which authorities can use to reinforce the harmful policies. This dilemma is particularly acute in SAR operations[17] and in working with detained migrants.[18]

In grappling with this dilemma, international humanitarian agencies adopt a range of different strategies to allow them to respond while minimizing the

[15] See 'Search and rescue' in Chapter 12 on the 2015–2016 European 'Migrant Crisis'.

[16] See 'Criminalization of support for migrants' in Chapter 14 on Law and International Humanitarianism.

[17] See 'Search and rescue' in Chapter 12 on the 2015–2016 European 'Migrant Crisis'.

[18] A whole book could be written to discuss the challenges and dilemmas facing humanitarian agencies working in places of immigration detention, where they are often additionally constrained from providing independent, impartial, and quality services. For reflections from within IRC and MSF, see Phillips (2019) and Kotsioni (2016), respectively.

extent to which they bolster harmful policies and avoiding becoming complicit in policies they oppose. Often, they seek to avoid normalizing their assistance role, in which they are effectively substituting for the state. In detention contexts, in particular, IRC and MSF have made difficult decisions to provide only minimal services and to refuse to undertake certain activities (e.g. refusing to repair sanitary infrastructures in detention facilities or provide tents for shelter)—even when these activities would have improved the well-being of detained migrants (Kotsioni 2016: 50–51; Phillips 2019: 15–19). With respect to SAR operations and immigration detention especially, but also in responding to migrants more generally, several humanitarian actors see advocacy as central (Cusumano 2018b: 393; Cuttitta 2018: 640–641; del Valle 2016: 27, 37; Kotsioni 2016: 50–51, 53–54; Phillips 2019: 5, 29–34). Advocacy can serve to reduce (perceptions of) complicity and may also serve to induce policy change, though it is not clear how effective it has been in doing so, which calls into question the appropriateness of speaking out in settings (e.g. immigration detention) where vocal criticism may be in tension with access (Phillips 2019: 30). Dialogue and advocacy can focus on demanding that duty bearers fulfil their duties (to rescue at sea, meet the basic needs of detained migrants, etc.) or can go further and demand more significant policy changes (e.g. increasing legal migration routes and ending, or reducing, immigrant detention). The ICRC favours confidential dialogue with detaining authorities to persuade them to improve the conditions of detention but at the global level additionally advocates publicly for states to limit their use of detention for migrants (Bradley 2023). In general, MSF has stopped short of opposing the practice of immigration detention per se, focusing instead on improving the conditions of detention, but in relation to its SAR operations, its advocacy goals are more expansive and include calling on European states to reform their border policies and increase safe and legal migration routes (del Valle 2016: 31, 37; Kotsioni 2016: 54).

Conclusions

Just as the parameters of humanitarian emergencies have shifted outward to encompass states of affairs that were not always deemed to require an international humanitarian response, so too have the goals of humanitarian action expanded. In protracted crises and urban humanitarian response, the scope of humanitarian ambition frequently expands beyond emergency response to include longer-term transformative goals, including development and peace-building. Those international humanitarian agencies working in Europe and to support migrants have mostly understood advocacy as central to this work and, in many cases, have not only advocated for states to comply with their obligations in international law or with existing policies but have also challenged those policies and, in some cases, the law itself.

16
Nexus Concept

The first World Humanitarian Summit (WHS) took place in May 2016, bringing together 'representatives of [UN] Member States, non-governmental organizations, civil society, populations affected by crises, the private sector and international organizations [. . .] to commit to collectively and individually reduce the unprecedented level of humanitarian need and suffering in the world' (UN Secretary-General 2016: 2). Starting from the fact that assessed humanitarian need was growing faster than funding for humanitarian response, it was widely agreed that humanitarian actors needed to address needs more efficiently. On the assumption that reducing needs would be more cost-effective than simply meeting them, a range of actors thus committed to addressing the causes of humanitarian crises, rather than just their consequences, and to bringing together humanitarian, development, human rights, and peace and security efforts. The WHS endorsed a New Way of Working (NWoW) which, among other things, seeks to transcend the humanitarian–development divide by embracing the so-called triple (humanitarian–development–peace) nexus. Later that same year, the Comprehensive Refugee Response Framework (CRRF), which seeks to operationalize the humanitarian–development nexus in refugee settings, was adopted by the United Nations General Assembly (UNGA) as part of the New York Declaration for Refugees and Migrants.

The nexus, the NWoW, and the CRRF are the latest in a series of concepts and agendas since the 1980s that have sought to better connect humanitarian action, development programming and—in some iterations—conflict resolution or peacebuilding initiatives. Underlying these concepts and agendas are multiple rationales. Emergencies are costly and disruptive of development, while development policy and institutions are often insensitive to the risk of crises and insufficiently focused on protecting vulnerable households against risk (Ross et al. 1994: 1). In some cases, humanitarian relief has actively undermined development, and in others inappropriate development programming has caused emergencies (Ross et al. 1994: 3). Thus, it is hoped that better-designed development aid could help mitigate against emergencies by reducing people's vulnerability to shocks and that better-designed

The Politics and Everyday Practice of International Humanitarianism. Miriam Bradley, Oxford University Press.
 DOI: 10.1093/oso/9780198840695.003.0016

humanitarian aid could contribute to development, or at least provide the basis for future development work (Macrae and Harmer 2004: 2). At the same time, underdevelopment often contributes to conflict onset, and relief programming can exacerbate and prolong conflict (Anderson 1999; Macrae and Harmer 2004: 2). Designed differently, it is thus hoped that development aid could help to prevent conflict and humanitarian relief to contribute to peacebuilding.

For all the potential synergies between different kinds of international aid, there are also important differences—and sometimes tensions—between them. As outlined in the introduction to this part of the book ('The Boundaries of Humanitarianism'), the humanitarian ethic differs from that of development or human rights work. The differences in ethic correspond to, among other things, differences in time horizons, principles, and relations with affected states (Harvey 2009: 22; Lie 2020: 5). Despite the fact that humanitarian agencies often respond to humanitarian crises over many years, their funding and programme cycles tend to be short-term and the legitimacy of their response linked to emergency claims.[1] Conversely, development work has longer time horizons, with changes in social policy, governance, and equality expected to take time, and its legitimacy is linked to participatory approaches. The principle of impartiality—according to which humanitarian aid must be provided in proportion to need, and need alone—is in marked contrast to many development initiatives, which may be subject to explicit economic or political conditions. These different kinds of international response also vary in terms of how they relate to the states affected by crisis.[2] Humanitarian agencies very often substitute for the state, sometimes eroding its capacity and its accountability to its population, while a primary goal of development actors is to strengthen state capacity, and human rights actors emphasize holding the state to account. Indeed, sometimes, donors have funded humanitarian relief rather than development in protracted crises precisely to avoid engaging with states that were belligerents in active conflict or that were perceived to be repressive, undemocratic, or massively corrupt (Macrae and Harmer 2004: 3).

This chapter surveys the most prominent of their precursors before exploring the NWoW and the CRRF. The structure of the chapter is loosely chronological, but the different agendas and initiatives have overlapped both temporally and conceptually. Broadly speaking, the dominant agenda has

[1] See 'Protracted "emergencies"' in Chapter 15 on the Shifting Parameters of Humanitarian Emergencies.

[2] For a discussion on the different ways humanitarian agencies relate to and engage with affected states, see 'International aid agencies and the governments of affected states' in Chapter 19 on Government and Civil Society in Affected States.

shifted from one which promoted efforts to avoid gaps between humanitarian response to crises and subsequent, separate development programming which started (or restarted) after the crisis was over to one which promotes humanitarian, development, and sometimes also peace programmes that are not only simultaneous but also pursue shared goals. However, this general trend obscures significant differences in terms of which actors were driving, embracing, or contesting different agendas, and the discussion in this chapter distinguishes the following actors:

- donors, including multilateral donors, such as the European Union (EU), and individual donor governments;
- mainstream development actors such as the Organisation for Economic Co-operation and Development (OECD), the United Nations Development Programme (UNDP), and the World Bank;
- single-mandate or 'purely' humanitarian agencies, who 'work only with an emergency humanitarian mission based in international humanitarian law and humanitarian principles', the biggest of which are the International Committee of the Red Cross (ICRC) and Médecins Sans Frontières (MSF) (Slim and Bradley 2013: 4); and
- multi-mandate agencies, who not only respond to humanitarian crises but also seek to address broader, longer-term issues of poverty, human development, and social justice; this category encompasses the majority of the largest international humanitarian actors, including United Nations (UN) agencies, such as the United Nations Refugee Agency (UNHCR) and the World Food Programme (WFP), and non-governmental organizations (NGOs) such as Islamic Relief, Oxfam, Save the Children, and World Vision (Slim and Bradley 2013: 4).

Whereas the shifting parameters of humanitarian emergencies discussed in Chapter 15 provoke questions about how far humanitarian actors can, and should, expand their own goals and activities into what is traditionally considered development territory, the different concepts and agendas introduced in this chapter are primarily about coordinating and integrating the work of humanitarian actors with that of development (and, to some degree, peace) actors. Coordinating with development actors is widely assumed to be desirable, and most debates about bridging the humanitarian–development divide focus not on the goal itself but rather on how to achieve it. By contrast, the integration of humanitarian action with peace-building goals has always raised concerns with humanitarian actors, although the level of resistance seems to have reduced over time.

Linking Relief, Rehabilitation, and Development (LRRD)

Drawing on disaster practices, the Linking Relief, Rehabilitation and Development (LRRD) agenda in the 1980s and 1990s represented an early attempt by humanitarian actors to tackle the humanitarian–development nexus (Zetter 2021: 1768). LRRD is an umbrella term which encompasses a multitude of ideas, with some progression over time—broadly speaking from a sequential model (the 'relief–development continuum') to a simultaneous model (the 'contiguum'). Sequential models assume that not all development values can be realized *in extremis* and that it is appropriate to introduce them progressively, after the emergency phase is over, while simultaneous models combine long- and short-term goals and seek to integrate emergency and development ethics (Slim and Bradley 2013: 6).

Underlying 'continuum' thinking was the idea of a linear process in which crises were temporary shocks to the 'normal' development path and necessitated relief, but that through rehabilitation, the status quo ante could be re-established and progress along the development path resumed (Macrae and Harmer 2004: 2; Mosel and Levine 2014: 3; Ross et al. 1994: 4). According to this thinking, LRRD was primarily about humanitarian actors applying exit strategies to prepare the ground for, and hand over to, development actors (IRRC 2019: 1053–1054; Mosel and Levine 2014: 1, 3). The expectation was that better development interventions could reduce the need for emergency relief, better relief interventions could contribute to development, and better rehabilitation efforts could help to link them (Ross et al. 1994: 1). Implicit in continuum thinking, then, was the idea that relief should serve not only as a palliative but also as a springboard for recovery and the development of livelihoods (Macrae and Harmer 2004: 2).

Through the 1990s, the continuum model was gradually replaced by the idea of a contiguum, which sought to adapt the linear model to conflict-related crises. Contiguum thinking recognized that different needs may exist simultaneously and that conflicts and displacement crises were often protracted and disasters often cyclical, not just temporary shocks to an otherwise progressive development path (Fan 2013: 1–2; IRRC 2019: 1054; Macrae and Harmer 2004: 2; Mosel and Levine 2014: 3). Operationally, this shift in thinking implied a move towards greater collaboration between humanitarian and development actors and greater coherence between humanitarian and development programming (IRRC 2019: 1054). In practice, however, the linear logic persisted, and substantive changes in aid responses were limited (Macrae and Harmer 2004: 3; Mosel and Levine 2014: 3). Mainstream development actors tended to wait for conflict to end before intervening, and in many conflict contexts, donors restricted their funding to emergency response, so it was largely left to the multi-mandated UN agencies and NGOs who were driving the LRRD agenda to expand what they did under the rubric of humanitarian assistance (Macrae and Harmer 2004: 3).

Beyond LRRD: The security–development nexus and whole-of-government approaches

By the early 2000s, political and security goals had firmly entered the equation, and donor governments and development actors were leading calls for the integration of different types of aid, while humanitarian actors were often more reticent. The use of aid in pursuit of political goals has a much longer history, but donor demands to integrate aid and security policy were intensified after 9/11, with development actors like the World Bank increasingly engaging in so-called failed and fragile states and donors increasingly expecting aid to contribute to counter-terror efforts (Macrae and Harmer 2004: 4).

Through the 1990s and 2000s, a number of influential publications promoted thinking about the links between aid, conflict, and security. The human security agenda set out by UNDP in its 1994 Human Development Report shifted emphasis from state security to individual security, with health, developmental, and environmental threats highlighted as well as more traditional security threats relating to violence (UNDP 1994). OECD and World Bank publications argued that underdevelopment, poor governance, and inequality were sources of conflict and that conflict and insecurity were the primary causes of poverty, generating a so-called poverty trap or conflict trap in which conflict exacerbates poverty and poverty exacerbates conflict in a vicious circle (see, e.g. Collier et al. 2003; OECD 1997, 2001). Furthermore, donor governments and multilateral institutions increasingly framed underdevelopment in the global South as a security problem for the global North, especially after 9/11. According to this thinking, poverty, underdevelopment, and state fragility were driving international migration and international terrorism (Waddell 2006: 537–538). While development was understood as necessary for reducing conflict and countering terrorism, success in these endeavours was also presumed to require more systematic linkage between 'hard' and 'soft' measures (Macrae and Harmer 2004: 4).

In policy terms, these ideas about the security–development nexus had two principal implications. First, mainstream development actors began to engage extensively in conflict-affected, 'fragile', and 'failed' states, seeking not only to contribute to economic development and poverty reduction but also to the reduction of violent conflict. Crises were no longer seen as temporary shocks, so development actors could not simply wait for the 'normal' development path to be re-established. Instead, development actors sought to sustain effective engagement in difficult environments, promoting political transformation in 'poorly performing' countries (Macrae and Harmer 2004: 6). Humanitarian and development programming were to be undertaken simultaneously as per the contiguum idea. Second, donors advocated whole-of-government approaches with an emphasis on coherence across different government departments (in both donor and partner countries) and integration of different policy agendas—specifically the 3 Ds of defence, diplomacy, and development. In this context, development was generally understood as including a wide range of civilian actors, with those driving the security–development nexus agenda not always cognizant of mandate differences and so undermining important differences between humanitarian and development aid (Lie 2020: 3–4). Stabilization efforts emerged as an important part of this agenda. US-led missions in Afghanistan and Iraq are the best known, but stabilization discourses and practices also took root at the UN (e.g. with respect to UN peacekeeping missions in the Democratic Republic of the Congo and Haiti) and were employed by national governments as part of counter-insurgency campaigns, for example, in Colombia and Sri Lanka (Collinson et al. 2010: 5–7).[3]

The integration of security, development, and humanitarian programming was very much a donor-driven agenda, with varying levels of support, concern, or resistance coming from different development and humanitarian actors. While the main multilateral development actors were largely on board, some voices within the development sector were concerned that poverty reduction was being undermined as development was instrumentalized in the 'war on terror' (Waddell 2006: 532, 538). In humanitarian circles, there was considerable and widespread—if not entirely consistent or universal—concern about the integration of civilian and military forms of intervention and the implications of this for the ability of aid agencies to reach affected populations and of the vulnerable to access protection and assistance (Collinson et al. 2010). When it came to contributing to stabilization efforts in Afghanistan, for example, different humanitarian agencies had different stances, and in many cases, their stances shifted over time. Immediately after the overthrow of the Taliban, when many believed peace was within grasp

[3] See 'National militaries of affected states' and 'International military forces' in Chapter 20 on Armed Actors.

and donors were promoting a development and reconstruction agenda, several humanitarian agencies were unconcerned about working as part of an integrated approach, but as the intensity of conflict increased, many sought to distance themselves.[4]

Disaster risk reduction and 'building back better'

In parallel, new thinking and buzzwords emerged with regard to disasters. While many of the events that trigger disasters—including earthquakes, droughts, and floods—may be impossible to prevent, at least in the short and medium term, the extent of disaster is determined by vulnerability and by local and national capacity to respond.[5] Disaster risk reduction (DRR) and the commitment to 'build back better' (BBB) are primarily aimed at reducing vulnerability and strengthening capacity—although DRR has mostly been used with a technical focus, and the 'better' in BBB commitments is ambiguous and leaves space for widely-varying interpretations.

In 1987, the UNGA proclaimed the 1990s as the International Decade of DRR, and the Yokohama Strategy and Plan of Action for a Safer World: Guidelines for Natural Disaster Prevention was adopted in 1994 (IDNDR 1994; IFRC 2007: 55;). These initiatives were, as their names suggest, centrally concerned with preventing disasters. Thus, the Yokohama text, for example, argues that a response-only approach is inefficient and calls for prevention, mitigation, and preparedness to be prioritized instead (IDNDR 1994). The DRR concept gained increased prominence in the aftermath of the 2004 Indian Ocean tsunami and the 2005 Hyogo Framework for Action and with increasing recognition of the likelihood of greater numbers of disasters linked to climate change (Mosel and Levine 2014: 5). Building on the 1994 Yokohama Strategy, the Hyogo Framework set out DRR priorities for 2005–2015 and the Sendai Framework did the same for 2015–2030.

DRR does not solely (or even primarily) take place as part of disaster response, and the DRR agenda is not, for the most part, driven or implemented by humanitarian actors. While mostly not driven by development actors either, Yokohama, Hyogo, and Sendai all called on donor governments to provide funding for disaster prevention and on development actors to incorporate DRR inro their programming. In line with a developmental ethic, there is significant emphasis on strengthening national capacities. However,

[4] See 'Integration and coordination' and 'Instrumentalization of aid by military actors' in Chapter 6 on Afghanistan, 2001–2014.

[5] See 'Crises in the contemporary world' in Chapter 1 on Humanitarian Emergencies.

and despite the fact that the DRR agenda is underpinned by recognition of the fact that societal conditions are a major determinant of disasters, DRR has tended to focus on technical solutions, with less attention to the political, economic, and social factors underlying vulnerability (Mosel and Levine 2014: 5). The Yokohama, Hyogo, and Sendai initiatives have little or nothing to say about conflict-affected countries, but some donors began to talk about DRR or disaster resilience in conflict settings in around 2012, and some recent research highlights the high incidence of disasters in conflict-affected countries and seeks to understand how, and to what extent, DRR programming is implemented in such contexts (Mena and Hilhorst 2021; Mosel and Levine 2014: 5).

BBB was the mantra of the international responses to the 2004 Indian Ocean tsunami and the 2010 earthquake in Haiti. The basic rationale is that a disaster can provide an opportunity for development, particularly when it attracts significant external assistance. With funding and international expertise, so the logic goes, it should be possible to do more than simply save lives—and, indeed, more than simply re-establish the status quo ante. While the terminology might be relatively new, the basic idea is not—and nor is it necessarily limited to disasters. Nonetheless, the BBB agenda gained particular prominence after the 2004 Indian Ocean tsunami and again after the 2010 earthquake in Haiti, and the language of BBB is used in the discussion of disaster response in the Sendai Framework.

In practice, there are three broad BBB approaches. First, in some contexts international agencies repackage existing programming as BBB activities. This was the case in Myanmar after Cyclone Nargis, for example, where some agencies explicitly used the language of BBB to describe activities that would previously have been classified as sustainable livelihoods programming (Fan 2013: 16). Second, in what is perhaps the most common approach adopted by international actors under the rubric of BBB, technical rather than political fixes are favoured, and 'better' is interpreted narrowly to mean something more like 'safer'. This implies activities such as constructing earthquake-resistant housing or distributing drought-resistant seeds. In this sense, BBB—or BBS, as some have argued would be a better name for it (Kennedy et al. 2008)—essentially implies incorporating DRR into disaster response. Third, BBB sometimes goes beyond rehabilitation of better/safer assets and infrastructure to include more transformative structural changes in economic, social, or political relations. This was the case with the nationally led tsunami response in Aceh, for example, where the Rehabilitation and Reconstruction Agency of Aceh-Nias saw an opportunity to pursue such goals as poverty reduction, gender equality, and peace (Fan 2013: 16–17).

BBB raises a number of issues for humanitarian actors, beginning with the question of who is to decide what 'better' means. Even if its application is limited to assets and infrastructure, different actors have different priorities, and different priorities imply different responses. In relation to the shelter response in Aceh after the 2004 Indian Ocean tsunami, for example, concerns about safety, durability, sustainability, modernity, and aesthetics pulled in different directions in terms of the materials for house building (Kennedy et al. 2008: 27). Furthermore, when resources are limited and the money for one earthquake-resistant home could build ten or twenty homes that meet people's immediate need for a roof over their heads but could be death traps if or when another earthquake strikes, it is not immediately obvious which option is 'better' (Fan 2013: 2). Moreover, BBB is arguably most meaningful when it goes beyond infrastructure and asset replacement to address social and political factors and focus on transforming power relations in society (Fan 2013: 25).

These more expansive goals do not align well with humanitarian principles, timelines, and skill sets. While the pursuit of transformative goals may be comfortable for so-called new humanitarians, it is less so for classical humanitarians.[6] There are concerns about BBB distracting attention and resources from urgent life-saving humanitarian priorities in an emergency and exploiting people's vulnerability in the aftermath of a disaster to drive social change (Fan 2013: 2, 26–27). With their top-down approaches, short time horizons, and tendency to work around rather than with the state, humanitarian agencies are generally also poorly equipped to address wider social and political issues, which most agree require longer-term, participatory approaches with significant state involvement. It is probably no coincidence that in Aceh, where BBB efforts contributed significantly to societal changes, they were led by national, not international, actors. Less ambitious goals are, by definition, more achievable, and humanitarian skill sets may be more suited to a technocratic approach focused on assets and infrastructure. However, responses to the 2004 Indian Ocean tsunami and the 2010 earthquake in Haiti highlight the importance of land tenure and market conditions and the poor capacity of international humanitarian agencies to address such issues, suggesting that the ability of humanitarians to achieve even a more limited version of BBB is questionable.[7]

[6] See 'Desirability of apolitical action' in Chapter 13 on Politics, Principles, and Humanitarian Action.

[7] See 'Recovery and rehabilitation' in Chapter 7 on the 2004 Indian Ocean Tsunami and 'Recovery and rehabilitation' in Chapter 9 on the Haiti Earthquake, 2010.

Resilience

The concept of resilience brings together thinking about disasters and DRR, thinking about protracted conflicts and displacement situations, and recognition of the effects of climate change on lives and livelihoods. As early as the 1990s, when the trend was for LRRD, resilience was deemed key, but resilience discourses and programming gained renewed prominence in the 2000s (Hilhorst 2018: 5; Ross et al. 1994: 3, 8). The sub-title of the 2005 Hyogo Framework, for example, is 'Building the Resilience of Nations and Communities to Disasters'. Resilience approaches have also been promoted in displacement crises, and building resilient systems is emphasized as a way to save lives continually throughout protracted conflict (Bailey and Barbelet 2014; ICRC 2016).

Resilience programming moves away from treating crises as exceptions, instead accepting them as normal and seeking to create conditions in which individuals, families, communities, cities, systems, and states can withstand or adapt to shocks and disaster (Hilhorst 2018: 6; Jaspars 2021; Slim and Bradley 2013: 7). It aims to bring together humanitarian and development actors, in pursuit of a single ethical goal, and in this sense is closely related to LRRD thinking (Mosel and Levine 2014: 4; Slim and Bradley 2013: 7). Whereas LRRD focuses on the links between different kinds of international aid, however, resilience discourses shift the focus to crisis-affected people, ascribing them agency to govern and enable their own survival (Hilhorst 2018: 6; IRRC 2019: 1054–1055; Mosel and Levine 2014: 5). In addition, the concept of resilience is seen as a way to bring climate change adaptation into mainstream development practice (Levine et al. 2012b: 1).

The rationales underlying resilience thinking involve questionable assumptions about the cost-effectiveness of resilience programming and the relationship between resilience and migration. Investment in building resilience, which takes into account people's ability to respond, adapt, and bounce back, is widely assumed to be more effective and cost-efficient than the direct provision of humanitarian relief (Fan 2013: 25; Hilhorst 2018: 5; ICRC 2016: 28). However, credible evidence of the impact of resilience programming on future humanitarian needs and costs is yet to be produced (Fan 2013: 25; Mosel and Levine 2014: 5). Resilience is also promoted by some actors, including the European Commission, as a means to reduce people's propensity to migrate (Jaspars 2021: 205). However, the logic underpinning this rationale may also be flawed in that the factors shaping migration decisions are multiple and complex, and it is not clear that people migrate to Europe due to lack of resilience—indeed, migration can itself be a way of adapting to or withstanding crisis (Jaspars 2021: 205).

While resilience has become something of a buzzword, it can mean a range of different things in practice, depending on whether it is primarily about withstanding shocks or adapting to them and whose resilience is being built. Resilience programming can be directed at people in crises as well as those vulnerable to crises, and there is a significant difference between building resilience *to* crisis (i.e. to withstand shocks and disasters) and building resilience *within* the context of crisis (i.e. to adapt to shocks and disasters). DRR approaches are primarily concerned with the former, but some humanitarian actors emphasize that they are focused on the latter. For example, the ICRC, with its focus on conflict and violence, would not seek to prevent displacement where that is a survival strategy but rather would seek to support displaced people to become resilient in their new situation (IRRC 2019: 1056). Resilience programming can also be directed at different actors or units of analysis, including individuals, communities, and systems—and the resilience of some actors may be in tension with the resilience of others. For example, a resilient community may obscure the vulnerability of individuals within it, and a resilient system is not necessarily a progressive system. A social system that is hierarchical, unequal, and exploitative can be highly resilient, and its resilience may thus increase rather than reduce suffering over time (Pain and Levine 2012: 8).

Critiques of resilience fall into two broad categories. First, there are those that question the adequacy of the conceptualization, theory, and evidence underpinning resilience frameworks (Levine et al. 2012b: 2). For example, the term 'resilience' is often used descriptively rather than analytically, and neither the drivers of resilience nor the individual-level outcomes are well theorized or understood (Pain and Levine 2012: 6–7). A second set of critiques highlights the ways in which resilience discourse and programming can serve to obscure the causes of crisis, shift responsibility to victims, and justify a lack of assistance for them. Resilience programming, much like classical humanitarianism,[8] does not seek to address the causes of crisis (Jaspars 2021: 201, 204). Unlike classical approaches, however, resilience thinking treats crises as normal. On the one hand, this reflects recognition of the empirical fact that many crises are protracted or cyclical. On the other hand, it sends a message that people must accept and

[8] On classical humanitarianism, see 'Desirability of apolitical action' in Chapter 13 on Politics, Principles, and Humanitarian Action.

adapt to crises and excuses national and international inaction to address their causes. Resilience approaches thus transfer responsibility for recovery to crisis-affected populations, overshadowing the responsibility of those who are creating humanitarian needs in the first place and the responsibility of the state to protect the population (Hilhorst 2018: 6; IRRC 2019: 1055; Jaspars 2021: 204–205). Resilience approaches may also reduce assistance to vulnerable populations. This can occur due to a shift of financial resources away from emergency response to resilience programming or because transferring responsibility to crisis-affected populations serves to justify not providing them with assistance (IRRC 2019: 1055; Jaspars 2021: 203–204, 206–207). A decision not to provide direct assistance can be framed in terms of avoiding aid dependency and ensuring the sustainability of humanitarian response (ICRC 2016: 24; Jaspars 2021: 203–204). Resilience-orientated approaches also tend to imply deep work on systems, which may jeopardize humanitarian principles, especially in armed conflict contexts, where the strengthening of political and economic structures may advantage one side against another (IRRC 2019: 1055; Slim and Bradley 2013: 7).

The New Way of Working and the Comprehensive Refugee Response Framework

Following the unanimous adoption of the Sustainable Development Goals (SDGs) by UN member states in 2015, the UN Secretary-General launched two initiatives which, among other things, sought to bring together different types of international aid to create more effective, efficient, and sustainable responses to humanitarian crises: the NWoW and the CRRF. The 2030 Agenda for Sustainable Development included a commitment to 'leave no one behind' and made specific references to humanitarian emergencies in what has been described as a common results framework in which humanitarian and development actors can work together (ICVA 2017a: 2). Less than a year later, at the WHS in Istanbul in May 2016, a range of different actors committed to the NWoW, which is grounded in thinking about the humanitarian–development–peace nexus. The NWoW and the triple nexus are part of an agenda that has been largely driven by the UN secretariat but with wide buy-in from both development and humanitarian actors. Then, at the UN High-Level Meeting on Addressing Large Movements of Refugees and Migrants in New York in September 2016, member states adopted the CRRF, which then became an integral part of the 2018 Global Compact on Refugees. The CRRF is intended to serve as a template for addressing the humanitarian–development nexus, which can be applied in concrete situations of large-scale movements of refugees.

The notion of the triple nexus has been embraced by a range of actors in recent years. On the humanitarian side, some principled agencies have cautiously adopted nexus thinking and appear broadly supportive of the SDGs. For example, the ICRC is engaged in long-term, 'deeper' interventions focused on structures and systems, concerned with sustainable impact, increasingly rejecting a clear distinction between relief and development programming and financing, and open to collaboration with peace and development actors (Carbonnier 2018b; ICRC 2016: 4, 20; IRRC 2019: 1060; Policinski and Kuzmanovic 2019: 970). At the same time, mainstream development actors are increasingly active in conflict and fragile settings, discussing the nexus concept, and committing to strengthening coherence between humanitarian, development, and peace efforts. UNDP is taking a lead in advancing the NWoW, and both the OECD and the World Bank are engaging explicitly with the NWoW and the triple nexus (ICVA 2017a: 4; OECD 2022; World Bank n.d.). Notwithstanding some complaints from NGOs that they have been excluded from many of the fora in which the NWoW is being discussed, it has brought a bigger range of actors to the table than previous efforts to achieve coherence across different kinds of international aid (ICVA 2017a: 7; IRRC 2019: 1054; Zetter 2021: 1769).

The NWoW also includes greater specification at the level of implementation, as compared with previous efforts to reduce tensions and increase synergies between humanitarian, development, and peace-building fields of practice. Implementation is to be structured around collective outcomes, where a collective outcome is a quantifiable and measurable result that development, humanitarian, and other relevant actors collectively agree they want to achieve over a three–five-year time frame, and to which each should contribute specific interventions, according to their comparative advantage (ICVA 2017a: 1; UN OCHA 2017: 7–8). Collective outcomes should reduce needs, risk, and vulnerability and increase resilience, and financing is linked to collective outcomes rather than specific interventions (ICVA 2017a: 1). Strong national and local ownership of collective outcomes is seem as central to sustainably reducing need, risk, and vulnerability (UN OCHA 2017: 7). At the same time, however, the NWoW underscores the need for humanitarian actors to adhere to the principles of humanitarian action and recognizes that, in some contexts, the pursuit of development goals may not be feasible and national priorities may be in tension with impartial humanitarian action (ICVA 2017a: 1).

The CRRF likewise builds on previous efforts to addressing the humanitarian–development nexus, incorporating a wider variety of actors and specifying a more concrete map or template for implementation. Each application of the CRRF in practice is intended to include multiple stakeholders—including the private-sector, faith-based groups and development and humanitarian actors—and may devolve significant power to local authorities and seek to promote refugee self-reliance and refugee participation. It aims for coherence between short-term emergency assistance and sustainable, resilience-building development that focuses on self-reliance for refugees and their host communities (Zetter 2021: 1769, 1771). The goals of the CRRF and the 2018 Global Compact on Refugees include taking pressure off host states and addressing refugee livelihoods needs in sustainable ways (Sharpe 2018: 707). Just as preventing migration is often a goal of resilience programming, so the humanitarian–development nexus in general, and the CRRF in particular, additionally have the goal of containing refugees in their regions of origin, even if discussion of these initiatives does not always state this so frankly (Zetter 2021: 1769).

It remains to be seen whether the nexus approach, as embodied in the NWoW and the CRRF, will achieve significantly more than its predecessors. Certainly, the CRRF is one of the most advanced statements of the humanitarian–development nexus with respect to displacement, and the operationalization of the humanitarian–development–peace nexus as part of the NWoW also provides a much more concrete framework for coordination and collaboration between different aid actors than previous efforts to transcend the divide(s) between them. However, insofar as humanitarian actors are committed to principled humanitarian action, the scope of their involvement in nexus programming is necessarily limited, especially in conflict contexts.

Even in displacement contexts, in practice, the understanding of development at the heart of the nexus approach is very narrow, focused largely on employment-generation strategies, requiring host states to open their labour markets to refugees, development actors to help with minimal infrastructure provision, and the private sector to create jobs and engage with refugees as consumers and producers (Zetter 2021: 1775). Cash transfer programming, which has been widely embraced by humanitarian actors, accounts for a small but growing proportion of humanitarian response[9] and is often seen as central to development-led response, turning recipients into consumers, and potentially also producers, through microfinance (Zetter 2021: 1772).

[9] See 'Cash and voucher assistance' in Chapter 22 on Material Assistance and Direct Service Provision.

This is development more as technocratic fixes or focused on the individual than as structural change.

The triple nexus approach in the NWoW is also limited both because the collective outcomes around which it is structured must be measurable and quantifiable and because many of the more transformative peace- and development-related goals would be in tension with humanitarian principles and approaches. Progress in achieving many of the most important humanitarian goals may not be easily quantifiable. Standard-setting regulation mechanisms for humanitarian action have been criticized for being overly technical and focused on quantitative measures in such a way that they risk undermining the humanitarian ethic, often coming at the expense of addressing more difficult ethical dilemmas and easily sidelining non-quantifiable aspects of humanitarian response such as solidarity and witnessing (Terry 2000). A similar criticism can be levelled at the collective outcomes approach of the NWoW; insisting on quantifiable outcomes not only limits the kind of goals that can be pursued under this approach but also risks diverting resources and attention from goals that are equally or more important but not (easily) quantifiable. Furthermore, the NWoW is explicit that it must be context-specific and always respect humanitarian principles, yet it is not clear how far principled humanitarian action can contribute to preventing and reducing, rather than simply meeting, need.

Finally, many of the risks and critiques identified with respect to previous attempts to integrate humanitarian, development, and conflict-prevention or reduction activities are equally applicable to these latest initiatives. For example, several actors have raised concerns that a focus on development and peace-building outcomes might reduce visibility for rights- and needs-based humanitarian priorities, with resources directed to long-term goals at the expense of meeting immediate needs (ICVA 2017a: 6; IRRC 2019: 1058). There is also a risk that the coordinated pursuit of collective outcomes will compromise impartial, neutral, and independent humanitarian action. Development is usually coordinated with national governments, and in armed conflict contexts those governments are often parties to conflict, such that working to support the government threatens humanitarian neutrality.[10] If humanitarian actors are associated with the government, this can also undermine their ability to engage with opposition groups, which can, in turn, undermine their access to territory and their ability to provide assistance impartially.

[10] On the different ways humanitarian agencies can, and do, engage with the governments, see 'International aid agencies and the governments of affected states' in Chapter 19 on Government and Civil Society in Affected States.

Conclusions

The fact that attempts to address the interconnectedness of humanitarian-, peace- and development-related needs and responses have gone through so many iterations suggests that success in such endeavours is elusive and the challenges persistent (Fan 2013: 1; Mosel and Levine 2014: 4). While the most recent iterations—in particular, the NWoW and the CRRF—have achieved wider buy-in and been more thoroughly institutionalized than many prior attempts, it remains to be seen how far they achieve their goals. Insofar as humanitarian actors retain their commitments to impartiality, neutrality, and independence, it is difficult to see how they can make a significant contribution to the kinds of structural change usually required for development and peace-building. On paper, these initiatives seek to reconcile principled humanitarian action with the pursuit of transformative goals, but in practice, it may be that the different actors involved will only be able to agree on limited development goals and technocratic means to achieve them.

III

AGENTS AND ACTORS

Introduction to Part III

Humanitarian Engagement with Other Actors

We cannot fully understand the politics and practice of international humanitarianism without taking into account a wide variety of actors. This third part of the book thus looks beyond the operational international agencies that are the main focus of the book as a whole and turns its attention to other actors with important impacts on humanitarian needs, responses, and outcomes: the media and celebrities (Chapter 17); donors (Chapter 18); governments and civil society in affected states (Chapter 19); and armed actors, including national and international militaries as well as non-state armed groups (Chapter 20). Each chapter examines the roles of these different types of actors in shaping humanitarian outcomes, their impact on international humanitarian response, and the ways in which international humanitarian agencies engage with them.

International media, celebrities, and donors can all influence the scale and focus of humanitarian response. Media and celebrities have an agenda-setting role, through which they help determine which humanitarian emergencies feature on the global agenda, and an agenda-framing role through which they help shape how those emergencies are understood and responded to. Their attention is consequential, but they do not always focus on the most severe emergencies, and they have a tendency to oversimplify and depoliticize those that they do focus on. A small number of donor governments provide a large majority of the funding for international humanitarian action and, hence, also play an important role in shaping the humanitarian agenda. While the level of need is a major factor in determining how they allocate their funding, it is often not the only factor. Through these different roles, the media, celebrities, and donors can have an impact on the size and nature of humanitarian responses around the world.

The governments of affected states and a range of armed and civilian actors also have important impacts on humanitarian response and outcomes. While a humanitarian emergency may mean that the state needs external assistance, it does not necessarily mean that the state cannot play a central role in the response. Civil society in affected states also plays a significant role in responding to humanitarian emergencies, yet—much like the state itself—is often sidelined by international responders. The localization agenda has emerged with the goal of transferring power and resources from international actors to national and local actors, but progress in implementation has so far been mixed. The conduct of armed actors—both state or non-state—is probably the most significant factor determining the ability of humanitarian agencies to reach those in need of protection and assistance. Armed actors can also have a direct effect on humanitarian outcomes—either negative, by increasing the insecurity of the population, or positive by providing relief assistance or protecting the population from threats posed by others.

The decisions and behaviour of these different actors, and hence their impact on humanitarian response and outcomes, are often treated as if they were determined independently of operational humanitarian agencies, but the chapters in this part of the book show that influence runs in both directions. For example, international media and celebrities are very often dependent on operational humanitarian agencies for information about what stories to cover and for logistical support with site visits. Sometimes, humanitarian agencies also produce news coverage directly and launch their own publicity campaigns, which can trigger further media coverage and celebrity engagement. Likewise, while donors have significant influence on operational humanitarian agencies, this relationship is not completely unidirectional. By lobbying donors directly or launching global campaigns, the agencies can have an impact on donor funding decisions.

Governments and armed actors in affected states may control access to territory, but different international humanitarian agencies gain different levels of access, presence, and acceptance as a result of different methods, reputations, and attitudes. Operational space should be understood not as something that is exogenously determined but rather as something that must be proactively created by humanitarian agencies. Humanitarian actors must engage not only with the civilian agencies of affected states but also—in armed conflict contexts at least—with a range of armed actors in order to get acceptance for their work and to negotiate access. Beyond gaining access and acceptance, humanitarian actors can work to persuade governments and armed actors to comply with humanitarian and human rights norms more broadly. Ignoring any of these actors, or trying to work around them, is generally not conducive to achieving the best possible humanitarian outcomes. Instead of asking whether or not they should engage with the different actors who have a bearing on humanitarian response and outcomes, international humanitarian agencies need to ask how they can best engage with them.

The ways that humanitarian agencies engage with governments, local and national civil society, and armed actors vary depending on the agency in question and on the context in which they are operating. Different ways of relating to the state correspond more to humanitarian, development, or human rights modus operandi, and the attitudes of humanitarian agencies to other actors is linked to where they position themselves in the boundary debates.[1] Those humanitarian actors who embrace a more expansive humanitarianism, with space for the pursuit of transformative goals and less concern for independence and neutrality, are sometimes more open to working closely with political and military actors, although this depends on the specifics of the context and the political and military actors involved. Taking the essence of the localization agenda seriously means not only channelling resources directly to local actors but also shifting power—including the power to define priorities and goals—from the international to the local level. Some analysts continue to criticize expanded goals and moves away from independence and neutrality, and many more continue to see a role for limited and strictly principled humanitarian response. However, there is a growing consensus that there is space—and a need—for multiple approaches to assisting people in humanitarian crises and that the most appropriate combination of approaches in any given contexts depends on the nature of the crisis and the needs arising from it.

[1] See the Introduction to Part II on 'The Boundaries of Humanitarianism'.

17
Media and Celebrities

The media and celebrities play significant roles in setting and framing the humanitarian agenda. Without their spotlighting of any given humanitarian emergency, the global public would mostly know little or nothing about it. Furthermore, the story they tell about any given emergency plays a part in how the wider public, and sometimes also politicians and policymakers, understand and respond to it. The way that media and celebrities tend to frame emergencies through simple, often depoliticizing, narratives that obscure complexity serves to inspire compassion, but those same frames call for a response grounded in charity and consumption rather than solidarity or justice.

Television-centred media coverage has facilitated celebrity humanitarian engagement and, more generally, the massive expansion of the humanitarian sector (Müller 2013a: 61–62). Between November 1984 and September 1986, celebrity-led Band Aid (and subsequently Live Aid) not only raised an enormous USD 110 million but also pressured Western governments to donate more and established itself as an independent operational charity to provide relief to victims of famine in Ethiopia.[1] Since Band Aid, there has been continuous high-profile celebrity engagement with humanitarian emergencies and responses (Müller 2013b: 470).

Celebrities performs site visits in the global South, and act as witnesses, fundraisers, activists, and diplomats in the global North (Richey and Budabin 2016: 2). The roles different celebrities play are often gendered. For example, Angelina Jolie and Madonna, like Audrey Hepburn, Mia Farrow, and others before them, are known for their engagement with children and women in crisis contexts and emphasize empathy, emotions, and their experience as mothers (Kapoor 2013: 17; Repo and Yrjölä 2011: 45, 49; Richey and Budabin 2016: 13). By contrast, Bono, Bob Geldof, George Clooney, and Ben Affleck are business-like in their engagement and operate at the centre of the world political stage (Kapoor 2013: 17; Repo and Yrjölä 2011: 45, 49; Richey and Budabin 2016: 14–15). All aim to mobilize compassion and donations or

[1] See 'Celebrity humanitarianism' in Chapter 3 on Drought and Famine in Ethiopia, 1983–1985.

The Politics and Everyday Practice of International Humanitarianism. Miriam Bradley, Oxford University Press.
 DOI: 10.1093/oso/9780198840695.003.0017

purchases to fund humanitarian response, and those operating on the world political stage often also advocate particular policy responses.

Setting the agenda

Media coverage and celebrity engagement can have a significant impact on the level of public attention and funds a cause attracts. The general public in distant countries will usually not know about a humanitarian emergency unless they are told about it by some intermediary, whether on traditional or social media. Celebrities can amplify the message about an emergency not only by increasing the volume of coverage but also by reaching segments of the public who may not be avid consumers of traditional news media. While they play an important role in raising awareness of particular emergencies, celebrities are not elected—this is 'undemocratic agenda setting' (Richey and Budabin 2016: 2). Furthermore, in focusing on particular causes, both the media and celebrities may inadvertently detract attention from other, equally important and urgent causes. To the extent that they are influential in setting the humanitarian agenda, then, it is important to understand what influences which emergencies they cover and which causes they promote.

Media coverage can be important in generating interest among public audiences in the global North, encouraging donations from individuals and driving responses from their governments. The so-called CNN effect refers to the idea that the media, especially television news, have decisive influence on political leaders and the foreign policy agendas of Western states (Robinson 1999). Real-time news coverage of distant suffering is expected to stimulate public demands to 'do something' which, in turn, places pressure on policymakers to offer some response to that suffering (Scott et al. 2022: 170). Dramatic media coverage from countries affected by the 2004 Indian Ocean tsunami was a factor driving massive Western public interest and unprecedented donations from individuals, with public concern additionally pushing donor governments to provide large-scale funding.[2] Celebrities often visit the sites and victims of humanitarian emergencies, providing exposure for their chosen causes—exposure which can also help to generate public awareness, concern, sympathy, and donations from the mass public in the global North (Richey and Budabin 2016: 3). They can thus be expected to reinforce or intensify any CNN effect, and both media coverage and celebrity engagement

[2] See 'Media coverage' and 'Funding for humanitarian response' in Chapter 7 on the 2004 Indian Ocean Tsunami.

were seen as instrumental in motivating donations for a large-scale response to famine in Ethiopia in the 1980s, for example.[3]

Western media coverage is not balanced or proportionate to the intensity of crises but may instead be driven by the level of economic or geopolitical interest in the global North, as well as geographic or cultural and social proximity (Franks 2006b: 281). Indeed, while the media might drive official policy on some issues, policymakers set the media agenda on others (Robinson 2000). In Afghanistan, for instance, media coverage and humanitarian funding followed the 2001 invasion by the United States and its allies, rather than the other way around. Official policies may also help to determine where celebrities focus their energy. As Alex de Waal has suggested, it may be no coincidence that so many celebrities converged on the Darfur issue, one of the few areas of bipartisan consensus in Washington, even as other crises were killing and displacing comparable numbers (de Waal 2008: 46–47).

Where governments lack a clear policy, news value may drive media coverage, and humanitarian emergencies that can be presented as clear cut and unambiguous stories have greater appeal for news reporters and audiences (Franks 2006b: 283). Celebrities amplify this tendency as they garner more positive coverage of themselves in transnational media, where their activism is relatively uncontroversial (Duvall 2015: 591). Sudden-onset disasters, such as earthquakes or tsunamis, have a double appeal in that they can be explained in a simple narrative, and they can often also be accompanied by dramatic visual imagery. Conflicts and famines with exemplary victims—the 'Biafran babies'[4] or starving Ethiopian children—and the possibility of presenting the story as one for which no one is to blame, or in which there are clear heroes and villains, may also appeal to media and celebrities. That the conflict in Darfur could appear, at first glance, to contain a clear moral narrative of (evil) perpetrators, (silent) victims, and (white) saviours undoubtedly contributed to its popularity as a cause for celebrities (de Waal 2008: 46–47).

Humanitarian agencies themselves can also influence which emergencies the media covers and celebrities engage with. There is something of a symbiotic relationship between aid agencies, the media, and celebrities, and it is not always clear how the agenda is set or who is driving it. Aid agencies need media and celebrities to bring attention to humanitarian emergencies and, indeed, to their own work in responding to emergencies (Franks 2010: 75–78). Celebrities benefit enormously from media coverage in general and

[3] See 'Media coverage' and 'Celebrity humanitarianism' in Chapter 3 on Drought and Famine in Ethiopia, 1983–1985.
[4] See 'Global public response' in Chapter 2 on the Nigerian Civil War, 1967–1970.

from coverage connecting them to humanitarian causes and agencies in particular. The media and celebrities very often rely on operational humanitarian agencies for information about what stories to cover and for practical assistance in covering those stories (Franks 2010: 79; Powers 2018: 46–47). In some contexts, humanitarian agencies produce news coverage directly and launch their own publicity campaigns, sometimes stimulating further coverage, as happened, for example, with the United Nation's 'four famines' campaign in 2017 (Scott et al. 2022: 176; see also Powers 2018: 48–50). All this means that United Nations (UN) aid agencies and non-governmental organizations (NGOs) play a role in determining which crises are covered in the media and which causes are taken up by celebrities.

Framing the agenda

The media and celebrities have significant power in determining not only which issues are brought to the attention of the global public but also how those issues are represented. More often than not, media and celebrity depictions simplify the complex reality of the humanitarian emergencies they are representing, albeit through different constructions.[5] These depictions can be effective in generating public concern and compassion, but they can also be depoliticizing and serve to perpetuate racial stereotypes and hierarchies.

Some emergencies are decontextualized by the media and celebrities in the global North, with humanitarian needs presented as the inevitable consequence of events for which no one is to blame. This is particularly true for emergencies linked to geophysical or climatological hazards, as we have seen in the case of drought in the Horn of Africa, which contributed to famines in Ethiopia in the 1980s, for example.[6] Very often, celebrity humanitarianism and media-inspired humanitarian agency fundraising campaigns rely on a simple narrative of 'natural' disaster (Müller 2013a: 62). A simple, straightforward narrative is preferred because it is more likely to engage audiences and attract funding (Franks 2010: 81). There appears to have been some change over time, with increasing—but still limited—nuance. Whereas in the early 1980s, for example, the media and Band/Live Aid celebrities depicted famine in Ethiopia as a natural disaster, some—but by no means all—of the media

[5] For more on this, and the role operational humanitarian agencies play in constructing humanitarian emergencies this way, see 'Constructing humanitarian emergencies' in Chapter 1 on Humanitarian Emergencies.

[6] See 'Media coverage' and 'Celebrity humanitarianism' in Chapter 3 on Drought and Famine in Ethiopia, 1983–1985.

coverage and celebrity responses to famine in Somalia in 2011 were slightly more nuanced (Müller 2013b: 476–478).

By contrast, representations of emergencies centred on armed conflict and violence tend to include context—but often a simplified misrepresentation of context, following one of two templates. In the first, the violence is presented as irrational and inexplicable. Contemporaneous Western media coverage of the 1994 Rwandan genocide exemplifies this kind of representation. Reflecting and reinforcing the perspectives of policymakers in the global North, mainstream media largely avoided calling the violence genocide and portrayed it as spontaneous and uncontrollable, the inevitable consequence of ancient tribal hatreds (Chari 2010; Melvern 2001).[7] In the second, perpetrators and victims are clearly distinguished in a black-and-white account of goodies and baddies (Dieter and Kumar 2008: 260). In this way, the highly complex conflict in Darfur has often been depicted as one in which the forces of evil (the Sudanese government and the 'Arab' militias it mobilized) were pitched against innocent, 'African' victims, lacking in agency, obscuring all nuance (Lanz 2009: 670; Pantuliano and O'Callaghan 2006: 1).

All these constructions are depoliticizing (but not, by any means, apolitical) and tend to depict helpless (often black, or at least non-white) victims and the need for/importance of (usually white) foreign saviours (Balaji 2011: 59; Chandler 2001: 690–691; Cloud 2014). On the Darfur conflict, for example, the dominant narrative suggests 'heroes [. . .] are Western aid workers, human rights activists, and the advocates themselves who courageously save lives and speak out against the atrocities' (Lanz 2009: 670). International humanitarian agencies are not only a factor in determining which emergencies the media cover but also have significant influence on how the media explain those emergencies (Franks 2010: 75). They have an obvious incentive to promote their own roles, so the centrality of the 'rescuer' is unsurprising, but the consequences of such depictions go beyond raising the funds that sustain humanitarian agencies and their work. These representations generate a racialized politics of pity, which may be instrumental in driving public interest and donations but, at the same time, separates and distinguishes those who suffer and are pitied from those who are able to pity and to help, either as donating audiences in the global North or as expatriate aid workers in the global South (Balaji 2011: 57, 58; Chouliaraki 2012: 2; Hutchison 2014: 3, 8).

Representations of Africa have been particularly problematic, with both mainstream media and celebrities mostly reducing the continent to a series

[7] See 'International inaction' in Chapter 5 on Genocide in Rwanda and Its Aftermath, 1994–1996.

of stereotypes. During the colonial period and the Cold War, Africa mattered to Western powers, and media coverage was more comprehensive than now, with serious, engaged reporting often by full-time, locally based African correspondents (Franks 2010: 72). Today, by contrast, African stories tend to be either ignored or misreported, with a focus on disasters and little engagement with the underlying politics (Franks 2010: 74–75). Celebrities tend to represent Africa as a place of perpetual suffering and with a particular place in the world system, dependent on charity (Kapoor 2013: 39; Müller 2013b: 471). Celebrity 'rescue fantasies' are not much different from the colonial 'white man's burden', and the ubiquity of images of suffering women or children (the 'Biafran babies' and the 'starving children of Ethiopia' being the archetypes) reflects and reinforces paternalistic–colonial relationships through which the African subject is treated as a victim or a child in need of Western help (Kapoor 2013: 39, 42).

More generally, celebrity 'representation' raises the question of whether or not their advocacy serves to reduce the voice and agency of the people suffering in the emergency in question. On the one hand, in their function as witnesses to crises in the global South, celebrities seek to represent the plight of 'others', who lack such easy access to international media themselves (Richey and Budabin 2016: 12). Very often, they present themselves as giving voice to the voiceless. To make their accounts resonate with their audiences, however, celebrities tend to rely on their own personal testimony, emphasizing their emotional reaction to the scenes they have witnessed and the conversations they have had rather than the experiences of those suffering themselves (Chouliaraki 2012: 6; Cloud 2014: 48–49). Because celebrity humanitarian constructs the global South, and Africa in particular, as passive, voiceless, and invisible, showing Africans as lacking in knowledge and agency, celebrities can ventriloquize and paternalize them (Kapoor 2013: 42). In speaking on behalf of the victims of humanitarian emergencies, celebrities both bring attention to their plight and, at the same time, reinforce the idea that they cannot speak for themselves.

Shaping (policy) responses

Just as their role in setting the humanitarian agenda helps to determine which humanitarian emergencies attract public attention and, to some degree, which attract large-scale funding and responses, the role of the media and celebrities in framing the agenda has implications for the kinds of response deemed appropriate.[8] Much as with the broader process of humanitarianization outlined in the introductory chapter on humanitarian emergencies, simplified narratives inspire responses focused on addressing symptoms rather than causes. At the same time, certain characteristics of the media, and especially of celebrity, inspire or demand individual, privatized responses centred on consumption that ultimately serve to reinforce rather than challenge the global economic and political order that creates and maintains the underlying conditions of humanitarian emergencies. In some cases, celebrities additionally advocate for specific political actions, and the influence of their advocacy can be out of all proportion to their expertise on the issues about which they are advocating.

Celebrity engagement in particular tends to promote individual initiative and philanthropy as a solution to structural problems, often calling on audiences to donate to international NGOs or UN humanitarian agencies or to buy products linked to the cause. With a small number of exceptions, North–South celebrity humanitarianism conforms to a neoliberal world view that prioritizes privatized, individual responses (Duvall 2015: 586). Furthermore, the idea of buying stuff to save lives makes addressing social concerns a lifestyle choice rather than an obligation. Audiences in the global North are offered the opportunity to buy a clean conscience through consumption (de Waal 2008: 47–48). As Ilan Kapoor has put it, watching 'charitainment' events such as Live 8 give the viewer not only enjoyment but also 'the illusion that something humanitarian is being done [. . .] Celebrities save us from having to change our world and allow us to be vicarious humanitarians without actually confronting global problems' (Kapoor 2013: 43). Through glamour and spectacle, the celebrity functions to absorb the shock, impact, and trauma of mass suffering in a humanitarian emergency and to offer a commodified conduit for charitable giving (Cloud 2014: 46).

The response that individual donations and purchases are expected to fund is thus one focused on alleviating suffering rather than challenging the global economic and political order. The representation of an emergency in terms of a simple narrative in which no one is culpable implies that little can be done to address its causes and, hence, favours an assistance model focused on alleviating symptoms. This is in the nature of the charitable ethos that characterizes humanitarianism more broadly, and some of the

[8] For consideration of the role that the media, celebrities, and other factors play in determining the level of funding for humanitarian emergencies, see 'In accordance with need?' and 'Other factors shaping funding decisions' in Chapter 18 on Donors and the Funding of Humanitarian Action.

more critical scholars of contemporary humanitarianism see it as an essential part of global governance arrangements aimed at securing a post-Cold War world order based on capitalism and liberalism (Duffield 2001, 2007). However, the framings favoured by media and celebrities that serve to generate compassion towards certain types of innocent victims may be particularly important in this regard, cementing the 'pathos of assistance' rather than promoting recognition of rights or addressing the conditions that give rise to humanitarian crises (Müller 2013b: 471). Furthermore, celebrity humanitarianism is arguably especially implicated in promoting and legitimizing neo-liberalism and capitalism, since celebrities both depend on and promote consumerism and mostly fail to recognize (or admit) that capitalism is driving the inequality they seek to address through humanitarian response (Kapoor 2013: 32).

In some cases, celebrities also explicitly lobby for particular policy responses—from humanitarian agencies Western states, and the political organs of the UN. While their fame and celebrity status does not derive from expertise on humanitarian crises or response, their elite status frequently sees them treated as experts (Richey and Budabin 2016: 5). On the one hand, the celebrity has 'become a "professional humanitarian" who explains to "us" what to demand from "our" respective governments' (Müller 2013b: 472). On the other hand, celebrities frequently have direct access to political elites. Bob Geldof and Bono, for example, have each secured one-to-one meetings with a rolodex of world leaders, among them Barack Obama, George W. Bush, Gordon Brown, Tony Blair, Stephen Harper, Gerhard Schröder, and Jacques Chirac (Dieter and Kumar 2008: 260; Kapoor 2013: 13–14). In several cases, 'celebrity humanitarians' have been called to high-level political meetings, with George Clooney speaking to the UN Security Council about Darfur and Ben Affleck addressing the US Congress on the Democratic Republic of Congo (Richey and Budabin 2016). Their policy influence seems vastly disproportionate to their level of expertise on the issues they are called to discuss, and the resultant policies can be damaging. In 2006, for example, when Clooney spoke at vast rallies calling for UN troops to 'save Darfur' and additionally addressed the Security Council, he made bombastic predictions—which were way off the mark—about what would happen without an immediate UN replacement for the African peacekeeping force which was about to end its mission in Darfur (de Waal 2008: 44–45). His advocacy served to pressurize the US government into an over-hasty effort to impose UN peacekeepers on the government of Sudan and arguably undermined slow-maturing efforts to negotiate an end to the war (de Waal 2008: 45).

While the media and celebrities wield a great deal of power in setting and framing the humanitarian agenda, and in (implicitly or explicitly) prescribing responses, they are unelected and largely unaccountable. With respect to traditional media, journalists have codes of professional ethics that relate specifically to how they report the news. Celebrities are neither professional humanitarians nor professional reporters, and yet they have significant power to shape dominant understanding of particular emergencies and even to influence high-level political and military responses. Their fame comes from acting or music, they are not democratically elected to public office, and they have no obvious mandate to play such an influential role in global politics (Dieter and Kumar 2008: 262). Indeed, the entrance of celebrity humanitarians into the elite world of politics depends on the goodwill and invitation of elite allies rather than the will of the wider public (Richey and Budabin 2016: 5). This means that policy is being driven by individuals who are neither experts nor elected officials.

Conclusions

Media coverage of humanitarian emergencies and celebrity engagement can be consequential for humanitarian funding and response. However, neither kind of actor pays equal attention to all emergencies and nor does their attention correspond to the level of need. To the extent that they influence funding and response decisions, therefore, they pose a problem for impartiality. Moreover, while the simple narratives favoured by media and celebrities in the global North may be especially effective in mobilizing empathy and donations, they are also problematic. Simplified representations of victims and saviours generate a racialized politics of pity, and Africa in particular is constructed as a place of suffering, in need of charity and guidance from the global North. These representations and the associated politics of pity call for responses primarily based on compassion and consumption that aim to reduce the suffering caused by humanitarian crises without trying to understand or address the causes of those crises. Where celebrities call for more muscular policy responses, their lack of expertise suggests that these will often be misguided: as Alex de Waal has put it with respect to Darfur, 'the moral hyperventilation of celebrities hasn't helped and probably has hindered' (de Waal 2008: 55).

18

Donors and the Funding of Humanitarian Action

Operational humanitarian agencies depend on voluntary donations to pay for their operations. Even the United Nations (UN) agencies like the United Nations Refugee Agency (UNHCR) and the World Food Programme (WFP) receive only a very small percentage of their funding from the UN regular budget and are dependent on voluntary contributions to be able to operate. This means that the donors, and their funding decisions, play a crucial role in determining the scale and focus of humanitarian action. It is not only the total amount of funding provided that is important but also what proportion of that funding is designated for which emergencies (or even specific locations affected by those emergencies) and for which particular sectors of humanitarian response (e.g. health, water, protection, etc.) it is earmarked.

Donors and their funding decisions are thus highly significant for compliance with the principles of humanitarian action.[1] For example, the principle of impartiality requires humanitarian assistance to be provided in proportion to need, which, in turn, requires funding in proportion to need. However, the amount of funding different emergencies receive is determined by a number of factors and not just by the level of need. Furthermore, funding often comes with restrictions about when, where, or on what it can be spent, or other conditions imposed by donors, and the policies of operational agencies are shaped by those restrictions and conditions. As such, the formulation and implementation of policy cannot be said to be independent of that donor, and the conditions may additionally prevent the operational agency from providing assistance impartially.

Concerns about the allocation of funding and the restrictions and conditions attached to it have led to a number of initiatives at the international level to improve the quantity, distribution, and quality of humanitarian financing. These initiatives include pooled funding mechanisms and commitments by donors to provide funding in particular ways, notably through the 2003

[1] See 'Principles of humanity, impartiality, neutrality, and independence' in Chapter 13 on Politics, Principles, and Humanitarian Action.

The Politics and Everyday Practice of International Humanitarianism. Miriam Bradley, Oxford University Press.
 DOI: 10.1093/oso/9780198840695.003.0018

principles of Good Humanitarian Donorship (GHD) and the Grand Bargain agreed by donors and aid agencies at the World Humanitarian Summit (WHS) in 2016.

Donors

Both public and private donors provide funding for humanitarian action. Public, or institutional, donors include both individual donor governments and inter-governmental organizations such as the European Union (EU). Private donors include individuals, foundations, companies, and national Red Cross and Red Crescent societies (Development Initiatives 2021: 52). Figure 18.1 shows how overall funding for international humanitarian assistance evolved from 2016 to 2020. Public donors consistently provide more than 80% of the total, and for the past three years, total annual funding for global humanitarian assistance has hovered around the USD 30 billion mark. To put these numbers in perspective, Italian military spending in 2020 was USD 28.9 billion—ten countries worldwide spent more than USD 45 billion each, and they collectively spent USD 1,482 billion (SIPRI 2021: 2). The global revenue of Coca-Cola, not a sector but a single firm, was USD 33 billion in 2020, while the combined revenues of the largest 500 companies in the United States was USD 14.2 trillion (Fortune 2021a: b). In short, while the sums for humanitarian assistance may appear large, they are a drop in the ocean relative to global defence spending or private-sector revenues.

Public donors are often divided into two further categories: so-called traditional donors, who are members of the Development Assistance Committee (DAC)[2] of the Organisation for Economic Co-operation and Development (OECD), and non-DAC/non-traditional or 'emerging' donors.[3] We have good knowledge of how much DAC donors are providing and where they are directing their funds because they tend to report their donations publicly, but not all non-DAC donors report their contributions consistently, leading to an undercounting of their contributions (Dreher et al. 2013: 402).

[2] As of 2 December 2021, there were thirty DAC members: Australia, Austria, Belgium, Canada, Czech Republic, Denmark, European Union, Finland, France, Germany, Greece, Hungary, Iceland, Ireland, Italy, Japan, Korea, Luxembourg, the Netherlands, New Zealand, Norway, Poland, Portugal, Slovak Republic, Slovenia, Spain, Sweden, Switzerland, the United Kingdom, and the United States.

[3] The category 'non-DAC' encompasses an enormous range of countries, of every income level and from every continent, and is not by any means a homogeneous category. To call some of them 'non-traditional' or 'emerging' is inaccurate, given that—to take just three examples—Russia/the Soviet Union was a major donor during the Cold War and China and India have a history of engagement dating at least to the 1950s (Dreher et al. 2013: 403). Moreover if we take into consideration non-financial contributions such as the hosting of refugees, non-DAC countries have long been of primary importance—and continue to be so.

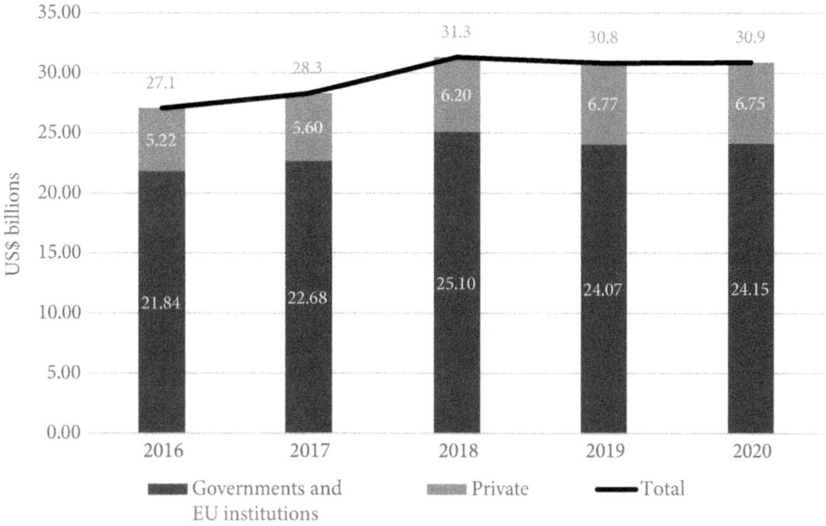

Figure 18.1 Funding for international humanitarian assistance, 2016–2020
Source: data taken from the Global Humanitarian Assistance Report 2021 (Development Initiatives 2021; figures for 2020 are preliminary estimates).

Yet, even with such undercounting, it is clear that since the 1990s there has been an upward trend in the number of governments funding humanitarian action and in the proportion of aid provided by non-DAC donor governments (Dreher et al. 2013: 403; Harmer and Cotterrell 2005: 7; Harmer and Martin 2010: 15, 17). Some non-DAC donors, notably Saudi Arabia, the United Arab Emirates (UAE), Kuwait, and Qatar, regularly provide more humanitarian funding than several DAC donors. Table 18.1 shows the figures for DAC and non-DAC governments donating more than USD 40 million in 2020.[4] When funding is considered as a percentage of gross domestic product (GDP), non-DAC donors appear especially generous—the countries on this list donating more than 0.1% of GDP included two DAC governments (Norway and Sweden) and three non-DAC (Saudi Arabia, the UAE, and Kuwait).

Just as the non-DAC component of funding from public donors is likely to be undercounted, so too is the diaspora component of funding from individuals, most of which does not flow through official channels. It is difficult to know how much funding is provided via remittances, but in many

[4] Bangladesh, Haiti, and Somalia are excluded because 2020 is an anomaly year for them in that they do not usually donate significant amounts and because most or all of the funding they provided was destined for a response within their own borders.

Table 18.1 Humanitarian funding provided by selected DAC and non-DAC donor governments in 2020

Donor government	Humanitarian funding (USD million)	Humanitarian funding (as percentage of gross domestic product)
United States	9,220.8	0.044
Germany	3,568.0	0.093
United Kingdom	2,118.1	0.077
Japan	1,216.4	0.024
Sweden	965.4	0.178
Saudi Arabia*	887.4	0.127
Norway	755.9	0.209
Canada	696.5	0.042
Switzerland	473.6	0.063
Netherlands	415.9	0.046
United Arab Emirates*	399.3	0.111
Denmark	315.1	0.088
Australia	264.3	0.020
Belgium	223.1	0.043
France	212.8	0.008
Italy	206.2	0.011
Ireland	144.6	0.034
Kuwait*	142.2	0.134
Finland	135.3	0.050
Korea, Rep.	105.7	0.006
Qatar*	70.7	0.049
Spain	60.1	0.005
Russian Federation*	59.9	0.004
New Zealand	57.0	0.027
Luxembourg	49.1	0.067
Austria	44.5	0.010
China*	43.3	0.000

Note: * = non-DAC donor governments.
Source: Data taken from OCHA Financial Tracking Service and World Bank World Development Indicators database as at 28 December 2021.

emergencies, this amount is significant, and may dwarf official humanitarian assistance. In 2017, for example, countries with recurrent humanitarian appeals are estimated to have received on average 4.5 times more per capita in the form of remittances than in official humanitarian assistance (Development Initiatives 2019: 44). In countries with large diasporas, remittances may account for an even larger share of international assistance.

Different types of donor favour particular channels for their humanitarian assistance. Western governments mainly fund multilateral organizations (mostly UN agencies) and international non-governmental organizations

(INGOs), while non-DAC governmental donors often provide a significant proportion of funding bilaterally, either government-to-government or via national RC societies (Harmer and Cotterrell 2005: 19–20; Harmer and Martin 2010: 21–22). Private donors generally favour INGOs, to whom they channelled 85% of their funding in 2019, for example, compared with just 12% to UN agencies (Development Initiatives 2021: 52). Between 2016 and 2020, the share of global humanitarian funding going directly to local or national organizations in affected states varied between 3.8 and 9.8% (Metcalfe-Hough et al. 2021: 53).[5] Diasporas mainly send remittances direct to family members or local non-governmental organizations (NGOs), and this kind of direct giving is also on the increase among distant publics, facilitated by crowdfunding and person-to-person giving platforms (Development Initiatives 2021: 60). There is some evidence to suggest that remittances are responsive to humanitarian emergencies, but at the household level, the receipt of remittances depends largely on historical migration patterns and so the recipients are not always those most affected by an emergency or with the greatest need (Bryant 2019: 6–9).

It is not only the preferred channels that differ according to the type of donor but also the countries and types of emergencies to which they tend to donate. Non-DAC donors have tended to provide humanitarian aid to states within their region, although there is some evidence that they are increasingly channelling donations outside their respective regions (Harmer and Martin 2010: 19). They often concentrate their donations in one or two high-profile crises per year, which means that, in some emergency contexts, the importance of their funding exceeds the average. For example, Saudi Arabia contributed 55% of funding to Bangladesh in response to Cyclone Sidr and floods in 2007, Kuwait was a major donor to the crisis in Gaza in 2009 (Harmer and Martin 2010: 19), and Saudi Arabia and Turkey together contributed around one-third of total humanitarian funding in response to the famine in Somalia in 2011. Private donors are usually thought to favour sudden-onset disasters, such as earthquakes, over slow-onset disasters, such as drought or famine, or armed conflict (Stoianova 2013: 1). Certainly, private donors provided record amounts of funding in response to the 2004 Indian Ocean tsunami[6] and the 2010 earthquake in Haiti[7], which would support this notion. However, private donors do not just give in response to particular

[5] The issue of direct funding for local actors is also discussed in 'Civil society actors' and 'Localization agenda' in Chapter 19 on Government and Civil Society in Affected States.

[6] See 'Funding for humanitarian response' in Chapter 7 on the 2004 Indian Ocean Tsunami.

[7] See 'Funding for humanitarian response' in Chapter 9 on the Haiti Earthquake, 2010.

appeals—individuals, for example, often give regularly to particular humanitarian agencies, allowing the recipient agencies to allocate the funds to the emergencies they choose.

While remittances may be significant sources of assistance for the victims of humanitarian emergencies, they differ in important ways from funds channelled through international humanitarian agencies or the governments of affected states—above all in that they are largely provided to kin or communities, in contrast to a humanitarian project largely defined by a commitment to the impartial provision of assistance to distant others.[8] Given these differences, and the difficulty in tracking remittances, they are mostly excluded from the discussion in the rest of this chapter.

In accordance with need?

Focusing on donations channelled through the major international humanitarian agencies, it is clear that the allocation of funding across different emergencies is not based solely on need. To put it another way, the principle of impartiality, which stipulates that humanitarian assistance should be allocated in proportion to need, does not function well across time or countries. This first point is evident from a cursory glance at Figure 18.1, which shows that following a reduction in overall humanitarian funding between 2018 and 2019, the global funding level stayed more or less constant in 2020, despite the obvious and sizeable additional humanitarian needs that year arising from the COVID-19 pandemic.[9] Disproportionate allocation of humanitarian funding across the globe is apparent from a comparison of different emergencies at similar points in time. For example, funding for the response to the 2004 Indian Ocean tsunami amounted to more than USD 7,000 per affected person, compared with just USD 3 for each person affected by floods in Bangladesh in the same year—a difference of a magnitude so enormous it cannot possibly be explained by differences in the scale of needs of those affected or differences in the costs of meeting those needs.

More generally, we can look to the percentage of different budgets that are funded, with the caveat that budgets themselves are not always a good representation of needs. In 2021, for example, UN-coordinated appeals were

[8] See 'Meaning of humanitarianism and the focus of this book' in Chapter 1 on Humanitarian Emergencies.

[9] A similar argument can be made by looking at how funding to particular emergencies varies across time. See, for example, the discussion on 'Funding for humanitarian response' in Chapter 6 on Afghanistan, 2001–2014.

48.3% funded, and this average conceals enormous disparity between different emergencies, from a Flash Appeal for Afghanistan, which was 135.8% funded, to a Regional Response Plan for the Democratic Republic of Congo (DRC), which received only 3.1% of requested funds (UN OCHA n.d.). However, funding appeals and budgets may be only loosely indicative of the level of need. In part, this is because it is often extremely difficult to get an accurate picture of the number of people affected by a humanitarian emergency, the ways they have been affected, and the needs they have as a result.[10] In addition, the operational humanitarian agencies appealing for funding are informed not only by the (usually limited) picture of needs but also by what they expect they can achieve and by what they know or believe about the likely funding willingness, preferences, and priorities of donors—especially their main institutional donors.[11] Thus, in the context of impending famine in Somalia, operational agencies requested insufficient funding through the Consolidated Appeals Process (CAP) launched in December 2010 due to their pessimism about the generosity of donors in this context, where much of the crisis was in territory controlled by Al-Shabaab, a group designated as terrorist by the United States.[12]

None of this is to say that need is not a factor in determining aid allocations, just that it is not the sole factor. Cross-national analysis of funding for sudden-onset 'natural disasters' suggests that both the needs arising from a disaster and the capacity of the affected country to respond affect the likelihood of receiving aid and the amount received (Fink and Redaelli 2011: 747, 750). Funding for humanitarian response in ongoing civil wars also appears to be sensitive to the level of need in affected countries (Narang 2016: 202–204). Around 70–90% of the humanitarian funding from major Western donors is allocated on a yearly cycle, set out in annual budgets and plans, and senior bureaucrats report allocating this in compliance with humanitarian and GHD principles, suggesting that the level of need is the most important factor (Scott et al. 2022: 179–181). However, allocating aid in accordance with need is open to distinct interpretations—at the most basic level, it can imply addressing the greatest amount of need overall or prioritizing those

[10] See 'Assessing needs, contexts, and capacities' in Chapter 21 on Needs assessment, Evaluations, and Response Decisions.

[11] In some cases, it is not only funding appeals that reflect expectations about capacity and donor demand but also needs assessments themselves. In responding to famine in Ethiopia, for example, UN agencies massively underestimated the need for food aid until at least mid-1984, presenting as an assessment of needs what was really their estimate of what they thought they could deliver, given their beliefs about the country's limited logistical capacity to distribute aid. See 'Underestimating the need for external assistance' in Chapter 3 on Drought and Famine in Ethiopia, 1983–1985.

[12] See 'Funding for humanitarian response' in Chapter 10 on Famine in Somalia, 2011–2012.

with greatest needs, and while sometimes these two approaches can be complementary, often there are trade-offs (Obrecht 2017: 13). Furthermore, many donors seek not only to meet immediate needs but also to reduce needs over the longer term, and here again, there are trade-offs, given limited resources (Obrecht 2017: 13).

Other factors shaping funding decisions

It is clear that the level of need is an important factor shaping funding decisions, but it is equally clear that, in many cases, there are disparities in the level of funding across time or across different emergencies that do not reflect differences in need. If need is not the sole factor determining the allocation of funding, how do donors choose whether, where, and what to fund? A variety of factors other than need drive funding decisions, and different factors may be more or less important for different types of donors and in different types of crisis.

Media coverage and celebrity engagement

As discussed in more detail in Chapter 17, media coverage and celebrity engagement play an important role in determining which emergencies attract significant levels of attention internationally, how those emergencies are understood, and what kinds of response are deemed appropriate. With respect to these latter two dimensions, it is not only the level of coverage but also the nature of that coverage—in particular, the ways that emergencies are framed—that is pertinent. While media coverage is widely deemed to be an important determinant of humanitarian funding allocations worldwide, the exact nature of its significance is debated. Several cross-country analyses show a correlation between the level of media coverage of a humanitarian emergency and the level of funding for humanitarian response, but it is not clear that this correlation reflects a causal relationship—rather, it could be that another factor (issue salience or political agendas) is driving both the media coverage and the funding (Scott et al. 2022: 169).

Comparing contemporaneous emergencies suggests that, *in some cases*, media coverage is decisive in generating funding for humanitarian response (Franks 2010: 77; Olsen et al. 2003: 114–116). Comparing the funds raised for particular appeals over time likewise suggests that media coverage can be decisive, as when a severely underfunded UN appeal in response to a severe food crisis in Niger in 2005 reached its target within days of the BBC covering the story with images of babies in feeding stations, distressed mothers, and anxious aid workers (Franks 2010: 76). Extensive media coverage is probably a necessary condition for private donors to give on a significant scale—without prominent media coverage, most members of the public will simply not have knowledge of humanitarian emergencies around the world. In some cases, as in the famine in Ethiopia in the 1980s and the Indian Ocean tsunami in 2004, media coverage seems to have stoked public interest, generating pressure on policymakers to respond, which, in turn, led to increases in public funding.[13]

However, media coverage (and celebrity engagement) is not a necessary condition for significant funding from donor governments, and sometimes it is foreign policy that drives media coverage rather than the other way around. Some humanitarian emergencies attract significant public funding despite little media attention (Olsen et al. 2003: 118). When planning their annual aid allocations, bureaucrats report sometimes deliberately seeking out so-called forgotten crises that have had little or no media coverage (Scott et al. 2022: 181–182). Even where an emergency attracts significant media coverage and significant funding, there is not necessarily a causal relationship between them. Media coverage sometimes reflects and reinforces, rather than influences, the foreign policy of powerful states, who are often also major donors. In Afghanistan after the US invasion at the end of 2001, for example, it appears that US security interests and foreign policy drove the massive increase in media coverage, and it is not clear that that coverage had any impact on funding or whether the funding was also a direct consequence of US and other donor interests (Olsen et al. 2003: 122).

Strategic interests of donor governments (and other powerful states)

As they provide the lion's share of humanitarian finance (see Figure 18.1), the funding decisions of donor governments are the most important determinant of the global distribution. Governments have strategic interests and a long history of using foreign aid in support of their foreign policy objectives. The humanitarian sector is thus highly vulnerable to the political interests of a small number of major donors among whom funding is concentrated (Bennett et al. 2016: 58).

[13] See 'Media coverage' in Chapter 3 on Drought and Famine in Ethiopia, 1983–1985 and 'Funding for humanitarian response' in Chapter 7 on the 2004 Indian Ocean Tsunami.

Several quantitative studies have sought to measure the effect of political and strategic interests, either on the funding decisions of individual donor governments or on global allocations of humanitarian funding, and have arrived at somewhat varied conclusions, but none have suggested that political interests do not matter. For example, analysis focusing on US humanitarian funding for responses to 'natural disasters' between 1964 and 1995 found that foreign policy was a more important determinant than humanitarian need in deciding which disasters to fund, and also important (although less so) in determining how much funding to provide (Drury et al. 2005). By contrast, a more recent study on US humanitarian funding across both 'natural' and conflict-related disasters in developing countries between 1989 and 2009 found that need is a more important driver of funding decisions than US self-interest (Kevlihan et al. 2014). Analysis of cross-national data generates similar results. Focusing on 'natural disasters' occurring between 1992 and 2004, one study found that both humanitarian need and political and strategic factors appear to be crucial determinants of humanitarian funding allocations (Fink and Redaelli 2011). Another, which focused on civil wars, suggests that aid allocations to ongoing civil wars are driven more by need that strategic interests, while aid allocations in the post-conflict phase are significantly driven by the perceived self-interest of donors (Narang 2016). All these studies find that political and strategic interests have an impact—the debate here is not about whether or not they matter but about whether they are, on average, a more or less important driver of humanitarian funding allocations than the level of need.

Even if political interests may be of less importance than need in determining the global distribution of aid, funding decisions for particular emergencies can sometimes be driven primarily by strategic considerations. Aid to Afghanistan from 2001 to 2014 provides a clear example of donor states using aid to support their political or military endeavours.[14] Most humanitarian funding to Afghanistan during this period came from the governments of countries contributing troops to the International Security Assistance Force (ISAF), and this funding was often provided explicitly for the provinces where international troops were present. Furthermore, a massive increase in humanitarian funding in 2002, followed by a big drop in 2003, did not correspond to changes in the level of humanitarian need but rather reflected

[14] See 'Funding for humanitarian response' and 'Instrumentalization of aid by military actors' in Chapter 6 on Afghanistan, 2001–2014.

changes in the narrative major donors wanted to present. Cutting humanitarian funding was the logical corollary of a post-conflict discourse. Similarly, in Somalia after the installation of the Transitional Federal Government (TFG), many UN and donor state diplomats sought to downplay the humanitarian crisis, which they saw as an embarrassment with the potential to delegitimize the entire government (Menkhaus 2010: S334).

Linked to the strategic interest of states are their counter-terror initiatives.[15] Counter-terror laws and regulations affect funding not only through their impact on donor funding decisions, as when the United States cut humanitarian funding for Somalia by 88% within two years of listing Al-Shabaab as a terrorist group, but also through their impact on the banking sector and its willingness to transfer funds to individuals and companies in, or associated with, countries deemed risky. Donors that once quietly accepted the risk of some aid diversion as a necessary evil in volatile environments have significantly lowered their risk threshold when it comes to designated groups (Pantuliano et al. 2011: 7). Islamic charities, and local NGOs in places like Palestine and Pakistan have come under greater scrutiny than others and have been particularly badly affected by funding cuts and the freezing of bank transactions (Pantuliano et al. 2011: 7). Counter-terror measures have also had an impact on remittances, with bank de-risking affecting transfers to and from Islamic organizations and accounts held by some money transfer firms, increasing the costs of remittances and pushing some underground, with concerns that some diaspora may have turned to individuals or 'briefcase NGOs', which were less likely to be compliant with counter-terror legislation (Bryant 2019: 4; Pantuliano et al. 2011: 8).

Where humanitarian emergencies occur in places where donor states have little political or strategic interest, they may provide humanitarian funding not to support their foreign policy engagement but to *avoid* any such engagement. This is the idea of the 'humanitarian alibi' whereby a (large-scale) humanitarian response is funded to conceal political or military inaction. Thus, in the Bosnian war, for example, Western leaders emphasized the centrality of the humanitarian mission, and for a long time, this allowed for a moral failure (to protect civilians or end the conflict) to be presented as an organizational triumph (operating an airlift and distributing large quantities of material assistance).[16]

[15] For a more detailed discussion from a legal perspective, see 'Counter-terror legislation' in Chapter 14 on Law and International Humanitarianism.

[16] See '"Humanitarian alibi"' in Chapter 4 on the Bosnian War, 1992–1995.

Prior relationships and humanitarian agency activism

Other factors, which are neither strictly needs-based nor necessarily strategic, may also guide donor allocation decisions. In particular, prior relationships between a donor government and any given country in crisis or the long-term presence of international agencies in a given country have been shown to be linked to funding decisions.

Some donors explicitly identify prior relationships with countries in crisis as a criteria for deciding which crises to fund (Obrecht 2017: 8). Cross-country studies have also shown that donor governments are more likely to provide humanitarian funding in response to 'natural disasters' in their former colonies and that global funding for post-conflict settings is biased to the former colonies of the permanent five members of the UN Security Council (Fink and Redaelli 2011; Narang 2016).

Operational humanitarian agencies can also have an impact on the level of funding, either by lobbying donor bureaucracies directly or launching global publicity campaigns, which generate public interest and pressurize policymakers to respond. This kind of activism can have both direct and indirect effects on funding. The presence of significant international humanitarian architecture in any given context ensures that needs are assessed, responses designed, and funding appeals launched, and in some cases, significant levels of funding are provided for humanitarian response even in the absence of media coverage or strategic interests on the part of donor states (Olsen et al. 2003: 119). It can also have an indirect effect in as far as the humanitarian lobby can influence media coverage, with journalists frequently depending on aid agencies as a source of information and logistical support or covering publicity campaigns launched by operational agencies.[17]

Independence, earmarking, and conditionality

It is not only the quantity of funding but also its quality—or the extent to which it is provided flexibly or with conditions attached—that affects humanitarian response. Donors (primarily governmental donors) mainly fund UN agencies and INGOs and often attach particular conditions to their funds. These conditions may vary from earmarking for a sector of activity (e.g. food, shelter or health care) or a particular part of the country to more specific

[17] See 'Setting the agenda' in Chapter 17 on Media and Celebrities.

conditions prohibiting, for example, working with individuals and NGOs thought to be associated with organizations listed as terrorist.

Earmarking is a common practice, by which official donors specify where and/or on what the funds they are contributing can be spent. For example, UNHCR distinguishes between tightly earmarked, earmarked, softly earmarked, and unearmarked donations (UNHCR 2021d: 7). Tightly earmarked funding is earmarked for a specific project or sector within a designated country or division. Earmarked funded is provided for a designated country or a designated emergency within a specific country but without restrictions as to project or sector. Softly earmarked funding can be spent across a range of countries and activities in a given region or situation or a specific theme or activity. Unearmarked funding is provided without restrictions on its use. Funding from official donors often has to be spent within a given financial year and is often short-term, requiring operational agencies to spend money quickly and limiting their ability to plan ahead or develop long-term programming. By contrast, multi-year funding is pledged for twenty-four months or more—and may or may not be earmarked (UNHCR 2021d: 7). Even if earmarked, multi-year funding is generally considered to be quality funding because it provides recipient operational agencies with predictable income, which allows them to plan ahead, and can thus contribute to efficiency and effectiveness (Development Initiatives 2021: 78; UNHCR 2021d: 7).

Just as strategic interests may drive allocation decisions across countries, they can also drive decisions about the allocation of funds across agencies, sectors, and regions within countries. In the context of a stabilization mission after the overthrow of the Taliban in 2001, for example, many donor governments (most of which were also contributing troops) sought to support that mission by earmarking funds for those parts of the country in which international troops were present and for activities aimed at bolstering the government or made funding conditional on agencies sharing information with the Afghan army.[18] A number of donors attach condition linked to counter-terrorism laws and policy objectives, requiring humanitarian agencies receiving those funds to ensure that they are not used to support terrorism or to vet local partners, vendors, and suppliers (Burniske et al. 2014: 6; Pantuliano et al. 2011: 5).

Needs and other factors can also play a role in earmarking and conditionality. It is not always clear what allocating aid in accordance with need means in practice—since no donor alone can tackle the entire humanitarian

[18] See 'Funding for humanitarian response' in Chapter 6 on Afghanistan, 2001–2014.

caseload, donors must prioritize, and this may involve focusing on certain aspects of need, certain crises or populations, or certain kinds of interventions (Obrecht 2017: 8, 12). There is thus a range of considerations that are not strictly needs-based but that reflect other broadly altruistic priorities. For example, some donor governments may have a particular commitment to supporting local humanitarian action, coordination mechanisms, development or peacebuilding efforts, resilience, gender equality, or cash-based programming (Obrecht 2017: 7, 8, 12, 13).

Some operational agencies receive much more flexible funding than others. To a significant degree, this depends on their funding model. Private voluntary contributions, especially those provided through regular giving from individuals rather than one-off donations in response to particular appeals, are seen as high-quality funding. Where provided in sufficient quantity, this kind of donation helps to ensure that humanitarian response is not limited by earmarking and conditionality imposed by official donors and can be used for sudden-onset emergencies and hidden crises (Stoianova 2013: 1, 25). The Médecins Sans Frontières (MSF) funding model, in which individual donors provide the vast majority of funding, is designed to ensure financial independence and the autonomy to make decisions without consideration of the preferences of donor governments. In 2020, for example, more than 7 million individual donors and private institutions (private companies and foundations) provided 97.2% of the €1.9 billion raised (MSF 2021). MSF sometimes also refuses funding from particular governments. In several conflict contexts, including Afghanistan, for example, MSF has refused any government funding. In protest at EU migration policies, in 2016 MSF announced that it would no longer accept funding from the EU or member state governments for any of its operations worldwide (MSF 2016). While 97% is exceptional, some other NGOs also receive significant funding from private sources. For example, 63% of World Vision International's operating budget comes from individuals, foundations, and corporations, and Oxfam receives well over half its funding from private sources, including trading revenue from its shops (Oxfam n.d.; World Vision International n.d.). UN agencies, by contrast, tend to depend heavily on official donors, and earmarked funding accounts for more than 80% of the total funding they receive, as shown in Figure 18.2. However, the proportion varies significantly across different UN agencies, with the United Nations Relief and Works Agency for Palestine Refugees in the Near East (UNRWA) and the Office for the Coordination for Humanitarian Affairs (OCHA) receiving more than half unearmarked or softly earmarked in 2020, for example (Development Initiatives 2021: 76).

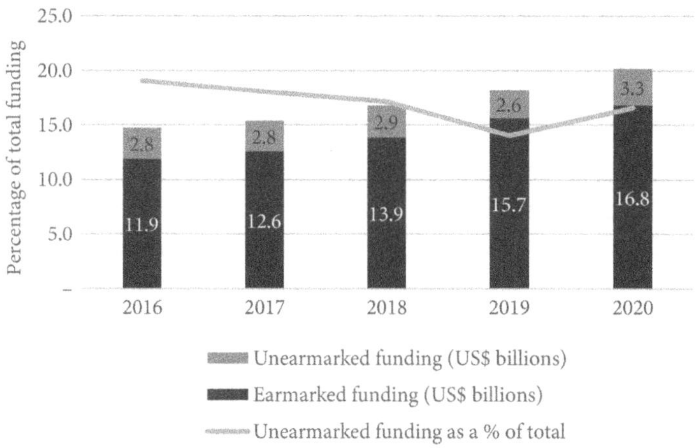

Figure 18.2 Resources received by UN agencies reported as earmarked and unearmarked
Source: Development Initiatives (2021: 75).

Initiatives to improve the allocation and quality of humanitarian financing

Since the early 2000s, a number of initiatives have been launched with the aim of making funding more responsive to needs. These take various forms, including principles and commitment that donor governments have signed up to and global and country-level mechanisms through which funding from different donors is pooled, and to which those donors delegate allocation decisions.

Good Humanitarian Donorship

In 2003, the main donor governments established twenty-three GHD objectives, principles, and good practices, and in 2005 these were adopted by the OECD DAC. Led by the Swedish government, the GHD initiative sought to address concerns that humanitarian assistance was not distributed across the globe in accordance with need and did not always meet an adequate range of needs beyond basic sustenance (Schaar 2008: 37). The resulting document defines humanitarian action to include both protection and the provision of material assistance; includes a commitment to the principles of humanity, impartiality, neutrality, and independence; and is concerned with not only where funds should be allocated but also what kinds of objectives should be pursued and, to a lesser degree, what activities should be undertaken (GHD n.d.). As one of the government representatives driving the initiative puts it, the vision was that 'aid should be given according to need, when and where it was required, in sufficient amounts and with appropriate quality, and it should include measures to prevent and prepare for emergencies, while also helping people rebuild their lives and livelihoods after a crisis' (Schaar 2008: 39). Thus, the goals went beyond making funding more responsive to urgent need in the 'emergency present' and encompassed commitments to prevention (before emergencies occur) and recovery and reconstruction (after emergency response).

The GHD initiative encompassed stipulations and recommendations for donors vis-à-vis allocation and quality of funding as well as suggestions for what they should demand or incentivize from the operational agencies they are funding. In terms of the allocation and quality of funding, for example, principle 6 commits donors to 'allocate humanitarian funding in proportion to needs and on the basis of needs assessments' and principles 12–14 recommend that donors increase the flexibility of their funding and contribute to inter-agency appeals (GHD n.d.). In terms of promoting particular practices by operational agencies, principles 7–9 address such issues as participation of beneficiaries, strengthening local and national capacity in affected states, and supporting recovery and long-term development (GHD n.d.). Principles 15–20 focus on promoting standards and enhancing implementation, and in 2018, a twenty-fourth principle promoting the use of cash transfers was added (GHD n.d.).

Pooled funding mechanisms

Funding mechanisms which pool donations, mainly from institutional funders, seek to improve the timeliness and the distribution of funding for humanitarian emergencies, with the aim of making it more responsive to need. Usually administered by the UN, these can be divided into mechanisms that operate at the global level and those that operate at the country level. By pooling funds, and investing a central body with the authority to distribute them, donors delegate some of the allocation decisions, with the goal of reducing the influence of non-needs-based factors in driving these decisions.

At the global level, the UN Central Emergency Response Fund (CERF), established by the UN General Assembly at the end of 2005, began with a USD 450 million annual funding target, increased to USD 1 billion from 2017.

The CERF collects financial contributions throughout the year—most of which come from governments, but foundations, companies, charities, and individuals can also contribute—into a central fund, to which operational humanitarian agencies can apply to finance emergency response (UN CERF 2016: 1). While a large proportion of CERF funds are provided by a small number of donors (with ten governments providing 87% of funding between 2006 and 2015), the CERF also attracts contributions from countries that *receive* CERF funds (Cosgrave 2009: 78; UN CERF 2016: 2). Funds can be distributed at the onset of a crisis through rapid-response grants, which can be approved in as little as forty-eight hours, with the aim of ensuring a timely response, or to underfunded crises through disbursements twice per year, with the aim of ensuring coverage of otherwise neglected situations (UN CERF 2016: 1). In this sense, the CERF aims to improve the distribution of humanitarian funding *across different humanitarian emergencies.*

While the CERF has undoubtedly helped to make humanitarian funding more responsive to need, its impact is limited in important ways. In some contexts, the CERF has played an important role, providing a significant proportion of funding for some emergencies (Cosgrave 2009: 78–79). However, the proportion of total donations channelled this way is relatively small, which necessarily restricts its impact on the global distribution. For the past ten years, annual contributions to the CERF have hovered around the USD 0.5 billion mark, with the exception of a record USD 831.4 million in 2019 (UN CERF 2021). Between 2016 and 2020, the CERF accounted for between 1.6 and 2.7% of humanitarian funding each year, a small proportion that can only compensate to a limited degree for insufficient responsiveness to needs in other donations. The CERF is also limited in its ability to compare needs (and hence identify underfunded crises) by its dependence on assessments by operational agencies, whose processes and methodologies are often not consistent or comparable (Obrecht 2017: 21).[19] A further limitation of the CERF is that it is a UN fund, and NGOs can only access CERF funding indirectly.

CERF funding has also had a number of indirect consequences, both positive and negative. Donor governments often highlight their funding to the CERF to domestic audiences to show that they have provided funding to emergencies they have not funded directly (Mowjee and Poole 2014: 9; Scott et al. 2022: 179). In some cases, this can help them to resist pressure from elected officials and the public responding to intense media coverage of a crisis (Scott et al. 2022: 179). For some donor governments, funding the

[19] On the lack of consistency and comparability of needs assessments, see 'Assessing needs, contexts, and capacities' in Chapter 21 on Needs Assessment, Evaluation, and Response Decisions.

CERF enables them to ensure—and explain to their constituents—that they are providing at least some funding to underfunded emergencies even as they focus their direct funding on other priorities (Mowjee and Poole 2014: 9–10). Where the CERF accounts for a large proportion of humanitarian funding for any given emergency, however, it can come at the expense of accountability to the donor community. For example, after the end of the war in Sri Lanka, the CERF compensated for underfunding that resulted from the reluctance of some key donors to foot the bill for a government unwilling to pay to address the massive human rights needs generated by its own deliberate military strategy (HPG 2010: 15).

The UN also administers regional- and country-based pooled funds. When a new humanitarian emergency occurs, or an existing situation deteriorates, the UN Emergency Relief Coordinator in the affected country can establish a country-based pooled fund (CBPF). The majority of contributions come from a small number of donor governments, with the top ten providing over 90% of the USD 956 million given to CBPFs globally in 2019 (UN OCHA 2020a: 2). While many of these donors increased their contributions in 2020, the United Kingdom—by far the largest donor in 2019—cut its funding, leading to an overall decrease in funding to CBPFs in 2020 (Development Initiatives 2021: 71; UN OCHA 2020a: 2).

Donors contribute to specific CBPFs, and each CBPF is locally managed by OCHA, which is responsible for allocating funds to specific sectors and operational agencies. These allocations are based on the country-level Humanitarian Response Plan (HRP) and made to UN agencies, national and international NGOs, and Red Cross & Red Crescent (RC) organizations (UN OCHA 2020a). The allocation process has been praised both for its transparency and for directing a significant proportion of funding to national actors (Obrecht 2017: 21). Worldwide, the proportion allocated to national actors was on an upward trend, from 24% in 2016 to 34% in 2020, mostly going directly to national NGOs, but with some going indirectly as subgrants to national NGOs and other national actors (Development Initiatives 2021: 72). However, the worldwide total masks significant differences across different CBPFs and across time (Development Initiatives 2021: 73).

While the CERF is intended to make funding more responsive to the distribution of need globally, CBPFs are intended to make funding more responsive to need within given humanitarian emergencies by providing unearmarked funding that can be used in the sectors that most need it and by the implementing agencies best able to use it. As with the CERF, the impact of CBPFs is necessarily limited due to the relatively small proportion of humanitarian funding they account for. Total CBPF contributions worldwide have risen fairly steadily from around USD 500 million in 2014 to USD 1 billion in 2021, with a dip in 2020 linked to the cut in funding by the United Kingdom (Development Initiatives 2021: 70). While the CBPFs collectively receive greater funding than the CERF, they still only account for around 3% of humanitarian funding each year. Moreover, it is not clear that in-country allocations of CBPFs are truly responsive to need. Some donors have raised concerns that HRPs are more the product of negotiation among implementing agencies than of evidence of needs and response effectiveness (Obrecht 2017: 21).

The Grand Bargain

At the WHS in 2016, eighteen donor governments and sixteen operational agencies agreed a 'Grand Bargain', which has since been signed by many more donors and aid agencies (Metcalfe-Hough et al. 2021: 17). Consisting of fifty-one commitments aimed at improving the efficiency and effectiveness of international humanitarian aid, part of the logic underpinning the Grand Bargain is that if donors improve the quality of their funding (e.g. by reducing earmarking and increasing multi-year financing) and operational agencies improve how they spend the funds they received (e.g. through increased use of cash-based programming and more joint needs assessments), aid delivery will be more efficient, generating efficiency savings of up to a billion dollars over five years (ICVA 2017b: 3–4; UN Secretary-General 2016: 14). It was hoped that the cost savings from efficiency gains would be large and could be used for the direct benefit of populations affected by humanitarian crises.

Grand Bargain workstreams 7 and 8 (subsequently combined into a single workstream, 7+8) are specifically concerned with the quality of funding, with a focus on increasing multi-year planning and funding and reducing earmarking. While progress has been made in some respects since 2016, many of the targets agreed in the Grand Bargain have not been met. Increases in the volume of multi-year funding and of unearmarked or softly earmarked funds did not always translate into equivalent increases in the proportion of funding provided in these ways (Metcalfe-Hough et al. 2021: 101–102, 105–106). Furthermore, while a core group of donor governments appear committed to providing flexible funding, some of the largest donors provide only a small percentage flexibly (Metcalfe-Hough et al. 2021, 95–97). Thus, in 2020, the proportion of total funding that was unearmarked was well below the target agreed in the Grand Bargain and even below 2016 levels (Development Initiatives 2021: 76). Part of the problem is that moving to more flexible funding is a political issue—which requires donor governments to convince their domestic audiences that the added value of flexibility outweighs any added risks—whereas the Grand Bargain has adopted a largely technical perspective (Metcalfe-Hough et al. 2021: 96).

Conclusions

For all the diversification in donorship in the first two decades of the twenty-first century, most of the biggest international humanitarian agencies remain heavily dependent on a relatively small number of official donors. There is some indication that the percentage of humanitarian funding provided by private donors is increasing, and insofar as private donors are making regular donations rather than responding to specific appeals, this funding is flexible. However, it still accounts for only a tiny proportion of humanitarian funding overall. Pooled funding mechanisms also offer certain kinds of flexibility, but they continue to account for only small percentages of humanitarian funding. In any case, pooled funding mechanisms are not a panacea. For example, the process through which CBPFs are allocated, based as it is on the HRPs, may not be any more responsive to needs than earmarked funding given direct by donor government to implementing agencies. Furthermore, while they may succeed in reducing the impact of donor political interests on funding allocations, they can have unintended political impacts. As seen in the case of the CERF in Sri Lanka, pooled funding can undermine accountability. In other contexts, the fact that these different pooled funding mechanisms are administered by the UN can pose a risk. In Somalia, for example, some NGOs questioned whether they should seek finding through such mechanisms, given the political stance of the UN in that context and its designation of Al-Shabaab as a terrorist group.[20] Grand Bargain commitments to increase the quality of humanitarian funding have not been met, but there are some signs that the largest donors are beginning to improve the quality of their funding, and this may mark a turning point.

[20] See 'Funding for humanitarian response' in Chapter 10 on Famine in Somalia, 2011–2012.

19

Government and Civil Society in Affected States

In principle, international humanitarian agencies respond to emergencies because the affected population needs external help. This may be because the capacity of the affected state is temporarily overwhelmed by the crisis or because it is more generally unable, or unwilling, to live up to its responsibility to protect and assist those on its territory. The international response can also have an impact—positive or negative—on the capacity and will of states and other responsible actors to fulfil their obligations. This chapter examines the role that the governments and civil society of affected states play in responding to humanitarian emergencies and the interaction between international humanitarian agencies and these different civilian actors in the places where they are operating.

International responders often sideline governments and civil society in affected states, despite the fact that, in many cases, these national and local actors can play a crucial role in humanitarian response. This can serve to reduce, rather than reinforce, the capacity of domestic actors. Alternative approaches emphasize working to support the state or to hold the state to account, but such approaches also have their downsides. While the most appropriate response from international humanitarian agencies would probably draw on a combination of these approaches, based on solid contextual understanding, in practice, responses are often shaped more by donor politics or ill-informed assumptions about the will and capacity of the affected state than by analysis of the context.

Primarily with respect to civil society, the so-called localization agenda promotes the idea that local and national, not international, actors should be at the forefront of humanitarian responses and that power and resources should be transferred to them. This agenda has been widely embraced in recent years and was given added impetus with the Grand Bargain agreed at the World Humanitarian Summit (WHS) in 2016. However, progress in implementation has been mixed, and the impact of localization on the quality and cost-effectiveness of humanitarian response remains unclear.

The Politics and Everyday Practice of International Humanitarianism. Miriam Bradley, Oxford University Press.
 DOI: 10.1093/oso/9780198840695.003.0019

State sovereignty, responsibility, and practice

In the 1990s, ideas about state sovereignty were reconfigured to stress responsibility, although the extent to which this reconfiguration was embraced globally is contested. Thinking about state sovereignty has traditionally emphasized the right of states to be free from outside intervention in their domestic affairs, but more recent perspectives build on the work of a group of scholars at the Brookings Institution, who conceptualized sovereignty as responsibility rather than right. According to these perspectives, a state's external sovereignty, which implies freedom from outside interference, is not absolute but is conditional on its internal sovereignty, which depends on it protecting and assisting its population and ensuring that basic needs are met. Among other things, these ideas provided the intellectual underpinning for the 1998 Guiding Principles on Internal Displacement[1] and for the responsibility to protect doctrine, initially set out in 2001.[2] These normative developments reiterate the primacy of the state but also set out duties and obligations of the wider international community.

In the context of a humanitarian crisis, the state has primary responsibility to protect the population from violence and to assist the victims of emergencies occurring on its territory. Where states are unable or unwilling to protect and assist the population, international humanitarian agencies have a role to play. In principle, it is the affected state that should 'call' a crisis and appeal for international aid (Harvey 2009: 6). International humanitarian agencies may also offer their services in the absence of any such appeal, but, in general, they cannot provide those services without the consent of the state.

For a number of reasons, states may resist calling for, or consenting to, international assistance. An international humanitarian response broadly implies that the state is not living up to its responsibilities, either through lack of will or lack of capacity, so it can be understood as a criticism of the state (Harvey 2009: 2; Kahn and Cunningham 2013: S146). States may fear appearing weak or incapable of managing their own affairs (Harvey 2009: 6; Kahn and Cunningham 2013: S146). Furthermore, states may mistrust the motivations of international actors, and for many, the predominantly Western identity of humanitarian actors recalls colonialism (Harvey 2009: 6; Kahn and Cunningham 2013: 147). Where a state lacks the will to protect and assist its own population, it may not want those parts of the population it is neglecting to be protected and assisted. In the Nigerian Civil War,

[1] See 'International displacement and migration law' in Chapter 14 on Law and International Humanitarianism.

[2] See 'International military forces' in Chapter 20 on Armed Actors.

for example, the national government imposed a blockade with the deliberate aim of starving the secessionists into submission.[3] In such situations, a humanitarian response may undermine government strategies. Where a state is actively targeting some of the population, it may also want to prevent international actors from witnessing the abuses it is perpetrating. This was the case in the final stages of the Sri Lankan Civil War, when the government sought to remove international witnesses from the Vanni, where it was shelling the civilian population.[4]

States should not arbitrarily withhold consent for international humanitarian assistance, but sometimes they do. The rule prohibiting the arbitrary withholding of consent is clear in international humanitarian law (IHL), but IHL only applies in armed conflict contexts. In other kinds of humanitarian emergency, there is an expectation that states will consent to external assistance, but the rule is less clear.[5] It bears emphasizing that the issue of arbitrary withholding of consent for relief operations only arises when the population is inadequately provided with essential supplies (Akande and Gillard 2016: 21). This means that where the state itself is meeting the needs of the population, it is under no obligation to accept international assistance. Thus, after the 2004 tsunami, for example, the Indian government stated that sufficient national resources were available for relief and that international efforts should focus on the worst-affected countries—though it subsequently accepted contributions towards longer-term rehabilitation, subject to certain restrictions and close coordination with the national authorities (Price and Bhatt 2009: 16–17). By contrast, when Cyclone Nargis hit Myanmar in May 2008, resulting in 140,000 deaths and affecting around 2.4 million people, the disaster clearly overwhelmed national capacity, yet the government was reluctant to accept international aid (Simm 2018: 133). After some negotiation, the Myanmar government did eventually accept international assistance, as states usually do.

States only rarely withhold consent altogether, but they sometimes impose restrictions on where international agencies can work and what activities they can undertake in such as way as to make quick, impartial, and independent operations virtually impossible (Bradley 2016a: 130–132). In armed conflict contexts, parties to conflict must allow and facilitate rapid and unimpeded passage once relief operations have been consented to. IHL sets out rules and recommendations regarding the entry of personnel, customs clearance, and the taxation of relief, among other things (Akande and Gillard 2016: 26–29).

[3] See 'Conflict dynamics' in Chapter 2 on the Nigerian Civil War, 1967–1970.
[4] See 'Manipulation of aid and aid agencies' in Chapter 8 on Sri Lanka, 2008–2009.
[5] See Chapter 14 on Law and International Humanitarianism.

In other kinds of humanitarian agencies, the legal framework is generally less clear, but there are nonetheless widely shared expectations that states will facilitate the provision of international assistance, where it is necessary and they have consented to it. Some restrictions are imposed for legitimate reasons, and under IHL the parties to conflict have the right to search relief consignments and convoys to check for weapons or military equipment, for example, and to require relief convoys to use prescribed routes at specific times.[6] Thus, during the Nigerian Civil War, when the government restricted aid into Biafra, refusing consent for air delivery of food unless transported on its own planes or those of the International Committee of the Red Cross (ICRC), ostensibly due to concerns that other planes also carried weapons, it was arguably acting reasonably.[7] However, these kinds of specifications must be applied in good faith and not in such a way that they make it impossible to provide relief items that are essential to the survival of the civilian population. In Sri Lanka in 2008–2009, when the government obstructed the provision of relief in bad faith and in such a way as to make it impossible for aid agencies to provide adequate relief in the Vanni, it was thus violating IHL.[8]

Where non-state actors control territory, as is often the case in civil wars (and sometimes in other situations of large-scale violence), their consent is necessary to operate on that territory, and such operations may be possible without the consent of the state. It is not clear that the consent of armed non-state actors is *legally* required for humanitarian relief operations, but where such actors control territory, their consent and acceptance is necessary in practice.[9] Conversely, where the territory they control can be accessed without crossing government-controlled areas (crossing the border from Turkey directly into rebel-held areas of Syria, for example), operations without the consent of the state may be practicable—and whether or not state consent is legally required in such situations is a matter of debate (Akande and Gillard 2016: 16–18; Gillard 2013: 364–367). In deciding whether or not to act without state consent in any given situation, humanitarian agencies are considering questions not only of legality but also of safety, potential impacts on their operations elsewhere in the country, and broader reputational risk. Thus the consortium of non-governmental organizations (NGOs) providing cross-border aid from Sudan to Ethiopia's northern regions in response to conflict and famine in the 1980s concealed their individual identities and worked anonymously under the umbrella of the Emergency Relief Desk

[6] See 'International humanitarian law' in Chapter 14 on Law and International Humanitarianism.
[7] See 'International humanitarian response' in Chapter 2 on the Nigerian Civil War, 1967–1970.
[8] See 'Manipulation of aid and aid agencies' in Chapter 8 on Sri Lanka, 2008–2009.
[9] See 'Non-state armed groups' in Chapter 20 on Armed Actors.

(ERD).[10] Elsewhere, international humanitarian agencies have sometimes sought to ensure that assistance could reach people in need throughout a country, with some operating exclusively in government-controlled territory and others exclusively in opposition areas.

The governments of affected states also have a role to play in the coordination and quality control of outside assistance (Harvey 2009: 8). Given that most states prize their (external) sovereignty very highly and humanitarians can only operate on their territory with their consent, it is unsurprising that governments expect that humanitarian actors will operate according to their rules (Kahn and Cunningham 2013: S145). A government's regulatory approach can serve to facilitate or to constrain the delivery of timely and effective assistance, and in many countries, much of the day-to-day work of international humanitarian agencies involves negotiating visas for expatriate staff, travel permits for different parts of the country, and access to particular populations (Harvey 2009: 9, 33). In some cases, regulations are applied in bad faith, as a means to pose obstacles to unwanted international assistance while claiming to facilitate it. In the final stages of the civil war in Sri Lanka, for example, the government frequently delayed and obstructed visas, authorizations to travel within the country, and import licences, using these bureaucratic obstacles together with threats in order to deter potential criticism.[11]

For refugee populations, governments in host states provide the policy framework. Even where the United Nations Refugee Agency (UNHCR) and other international agencies take responsibility for the actual provision of assistance, host governments are responsible for admitting and recognizing refugees on their territory, respecting the principle of non-refoulement, and providing security to refugees and humanitarian actors (Slaughter and Crisp 2009: 1–2). They also determine whether refugees will be settled in camps, whether they will be allowed to work, and what degree of self-sufficiency they will be permitted to develop (Harvey 2009: 10). Many states fear that by offering the conditions for integration, they will discourage refugees from returning. (Somewhat counter-intuitively, the opposite may be true, as the capacities developed for and through integration also prepare people for return; see Bradley 2017: 25.) Due to such fears, humanitarian agencies sometimes have strict limits imposed upon them by the governments of the countries in which they are operating. In many contexts, host governments

[10] See 'Non-cooperation and access restrictions imposed by the government' in Chapter 3 on Drought and Famine in Ethiopia, 1983–1985.

[11] See 'Manipulation of aid and aid agencies' in Chapter 8 on Sri Lanka, 2008–2009.

insist that refugees live in camps and/or impose limits on their rights to movement, work, and status, and constraints on the humanitarian actors seeking to protect and assist them (see, e.g. Brun 2016: 397–398; Lindley 2011: 20–21, 37; Slaughter and Crisp 2009: 7).

International aid agencies and the governments of affected states

Beyond getting consent for their operations and adhering to domestic rules and regulations, there is scope for significant variation in the ways that international humanitarian agencies interact with the governments of affected states. It is helpful to distinguish three main approaches adopted by international agencies: (i) substituting for the state in question, often bypassing it in the process and sometimes setting up parallel structures; (ii) working to support that state through such activities as upgrading infrastructure, funding government posts, and providing advice and capacity-building; and (iii) seeking to hold that state to account for its own obligations and responsibilities, either directly or indirectly. These three approaches can be combined, but, in many cases, tensions arise—substituting for a state can serve to undermine its capacities, complicating capacity-building, for example, and working closely with a state to support its work in particular areas can make it difficult to hold that same state to account when it fails to meet its own obligations and responsibilities.

Substituting for the state

Probably the most common mode of action for international humanitarian agencies is substituting for the state. During the Cold War, both development and humanitarian aid were often provided bilaterally, with donor governments directly funding the governments of affected states (Harvey 2009: 1, 5). However, with the Washington consensus came suspicion of the state and a tendency to sub-contract public functions to NGOs. In the humanitarian sector, this translated to a wholesale shift to funding UN agencies, international governmental organizations (INGOs) and the Red Cross and Red Crescent (RC) Movement in a model of humanitarian assistance which implicitly assumes that governments are either too weak or too corrupt to manage large-scale humanitarian funding (Duffield 1997: 527, 532; Harvey 2009: 1, 6).

Substitution can mean simply taking on tasks that would normally be the responsibility of the state but often additionally means doing so in such a way as to exclude the state from those tasks, as when coordination meetings are held in English or in locations that are difficult for nationals to access.[12] The use of jargon and acronyms can also serve to exclude those who are not (Western-trained) professional humanitarians (Bennett et al. 2016: 64). Sometimes, substitution is a response to lack of capacity in the affected state. However, humanitarian emergencies do not necessarily completely overwhelm the capacity of a state to act, even where the affected population requires—or can benefit from—international assistance (Harvey 2009: 2). Sometimes, substitution is a reaction to a lack of will on the part of the affected state to provide an adequate response. Refugee-hosting states in particular often prefer to leave UNHCR and other international agencies to assist refugees (Slaughter and Crisp 2009: 7–8). In many cases, however, the state has both the will and the capacity to play a central role, but international humanitarian agencies are generally poor at assessing capacities and tend to disregard local capacity and assume that affected states are too corrupt to deliver aid effectively (Harvey 2009: 16, 29).[13]

Substituting for the state can have unintended consequences and sometimes serves to reduce the capacity and accountability of affected states. State capacity may be reduced due to inactivity where international actors have taken on functions normally carried out by state officials or due to international agencies poaching government staff or paying much higher salaries than the government (Harvey 2009: 27). Substitution can also allow states to abdicate responsibility and undermine relationships or mechanisms of accountability (de Waal 1997; Harvey 2009: 29).

Substitution can thus pose serious dilemmas for international humanitarian agencies—where they compensate for the limited will or capacity of states, they can help to ensure people's needs are met but risk further undermining that will or capacity and even facilitating abuses. In Sri Lanka, for example, in the aftermath of the civil war when the government set up what were essentially internment camps for internally displaced persons (IDPs), the humanitarian agencies which provided shelter, food, water, sanitation, and other services for the IDPs in those camps were facilitating a system in which people were deprived of their liberty and forced to live in appalling conditions.[14]

[12] See, e.g. 'Coordination, competition, ownership, and accountability' in Chapter 7 on the 2004 Indian Ocean Tsunami and 'Coordination and ownership' in Chapter 9 on the Haiti Earthquake, 2010.

[13] See 'Assessing needs, contexts, and capacities' in Chapter 21 on Needs Assessment, Evaluation, and Response Decisions.

[14] See 'International humanitarian response' in Chapter 8 on Sri Lanka, 2008–2009.

Supporting the state

Working with, through, or in support of the state can take different forms, including capacity-building and channelling assistance through the government. The post-Washington consensus has led to renewed commitment to centring the state in *development*, but humanitarian aid is still mostly not aimed at bolstering the state. There are some exceptions to this general rule and some indications that the balance is shifting, albeit slowly. Non-Development Assistance Committee (DAC) donor governments tend to channel more of their funding directly to governments or national RC societies in affected states, so it may be that, with increases in the proportion of humanitarian funding provided by such donors, we will see an increase in bilateral humanitarian aid in coming years.[15] Among operational humanitarian agencies, there is also increasing concern to work with, and to seek to strengthen, affected states. In part, this is a response to the criticisms frequently levelled at international responses centred on substitution, but it is also closely linked to the protraction and urbanization of humanitarian emergencies and response[16] and to efforts to integrate humanitarian and development thinking in international response[17] as well as demands from assertive governments of affected states themselves (Harvey 2009: 34; Hilhorst 2018: 6).

When working to deliver basic services (e.g. health care or water supply) rather than commodities (e.g. food aid or cash), international humanitarian agencies have always tended to work with the state, often with technical line ministries in sectors like health, education, and water (Harvey 2009: 1). Even agencies with exclusively humanitarian mandates, such as the ICRC and Médecins Sans Frontières (MSF), who do not explicitly engage in state-building processes, continuously work with parts of the state. For example, ICRC hospitals often go beyond treating patients directly and additionally train medical personnel and secure their government salaries, and the ICRC sometimes works to upgrade government hospitals, all with the goal of expanding the quality and quantity of health-care services (ICRC 2016: 27). The more numerous multi-mandate agencies often have state-building as a central goal of their development work, and while they may try to keep their humanitarian work separate, they do not necessarily do so (Harvey 2009: 22). In contexts where donors or the political organs of the UN seek to support fledgling regimes, they may pressurize operational humanitarian agencies to

[15] See 'Donors' in Chapter 18 on Donors and the Funding of Humanitarian Action.
[16] See Chapter 15 on the Shifting Parameters of Humanitarian Emergencies.
[17] See Chapter 16 on the Nexus Concept.

support that government with the goal of increasing its strength or domestic legitimacy. This occurred with efforts to bolster the Karzai government in Afghanistan and the Transitional Federal Government (TFG) in Somalia.[18]

Like substitution, working to support and strengthen the state carries its own set of problems. Where governments are corrupt, support for the state may translate into support for corruption. While the levels of corruption among the governments of states affected by humanitarian emergencies may be overstated, some governments are corrupt and seek to control humanitarian aid to enrich themselves. Several of the case studies in the first part of this book provide examples of aid being diverted and manipulated by governments,[19] and if aid thus becomes embedded in systems of patronage, it can contribute to cultures of corruption (Harvey 2009: 29).

From a principled perspective, supporting the state can undermine the impartiality, neutrality, and independence of humanitarian action.[20] Working with and through the state may mean that planning and funding become contingent on, or subsumed under, processes led by the state, potentially reducing the autonomy of humanitarian agencies to take programming decisions independently. Working to support the state also risks jeopardizing the principle of neutrality, above all in armed conflict contexts but also in non-conflict contexts where supporting the state means supporting the regime in power. In such contexts, making strengthening the state an explicit goal of international assistance, even if only a secondary goal, implies non-neutral intentions. Even where strengthening the state is not a goal of humanitarian agencies but rather an incidental consequence of, say, expanding the quality and quantity of health-care services, it is likely to be non-neutral in outcome. Furthermore, if state-building is a goal that comes to define or shape what assistance is provided, and to whom, then that assistance is—by definition—not being provided in accordance with need alone and, hence, cannot be said to be impartial.

In practice, however, the degree to which humanitarian agencies and, especially, donor governments are prepared to work to support the governments of affected states may depend less on concerns about will, capacity, corruption, or humanitarian principles and more on how governments are perceived or favoured by powerful international actors (Harvey 2009: 15).

[18] See, respectively, 'Instrumentalization of aid by military actors' in Chapter 6 on Afghanistan, 2001–2014, and 'Diversion, manipulation, and instrumentalization of aid' in Chapter 10 on Famine in Somalia, 2011–2012.

[19] See, e.g., 'Use of food aid for non-humanitarian ends' in Chapter 3 on Drought and Famine in Ethiopia, 1983–1985, 'Manipulation of aid and aid agencies' in Chapter 8 on Sri Lanka, 2008–2009, and 'Diversion, manipulation, and instrumentalization of aid' in Chapter 10 on Famine in Somalia, 2011–2012.

[20] For a discussion of these principles, see Chapter 13 on Politics, Principles, and Humanitarian Action.

Both the Karzai government in Afghanistan and the TFG in Somalia were widely known to be corrupt, for example, yet donors in Afghanistan and the political backers of the TFG in the UN sought to use humanitarian assistance to bolster these governments and increase their legitimacy in the eyes of their populations.

Holding the state to account

The third way that international humanitarian agencies can engage with the state in the countries in which they operate is through direct or indirect efforts to hold the state to account. Such efforts can be aimed at protecting the civilian population from attack, or from other abusive policies, by the state. They can also be aimed at getting the state to consent to, and facilitate, external assistance, or at challenging access restrictions imposed by the state on humanitarian agencies.

Public criticism and legal action are among the most obvious ways to hold states to account for abuses. International human rights organizations like Amnesty International and Human Rights Watch conduct research on human rights violations and then 'name and shame' the perpetrators through public advocacy. The pursuit of accountability is likewise part of the purpose of international justice mechanisms, including the International Court of Justice (ICJ), which adjudicates inter-state disputes, and the International Criminal Court (ICC), which prosecutes individuals, including state officials. Humanitarian agencies can also seek to hold a state to account through naming and shaming or through support for international prosecutions.[21] However, many are cautious about these kinds of activities, concerned that they may imply compromising the principle of neutrality and that they may result in losing access if authorities are named, shamed, or prosecuted, and then retaliate against those they deem responsible. However, the impact on neutrality is not so straightforward in practice,[22] and, in some situations, naming and shaming authorities who pose obstacles to

[21] For more on 'naming and shaming', see 'Public advocacy and criticism of authorities' in Chapter 23 on Dialogue, Negotiation, and Advocacy, and on support for international prosecutions, see 'International justice' in Chapter 14 on Law and International Humanitarianism.

[22] See 'Interpretation and implementation of these principles in practice' in Chapter 13 on Politics, Principles, and Humanitarian Action.

humanitarian response might pressurize them to extend access rather than limit it.[23]

Where public advocacy is expected to be ineffective or counterproductive, or to have unwanted side effects in terms of principles, reputation, and access, there is a variety of other activities that humanitarian agencies can undertake to help increase the accountability of states for their responsibilities to protect and assist the population. For example, they sometimes pass information to other states through dialogue with diplomatic representatives or to other actors who can put pressure on the affected state, such as human rights organizations.[24] Similarly, they can work indirectly to hold the state—or parts of it—to account through legal mechanisms. International humanitarian agencies thus sometimes provide domestic actors with the information, tools, and resources they need to hold authorities to account for any failure to fulfil their duties and obligations. For example, UNHCR and the Norwegian Refugee Council run programmes focused on providing legal assistance to people affected by displacement to enable them to claim and exercise their rights (Bradley 2016a: 138).

As with substitution and support, efforts to hold states accountable carry certain risks and can be in tension with the other two approaches. As indicated, more adversarial approaches may have negative consequences for access, operational space, and neutrality, but this is not necessarily the case, and concerns about the impact on neutrality in particular seem overstated (but not entirely unfounded). The corollary of substitution activities undermining state accountability is that efforts to hold states accountable for their failures often require humanitarian actors to withhold direct assistance to those in need. For example, in responding to the so-called migration crisis and other humanitarian needs in Europe, most humanitarian agencies judged that the EU and its member states had the capacity both to fulfil their search-and-rescue (SAR) responsibilities and to protect and assist migrants and other marginalized groups on their territory and that they lacked only the will. Some agencies thus sought to minimize the extent to which they substituted for states and instead engaged in more extensive advocacy than normal.[25] In such situations, humanitarian agencies are choosing between putting significant pressure on states to change the policies that are the cause

[23] See 'Public advocacy and criticism of authorities' in Chapter 23 on Dialogue, Negotiation, and Advocacy.

[24] See 'Sharing information with third parties' in Chapter 23 on Dialogue, Negotiation, and Advocacy.

[25] See 'Humanitarian response' in Chapter 12 on the 2015–2016 European 'Migrant Crisis', and 'Europeanization' and 'People on the Move' in Chapter 15 on the Shifting Parameters of Humanitarian Emergencies.

of suffering, on the one hand, and doing all they can to relieve that suffering directly on the other.

Combining different ways of relating to the state

Clearly, no single mode of relating to the state is appropriate all of the time, and in practice, humanitarian agencies can, and do, adopt different approaches—and combinations of approaches—in different contexts. There is a certain logic to adapting the approach according to the will and ability of the state in question. Thus, a humanitarian agency might seek to hold able but unwilling states to account; to support willing but unable states, perhaps partially substituting for them until their capacity is strengthened; and to combine all three approaches in states that lack both will and ability (Bradley 2016a: 135; Harvey 2009: 16). However, this logic may be overly simplistic.

Where a state is failing to fulfil its responsibilities, it can be difficult to discern how much this is due to insufficient will and how much due to insufficient capacity, and it is not clear that most humanitarian agencies have the skills to distinguish. Even if they were able to distinguish lack of will from lack of capacity, will and capacity are not necessarily fully present or fully absent. In practice, almost all governments have some capacity and some will to protect and assist their populations, even if at inadequate levels. Furthermore, capacity and will often vary across different levels of government. It is thus important to distinguish the different branches, levels, and sectors of government. Where engaging with central government departments is deemed unwise or difficult, there may still be the option to work with local government actors or technical line ministries (Harvey 2009: 31). In some cases, humanitarian agencies work to support the judiciary in a country through such activities as providing advice or financial support to oversight mechanisms, so indirectly helping to hold the executive to account (Bradley 2016a: 138).

In addition, the apparently logical approach is not always the only or best way of increasing the will or capacity of the state. For example, increasing the will of a state (or some parts of a state) to ensure that its population is protected and assisted does not necessarily or only imply holding the state to account. Behaviour change can be encouraged by carrots as well as sticks, and support to the state may be used in conjunction with dialogue and made conditional on certain conduct. Working closely with the state can also enable international actors to observe state conduct from the inside, providing them with the understanding necessary to inform other accountability measures. Assistance does not have to be completely state-led or completely state-avoiding, and humanitarian agencies channelling humanitarian assistance through governments can invest in oversight measures to monitor and audit state-led responses (Harvey 2009: 29). At the same time, there can be tensions between different goals and the activities undertaken in pursuit of them—as when substituting for, or supporting, the state serves to undermine its capacity and accountability.

Civil society actors

Governments are not the only actors in states affected by humanitarian crises that are frequently sidelined by international responders. Local and national civil society actors play an important role in responding to crises, and they have a number of advantages over international humanitarian agencies. However, their role has often been under-acknowledged by international actors, and they face much greater difficulties in accessing international sources of funding than do the bigger and well-known international NGOs.

A range of different civil society actors in states directly affected by humanitarian crises, and in states hosting refugees, are essential players in humanitarian response. These actors include national RC societies, faith-based organizations, formal local and national NGOs, and refugee-led organizations. In the crucial hours immediately following sudden-onset disasters in particular, local actors are often the only, or the most important, responders. In the aftermath of the 2004 Indian Ocean tsunami, for example, local communities were the first to respond, quickly followed by national and regional (state and non-state) actors from relatively unaffected areas nearby.[26] Following the 2010 earthquake in Haiti, and despite the personal suffering and trauma of survivors, Haitians were the first to respond, with the local staff of INGOs already operating in the country especially effective in the initial response.[27] The Ebola outbreak which hit West Africa in 2014 was not such a sudden event, but delays in mounting a large-scale international response nevertheless meant that local actors, both governmental and civil society, were the first to respond, implementing protective and isolation initiatives.[28] In contexts of conflict and violence, local actors are often active in areas where international humanitarian agencies have little or no presence due to government restrictions or insecurity (Svoboda et al. 2018: 8). In many countries,

[26] See 'Tsunami and its immediate aftermath' in Chapter 7 on the 2004 Indian Ocean Tsunami.
[27] See 'Background and context' in Chapter 9 on the Haiti Earthquake, 2010.
[28] See 'Treatment of Ebola patients' in Chapter 11 on Ebola in West Africa, 2014–2015.

national RC societies play a central role, sometimes linked to their close relationship with their respective government and sometimes linked to the unique relationship they have with the rest of the RC Movement.

Local and national actors often have a number of advantages over international responders, stemming from their flexibility and their superior local knowledge and understanding. Local knowledge, understanding, and input can have an important impact on the quality and effectiveness of responses to humanitarian crises, with organizations embedded in affected communities well placed to identify and respond to the needs of those communities (Svoboda et al. 2018: 11). In the context of the Ebola outbreak in West Africa in 2014–2015, for example, international responders initially imposed top-down approaches to infection control and patient care, which were ill suited to the needs and social conditions of many communities, and protocols were more effective when adapted based on local understanding.[29] Familiarity with the context can also be an advantage in getting access in insecure environments. In Syria, for example, local organizations have drawn on their in-depth understanding of different armed groups in access negotiations, adapting language and behaviour at checkpoints depending on the armed group in charge (Svoboda et al. 2018: 11). Membership of affected communities and continued presence through periods of high insecurity, when international actors are more likely to withdraw, have also been shown to increase trust in the most locally based organizations (Svoboda et al. 2018: 11).

Despite the advantages and the importance of local and national NGOs, they receive only a small percentage of international funding for humanitarian response and are frequently sidelined by UN agencies and international NGOs. For the most part, they either receive no international funding at all or they receive it from INGOs, who effectively sub-contract certain tasks to them. While called 'local partners', the term 'partner' implies a level of equality that seldom characterizes these relationships. The tendency to use English as the language of coordination meetings and other obstacles that serve to exclude governmental actors in affected states often also serve to exclude local and national NGOs (Bennett et al. 2016: 64; Svoboda et al. 2018: 10). The formal humanitarian system has mostly failed to connect meaningfully with national and local actors and lacks the incentives for such engagement (Bennett et al. 2016: 5). Indeed, given the 'corporatization' and funding structures of the humanitarian sector, which tend to mean short-term funding for

[29] See 'International humanitarian response' in Chapter 11 on Ebola in West Africa, 2014–2015.

specific activities and tend to reward outputs rather than outcomes, the established international agencies are incentivized to pursue funds even when other actors are better placed to respond (Bennett et al. 2016: 59–60).

Localization agenda

Against the background of international actors holding all the power and local and national NGOs struggling to attract international funding and respect, the 'localization agenda' has emerged. The basic idea is that, instead of taking over themselves, international humanitarian actors should support the efforts and capacities of crisis-affected people and the local actors assisting them (Schenkenberg 2016: 5). Thus, funds should be rechannelled to local and national actors in countries affected by crisis, local and national partners of international actors should be more systematically involved in the development and implementation of projects, and the work of local and national actors should be properly acknowledged and given greater public visibility (Roepstorff 2020: 286). 'Upending humanitarianism' (Fast 2017) or 'Turning the humanitarian system on its head' (Gingerich and Cohen 2015) can be seen as a goal in and of itself insofar as it serves to transform the colonial mindset. However, localization is also widely expected to lead to higher-quality, faster, more cost-effective and sustainable responses, promoting resilience and contributing to developmental and peace-building goals (Barbelet et al. 2021: 20; Fan 2013: 2; Hilhorst 2018: 6; Metcalfe-Hough et al. 2021: 56; Roepstorff 2020: 286). The agenda is at least partly driven by a perceived need to do more with less, as humanitarian needs outpace funding increases, concerns about the protracted nature of crises, and a desire not only to meet but also to reduce humanitarian needs.[30]

Efforts to transfer power and financial resources to actors within affected states predate the WHS in 2016, but they were given significant added impetus with the Grand Bargain commitments made by donors and operational agencies at the Summit. Specifically, Grand Bargain Core Commitment 2.4 set a global aggregated target of at least 25% of humanitarian funding going to local and national responders as directly as possible by 2020. This is expected to improve outcomes for affected people and reduce transaction costs. By 2020, however, only thirteen out of fifty-three grant-giving signatories reported that they had met or exceeded this 25% target, and only 4.7% of global humanitarian funds, representing USD 1.3 billion, went to

[30] See 'Protracted "emergencies"' in Chapter 15 on the Shifting Parameters of Humanitarian Emergencies, and Chapter 16 on the Nexus Concept.

local and national responders (Metcalfe-Hough et al. 2021: 52). Between 2016 and 2020, the share of global humanitarian funding going directly to local or national actors fluctuated between 3.8% and 9.8%, with no discernible trend (Metcalfe-Hough et al. 2021: 53). Counter-terror legislation and policy pose an obstacle to meeting the 25% target since smaller local and national organizations are often unable to meet stringent due diligence requirements in places where groups designated as terrorist are in operation (Austin and Chessex 2018: 5).[31] There were, however, more positive trends with respect to pooled funds, with the percentage of country-based pooled funds (CBPFs) and Central Emergency Response Fund (CERFs) allocations to national and local responders increasing since 2016 (Metcalfe-Hough et al. 2021: 54).[32]

Despite only slow progress on meeting the quantitative targets for directing funding to local and national responders, the Grand Bargain appears to have helped to move the localization agenda forward in other ways. It has been credited with driving a system-wide normative shift such that the appropriateness of localization is no longer in question, even if implementation of the norm has been slower (Metcalfe-Hough et al. 2021: 46). Progress has been made on enabling government and civil society actors in affected states to engage in international coordinating mechanisms, with more than half of clusters indicating that a national/local language was spoken at meetings, for example (Metcalfe-Hough et al. 2021: 47). However, it remains unclear how far decision-making power has shifted from international actors towards local and national actors (Barbelet et al. 2021: 24). Moreover, there is currently little evidence to show how such a shift in power affects humanitarian outcomes. Research and evaluations have largely focused on assessing how far output goals, rather than outcomes, have been achieved.[33] Thus, progress is tracked against high-level commitments regarding the percentage of funding going to local actors, for example, but there is much less focus on the effectiveness or efficiency of more localized humanitarian response (Barbelet et al. 2021: 23).

The discourse of localization has been criticized both for a lack of clarity in terms of who is local/international and for reifying the distinction between the two. It is not always clear which kinds of actors are included

[31] On the impact of counter-terror measures on humanitarian action more broadly, see 'Counter-terror legislation' in Chapter 14 on Law and International Humanitarianism.

[32] For an explanation of pooled funding mechanisms, see 'Pooled funding mechanisms' in Chapter 18 on Donors and the Funding of Humanitarian Action.

[33] For a discussion of evaluations in the humanitarian sector and the wider tendency to focus on measuring outputs rather than outcomes, see 'Monitoring and evaluation' in Chapter 21 on Needs Assessment, Evaluation, and Response Decisions.

in the localization agenda or what exactly is meant by local. Some interpretations include internationally affiliated organizations, such as national RC societies or local offices of INGOs, but other interpretations exclude these (Roepstorff 2020: 290). NGOs based within the affected state are consistently included, whereas less formal entities and governmental actors are sometimes, but not always, included (Schenkenberg 2016: 9). The individuals, families, and networks that are most directly affected by crises and often among the most important responders are not always part of the kinds of formal organizations the aid system is set up to support (Schenkenberg 2016: 10). In refugee settings, actors which are national within the host state are not local to the refugees, while refugee-led organizations are not necessarily considered local in terms of the localization agenda (Barbelet et al. 2021: 22). Diaspora organizations, run by expatriates to provide assistance in their home countries, have seen significant growth, with Syrian diaspora organizations in particular playing an important role in the Syrian war (Knox Clarke et al. 2018: 106). While they may be established elsewhere in the world, many identify as local (Svoboda et al. 2018: 2). Lack of clarity on which actors are included makes assessing progress difficult insofar as progress is measured with quantitative indicators, but such ambiguity has the advantage of allowing for context-specific understanding and goals. More problematically, the localization discourse and agenda risks reinforcing the dichotomy between international and local or national actors and, as such, risks perpetuating the issues it seeks to address (Roepstorff 2020).

A second set of critiques calls into question the very goal of transferring power to local actors. These critiques highlight the risk that the localization agenda can be used by regimes seeking to assert their external sovereignty and by government or opposition forces seeking to prevent or reduce an international presence to bear witness to their IHL violations or human rights abuses (Schenkenberg 2016: 5). Furthermore, they raise a concern that local and national actors may struggle to adhere to the principles of humanitarian action (ICRC 2016: 22; Schenkenberg 2016: 5–6). They may not adhere to the principles of impartiality and neutrality for a variety of reasons—as a conscious choice to privilege a particular group, as part of an unconscious repetition of internalized norms of exclusion of a particular group, or as a response to a fear of reprisals if they do not privilege (or exclude) a particular group (Austin and Chessex 2018: 18–19; Schenkenberg 2016: 15). Local organizations often adopt a more solidarist approach and do not always separate out the provision of relief from developmental, human rights, or peacebuilding activities—and, in some cases, groups which began with explicitly and exclusively political goals shifted to a humanitarian focus out of necessity

(Bennett et al. 2016: 49; Svoboda et al. 2018: 9). An activist past can be perceived by those controlling territory as undermining the neutrality of the local actor in question, and this has sometimes led to denial of permission to access that territory (Svoboda et al. 2018: 11).

While these critiques raise important concerns, *international* humanitarian agencies do not always adhere to the principles of humanitarian action either, and local actors may sometimes be better, rather than worse, placed to ensure—or at least increase—the impartiality of humanitarian response. Their contextual understanding may enable them to identify traditionally excluded groups, with identification being a necessary, if insufficient, condition for receiving assistance.[34] Elsewhere, they may be able to access areas which international actors cannot, as was the case with the White Helmets in the Syrian Civil War, known for rescuing people from bombed buildings across opposition-controlled territory. Ed Schenkenberg sees the fact that the Syrian Arab Red Crescent was working mainly in government-controlled territory and the White Helmets mainly in opposition-controlled territory as an illustration of some inherent limitations of local humanitarian actors in armed conflict contexts (Schenkenberg 2016: 15). However, this kind of limitation is not unique to local actors. International humanitarian agencies were largely limited to operating in government-controlled parts of Syria (Martínez and Eng 2016). In light of the limitations of international actors in that context, then, the ability of the White Helmets to operate in opposition-held territory contributed to a response that met needs across the country.

Some of the more thoughtful analyses point to complementarity, recognizing that there is no single response model and that different actors and approaches have a comparative advantage, depending on the context (Austin and Chessex 2018; Bennett et al. 2016: 6). It is useful not only to distinguish between disasters and armed conflicts but also to identify more specific characteristics of different crises and the actors operating within them to determine what division of labour will yield the best outcomes for the affected population (ICRC 2016: 22; Schenkenberg 2016: 16). In insecure contexts, local actors may be able to access more areas than international agencies, but it is not always safer for them to do so and relying on local partners to deliver aid in insecure environments may imply transferring risk to them, which raises ethical concerns. Moreover, access negotiations take place at multiple levels, from the United Nations Security Council (UNSC) to the checkpoint and the affected community, and some organizations are better

[34] See 'Impartiality' in Chapter 10 on Famine in Somalia, 2011–2012.

able to negotiate with more leverage at one end of the spectrum while other organizations are stronger at the other end (Svoboda et al. 2018: iv). The UN Secretary-General's call at the WHS in 2016 to make humanitarian action 'as local as possible, as international as necessary' encapsulates this vision for complementarity at the same time as it recognizes that power and resources are currently excessively concentrated in the hands of international responders.

Finally, there is a set of critiques which focuses on how the localization agenda is being implemented in practice, arguing that current practice does not go far enough in transferring power. First, despite the formal commitments, local actors are often still relegated to roles as implementing partners rather than equals or leaders, without significant involvement in decision-making processes or control over financial resources (Svoboda et al. 2018: 8). In this sense, the agenda has not been sufficiently implemented. Second, implementation of the localization agenda is largely led by international agencies and focused on building capacity through technical fixes and the transfer of resources, with the implicit goal of making local and national actors look and function more like international actors (Bennett et al. 2016: 57; Fast 2017: 2–3). The problem identified here is not so much insufficient but rather misguided implementation. This kind of capacity-building (which can include, for example, building capacities related to evaluation, impact, and evidence) might help local actors to access international funding sources but it might also compromise their flexibility, agility, and situatedness—the very characteristics that give them many advantages over international actors in terms of understanding and responding to needs (Bennett et al. 2016: 57; Fast 2017: 3; Svoboda et al. 2018: iii). Letting go of power and control means giving up claims to control the definition of humanitarianism and with it the insistence that humanitarian action should necessarily be neutral, independent, and limited—remaking humanitarian action to encompass more developmental or solidarist approaches (Bennett et al. 2016: 6; Fast 2017: 5).

Conclusions

The ways that international humanitarian agencies work in relation to governmental and civil society actors in affected states have important impacts on humanitarian response and outcomes, and on the strength of national and local actors over the long term. Each different way of relating to the state reflects a different ethos—substitution reflects the humanitarian ethos, while working with and supporting the state is more often associated with development work and holding the state to account is more commonly the modus operandi of human rights organizations. Similarly, different ways of working with local and national civil society actors can fit with a more strictly principled humanitarian ethos and approach or can embrace more solidarist approaches, which seek to address not only the symptoms but also the causes of suffering. Each approach is subject to critique, and the challenge for the humanitarian sector is to strike the best balance in any given context. Notwithstanding the importance of adapting the approach according to context, there is now broad—albeit not universal—agreement that, in general, humanitarian response should be more localized, and the debate focuses primarily on what that means in practice and how to achieve it (Fast 2017: 2).

20
Armed Actors

Invariably in armed conflicts, and frequently in many other kinds of humanitarian emergency, international humanitarian agencies operate in the same spaces as armed actors—which can include international military forces, the national militaries of affected states, and non-state armed groups. Some interaction between humanitarian and armed actors is often therefore inevitable, but the nature and extent of interaction varies enormously depending on the humanitarian actor in question and the type of armed actor and context.

Humanitarian engagement with armed actors is often important, both because armed actors affect humanitarian outcomes directly and because they affects the ability of humanitarian agencies to operate and to address humanitarian need. Whether national or international, state or non-state, armed actors may pose a violent threat to the civilian population (or part of it), or they may protect that population (or part of it) from other violent threats. They may block, facilitate, or force displacement, and they may block, facilitate, or provide assistance. Accordingly, humanitarian agencies need to engage with armed actors, either directly or indirectly, to gain acceptance for their work and access to affected populations. They also need to engage if they seek to shape the conduct of armed actors towards the civilian population, notably in terms of security and protection. Many humanitarian agencies and their staff, however, are uncertain as to whether, and how, they should engage with armed actors.

Different humanitarian agencies interact with armed actors in very different ways. From a classical International Committee of the Red Cross (ICRC) or Médecins Sans Frontières (MSF) perspective,[1] humanitarian action and military action represent two completely separate spheres of action—but, in principle, they engage in dialogue with all kinds of armed actors to negotiate access and, in some cases, to try to persuade those actors to comply with a wider range of obligations in international humanitarian law (IHL)

[1] For a general explanation of how the major international humanitarian agencies position themselves with respect to principled humanitarian action, see 'Perspectives of the largest international humanitarian agencies' in Chapter 13 on Politics, Principles, and Humanitarian Action.

The Politics and Everyday Practice of International Humanitarianism. Miriam Bradley, Oxford University Press.
 DOI: 10.1093/oso/9780198840695.003.0020

and international human rights law (IHRL). This means that they endeavour to build relationships and engage with armed actors at the same time as they seek to avoid supporting those actors' political or military goals or becoming dependent on those actors. Other agencies tend to be more comfortable engaging with national and international military forces than armed non-state actors. Indeed, in several contexts, some agencies have been criticized for working too closely with international forces, sometimes compromising their commitments to principled humanitarian action as a result.[2] By contrast, many humanitarian agencies are reticent to engage with armed non-state actors, and there is broad consensus that positive humanitarian outcomes require more rather than less engagement.

National militaries of affected states

As parties to either international or non-international armed conflicts, the armed forces of affected states can pose a direct threat to civilians. In the Nigerian Civil War, for example, the national military imposed a siege on the Biafran enclave, causing severe and widespread famine among the civilian population (see Chapter 2). During the civil war in Ethiopia, government forces adopted brutal counter-insurgency tactics, including scorched earth policies and the forcible relocation of famine victims (see Chapter 3). The 1994 Rwandan Genocide was state-led, with the armed forces playing a central role (see Chapter 5). In the final stages of the Sri Lankan Civil War, the military shelled civilian targets and then confined the displaced Tamil population to internment camps (see Chapter 8). At the same time, national militaries may serve to protect the civilian population (or some parts of it) from violent threats posed by other parties to conflict. This is most obvious in international armed conflict, where the military is expected to use force to protect the country (and its population) from external attacks, but in non-international armed conflicts, the armed forces may also protect people from attacks by opposition forces.

National militaries are often involved in responding to emergencies on their own territory, particularly emergencies other than armed conflict. For example, the Indian army established relief camps for 23,000 people and provided emergency medical care following the 2001 earthquake in Gujarat, deployed 8,300 troops for rescue and relief operations in tsunami-affected parts of India (and also in neighbouring Sri Lanka) in 2004, and also

[2] See, e.g. 'International humanitarian response' in Chapter 6 on Afghanistan, 2001–2014.

contributed to the response to floods in Bihar in 2008 (Harvey 2009: 12). The Pakistan military played a central and reportedly highly effective role responding to the 2005 earthquake, providing food, shelter, and livelihood and medical services (Harvey 2009: 12). Military support for the domestic response can also go beyond the provision of relief and does not necessarily have a positive impact on humanitarian outcomes. For example, when faced with an Ebola outbreak in 2014, the governments of Liberia and Sierra Leone used the military to enforce quarantines and restrictions on movement.[3] In many contexts of large-scale criminal violence that fall short of the IHL thresholds for armed conflict, as in Brazil, Mexico, and Central America, governments have also used the military or military police for law enforcement purposes (Jütersonke et al. 2009; Kalmanovitz and Bradley forthcoming; Sampaio 2019).

Where the affected state is a party to armed conflict, different international humanitarian agencies engage with the national military to varying degrees. The day-to-day work of the ICRC involves providing training in IHL and engaging in a dialogue on operational space and the protection of civilians (Bradley 2016a: 162–163, 171–173). It is precisely because the national militaries can be parties to conflict that the ICRC seeks to increase their understanding of IHL and persuade them to comply with it (Bradley 2016a: 133). Some other humanitarian agencies also engage in training, dialogue, and advocacy with respect to national militaries but most do not do so as consistently as the ICRC (see, e.g., Thompson 2008: 21). By contrast, some other humanitarian actors are reticent to engage with parties to conflict beyond the bare minimum, which may include negotiating access and only rarely involves substantive protection dialogue (Bradley 2016a: 134). Some do not engage directly even to negotiate access, instead leaving this task to the Office for the Coordination of Humanitarian Affairs (OCHA) as part of its coordination role.[4] Despite the reticence of many agencies to engage with any parties to conflict, most seem to engage more with the national militaries of affected states than they do with other armed actors (Bradley 2016a: 143–146; Thompson 2008: 13).

The nature of engagement, and the consequences of that engagement, depend not only on the agency in question but also on the context. In civil wars, working with the national military raises concerns about neutrality and about perceptions of neutrality. To the extent that such work serves to strengthen the military, whether materially or by bolstering their legitimacy

[3] See 'Militarization and securitization' in Chapter 11 on Ebola in West Africa, 2014–2015.
[4] See 'Access negotiations' in Chapter 23 on Dialogue, Negotiation, and Advocacy.

in the eyes of the population, it may be non-neutral in terms of outcomes.[5] To the extent that humanitarian agencies are thus perceived by other parties to conflict as taking sides, it may reduce their ability to engage with those other parties. In the context of disasters linked to natural phenomena, humanitarian agencies are generally more open to working closely with the national military. In responding to the 2005 earthquake in Pakistan, for example, while the ICRC and MSF sought to keep their own work separate from the military's relief efforts, most other agencies put aside their concerns about independence and neutrality and worked together with the military—and even the ICRC coordinated with the army on some aspects of the response (Harvey 2009: 12).

Non-state armed groups

Almost all armed conflicts in the contemporary world are non-international and involve non-state armed groups (hereafter, armed groups), which can pose a direct threat to the population. Violence from a clash between two or more parties to conflict can spill over to affect civilians (the idea of being 'caught in the crossfire'), or civilians can be directly targeted by armed actors. In civil wars, clashes between parties to conflict necessarily involve at least one armed group, and long-term trends suggest that worldwide, armed groups are responsible for a greater proportion of civilian fatalities from direct targeting in conflict than are government forces (Eck and Hultman 2007; UCDP 2012). Furthermore, direct conflict deaths now account for only a fraction of violent deaths worldwide, and organized violence *outside* armed conflict—in particular that perpetrated by criminal groups—has significant effects on the wider population (Krause 2016; Mc Evoy and Hideg 2017). Armed non-state actors, whether insurgent groups in civil wars, paramilitaries supporting (but distinct from) the state, other militias, gangs, or organized crime groups, are thus a major cause of civilian insecurity.[6]

In addition to posing a direct threat to civilians, armed groups can also pose an indirect threat by restricting the ability of the population to access life-saving assistance. They may seek to obstruct humanitarian access for a variety of reasons, including strategic advantage, control of resources, and suspicion of humanitarian actors (Jackson 2012a: 2). In many contexts, armed groups

[5] On neutrality of intent and outcome, see 'Feasibility of apolitical action' in Chapter 13 on Politics, Principles, and Humanitarian Action.

[6] For examples from the case studies in the first part of this book, see Chapter 4 on the Bosnian War, 1992–1995, Chapter 6 on Afghanistan, 2001–2014, and Chapter 8 on Sri Lanka, 2008–2009.

exercise some control over territory and, with it, control over access to that territory for international humanitarian agencies. In Somalia, for example, Al-Shabaab controlled those parts of the country worst affected by famine in 2011 and expelled some agencies, while imposing tight restrictions on those it allowed to operate.[7] Even where an armed group's control over territory is limited, the fact that its members are armed enables them to exercise coercion over unarmed actors, and this can include roadblocks, where they prevent the passage of humanitarian workers and supplies or where they steal or 'tax' such supplies. Armed actors sometimes attack aid workers and agencies, and some research suggests that armed groups are the main perpetrators of such attacks (Stoddard et al. 2017). Gangs and other armed criminal groups may pose similar obstacles and threats to aid workers and operations, something that is of growing concern as many international humanitarian agencies seek to respond to contexts of urban violence (Stoddard et al. 2021).[8]

Conversely, armed groups can also protect and assist populations. They can protect by actively defending the population (or some parts of it) from violence perpetrated from other actors or by exercising restraint towards that population in the conduct of hostilities (ICRC 2018). In many cases, insurgent groups in civil wars are also providers of basic services, and in some cases, criminal groups in contexts of large-scale violence outside armed conflict also play this role (Lessing 2021; Stewart 2018).

As armed groups are important determinants of humanitarian outcomes, aid agencies need to engage with them. Where armed groups control territory, humanitarian agencies need armed non-state actors to consent to, and accept, humanitarian operations on that territory. While it is not clear that the consent of armed non-state actors is *legally* required for humanitarian relief operations, operating without their consent and acceptance is likely to imply security risks beyond the risk threshold of most international humanitarian agencies. Getting their consent and acceptance requires either direct or indirect engagement with armed groups in control of territory. Engagement with armed groups is also widely understood to be a necessary condition for improved protection outcomes (ADH 2011; Jackson 2012a; Mc Hugh and Bessler 2006; UN Security Council 2009, para. 40; Zeender 2005).

The extent and nature of humanitarian engagement with armed groups varies, depending on the agency and on the type of armed group as well as on the specifics of any given context. Organizations such as the ICRC,

[7] See 'Background' and 'International humanitarian response' in Chapter 10 on Famine in Somalia, 2011–2012.

[8] On the incorporation of contexts of urban violence into humanitarian mandates, see 'Urbanization' in Chapter 15 on the Shifting Parameters of Humanitarian Emergencies.

Geneva Call, and the Centre for Humanitarian Dialogue undertake practical efforts based on the assumption that engaging with groups is the best way to alter their behaviour to better conform to international law and human rights norms. By contrast, the majority of international humanitarian agencies do not engage strategically with armed groups (Jackson 2012a: 1). Even the ICRC is more reticent when it comes to engaging with criminal groups that are not party to armed conflict as per IHL (Bradley 2020). In territory where armed groups of any kind operate, occasional contact tends to be unavoidable, so humanitarian agencies operating in those areas have some level of interaction, even if they are not keen to publicize it. Occasionally, international humanitarian agencies contract armed groups to provide armed escorts or other kinds of protection. Even the ICRC and MSF, who seek to avoid using any kind of armed escort in an effort to ensure their neutrality and independence, have, in some contexts, opted to purchase armed protection when they deemed that was the only way they could operate. For example, all international humanitarian agencies—including the ICRC—operating in southern Somalia in the early 1990s were forced to buy armed protection from local clan militias (Menkhaus 2010: S323–S324).

The engagement of the sector as a whole appears to have varied over time, generally moving from greater engagement during the Cold War to less engagement today (Jackson 2012a: 1). This is difficult to measure systematically, in part because humanitarian actors are often reticent to acknowledge publicly their interactions and engagement with armed groups (Jackson and Davey 2014: 1; Svoboda et al. 2018). However, there are several examples of extensive engagement with non-state armed groups in the Cold War, including working closely with Biafran separatists in the Nigerian Civil War and with the relief wings of the Eritrean People's Liberation Front (EPLF) and Tigrayan People's Liberation Front (TPLF) in Ethiopia from 1983 to 1990.[9] Throughout the 1990s, the United Nations (UN) took the lead in negotiating with armed groups on behalf of the wider humanitarian community, as exemplified by Operational Lifeline Sudan, which operated to provide assistance in both government- and opposition-controlled territory from 1989 to 2005 (Jackson and Davey 2014: 24–26). However, UN leadership of humanitarian negotiations has diminished significantly since 9/11 (Jackson 2012a: 3). On the one hand, many humanitarian actors feel that the UN has become too politicized and is no longer an appropriate body to lead engagement with armed groups (Jackson 2012a: 3). UN humanitarian agencies frequently

[9] See 'Political impacts: Supporting the secessionists and prolonging the war' in Chapter 2 on the Nigerian Civil War, 1967–1970 and 'Non-cooperation and access restrictions imposed by the government' in Chapter 3 on Drought and Famine in Ethiopia, 1983–1985.

show deference to the government even as they profess their neutrality, as, for example, in Sierra Leone in the early 1990s and Sri Lanka in 2008–2009 (Internal Review Panel 2012; Keen 1998a: 321, 2014). Where the wider UN is engaged in a state-building project, and especially where this involves international military forces, the UN may be seen to be taking sides and hence lacking in credibility as an honest broker for access negotiations.[10] On the other hand, in many contexts this leaves a lacuna since operational agencies are often ill-equipped to negotiate access directly, and their efforts tend to be fragmented, which weakens their ability to negotiate effectively (Jackson and Davey 2014: 27–28; Muggah 2010: S457; Svoboda et al. 2018: 6).

There is wide agreement that humanitarian engagement with armed non-state actors is insufficient in current practice. The lack of senior-level engagement has been found, in some contexts, to undermine efforts to get the security guarantees necessary for accessing some populations in need and has limited the opportunities to undertake advocacy on issues relating to the protection of civilians (ADH 2011: 4; Jackson 2012a: 3). In Afghanistan, for example, most international humanitarian agencies did not engage with the Taliban systematically or at a senior level, and this was seen to contribute to limited access and high levels of aid worker insecurity.[11] In Somalia, weak or non-existent relationships with Al-Shabaab limited the scale and speed at which humanitarian agencies could scale up programming in response to famine in 2011.[12] Apart from limiting access to those in need of protection and assistance, when senior expatriate staff of humanitarian agencies fail to engage with armed groups, they may inadvertently transfer security risks to local staff or community leaders, who end up operating as unofficial liaisons for international actors.[13]

Respect for state sovereignty and, in some cases, uncertainty over the political, legal, and ethical implications of engagement, help to explain the limited extent and effectiveness of interaction between international humanitarian agencies and armed non-state actors. UN agencies in particular tend to be very cautious about doing anything which might be seen as antagonistic to states, and in many contexts, governments prefer that external actors avoid interacting with armed groups. Such preferences are sometimes formalized, with governments of affected states imposing legal and bureaucratic constraints on international humanitarian agencies. In 2005, for example, the Colombian government banned all organizations other

[10] See, e.g., 'Integration and coordination' in Chapter 6 on Afghanistan, 2001–2014.
[11] See 'International humanitarian response' in Chapter 6 on Afghanistan, 2001–2014.
[12] See 'International humanitarian response' in Chapter 10 on Famine in Somalia, 2011–2012.
[13] See 'Aid agency responses to these challenges' in Chapter 6 on Afghanistan, 2001–2014.

than the ICRC and, at times, the Catholic Church, from interacting with armed groups (Bradley 2016b: 90). In 2008, the government of Afghanistan similarly banned all contact between international organizations and 'Anti-Government elements' (Mc Hugh and Singh 2013: 4). Governments often also restrict access to territory controlled by armed groups through bureaucratic obstructions (including complex procedures for visas and in-country travel permits), which can effectively seal off parts of the country, without the same level of negative publicity or political pressure that an outright ban might provoke (Jackson and Davey 2014: 9). In the context of UN integrated missions, mission heads have sometimes limited humanitarian engagement with armed groups when such engagement was deemed detrimental to the political objectives of the respective missions, and there seems to be some confusion among operational agencies as to whether UN policies prohibit engagement with armed groups (Jackson and Davey 2014: 27).

Counter-terror policy also serves to restrict, and sometimes criminalize, engagement with members of groups designated terrorist organizations.[14] Operational humanitarian agencies sometimes also limit their engagement with armed groups in line with the interests of their funders. A small number of mainly Western donor governments provide the vast majority of funding for the major aid agencies, and heavy financial dependence on a few donors inevitably enables those donors to exert influence over where these agencies work and what work they do.[15] In some cases, this influence relates to engagement with armed groups, particularly those designated as terrorist organizations. Where donors are also participants in a conflict, or supporting one side or another, they may restrict their funding in such a way as they expect to advantage themselves or their allies, and this may include limiting funding to areas under the control of armed non-state actors. In Somalia, for example, after declaring Al-Shabaab a terrorist group in 2008, the United States suspended funding for humanitarian assistance in areas under Al-Shabaab control.[16] Even where engagement has not been criminalized and humanitarian agencies are not coming under direct pressure from donors, fear or uncertainty over the legal and financial consequences of engaging with non-state armed groups sometimes results in self-censorship by operational agencies.

Externally imposed legal and financial constraints go a long way to explaining humanitarian agency engagement (or non-engagement) with different

[14] See 'Counter-terror legislation' in Chapter 14 on Law and International Humanitarianism.

[15] See Chapter 18 on Donors and the Funding of Humanitarian Action.

[16] See 'Background' and 'Funding for humanitarian response' in Chapter 10 on Famine in Somalia, 2011–2012.

types of armed actors, but factors internal to humanitarian agencies also play a role. Some agencies explicitly cite their commitment to the humanitarian principle of neutrality as reason to avoid engagement with armed groups (Bradley 2016a: 60). However, engagement need not imply lending support. Notably, it is those agencies most committed to neutral humanitarian action—the ICRC and MSF—that engage most consistently and with a clear idea of what kinds of engagement are, and are not, acceptable to them. Neutrality implies treating non-state armed groups in the same way as other parties to conflict, but many humanitarian agencies treat armed groups differently from national and international military forces. There seems to be an ideational factor at play here, with some agencies limiting their engagement because they view armed groups as inherently illegitimate and dangerous, and this perspective appears to have become more pronounced over time.

International military forces

In the context of humanitarian crises, international military forces can be present and active in a variety of capacities and roles. Whether part of a multilateral operation, as in the North Atlantic Treaty Alliance (NATO) or UN peacekeeping missions, or intervening unilaterally, an international force may be a party to armed conflict or may explicitly take sides in an armed conflict. For instance, the International Security Assistance Force (ISAF), the NATO-led force in Afghanistan and the United Nations Organization Stabilization Mission in the Democratic Republic of the Congo (MONUSCO) were both mandated to support government forces against armed groups. Sometimes, international forces operate in contexts not characterized by armed conflict per se, as when the Indian military carried out rescue and relief operations in Sri Lanka in the immediate aftermath of the 2004 Indian Ocean tsunami, or when the UK and US militaries played a role responding to the Ebola outbreak in Sierra Leone and Liberia, respectively.[17]

International military forces can have an explicit mandate for the protection of the population, although a protection mandate is no guarantee of protective action or outcome. Some missions are justified precisely on the grounds of civilian protection, as envisaged by the responsibility to protect doctrine, and others are tasked primarily with peacekeeping or peacebuilding but are additionally mandated to protect civilians. Since United Nations Security Council (UNSC) Resolution 1270 on Sierra Leone in 1999,

[17] See 'Tsunami and its immediate aftermath' in Chapter 7 on the 2004 Indian Ocean Tsunami and 'Militarization and securitization' in Chapter 11 on Ebola in West Africa, 2014–2015.

when UN peacekeepers were, for the first time, explicitly mandated to protect civilians, protection language has been central in nearly every peacekeeping mandate (Mamiya 2016: 63; UN OCHA 2019: 10, 11). This represents a significant change from the situation in the Bosnian war, where the United Nations Protection Force in Yugoslavia (UNPROFOR) was limited to facilitating the delivery of humanitarian assistance and lacked both the mandate and the strength to protect even those Bosnians concentrated in UN-designated safe areas.[18] However, UN peacekeepers have been much criticized for their reticence to use force to protect civilians in practice, reticence which is linked to insufficient capabilities, conflicting directives, and ambiguity as to exactly what the protection of civilians entails (Paddon Rhoads and Welsh 2019: 610, 613–614). The responsibility to protect, set out by the International Commission on Intervention and State Sovereignty in 2001 and institutionalized as a political principle at the 2005 world summit, sought to clarify, among other things, the conditions under which the international community should intervene where a state is manifestly failing to protect its own population (ICISS 2001; Paddon Rhoads and Welsh 2019: 598, 601). However, the principle remains contested and has not been consistently implemented.

Through a range of different mechanisms, international military forces can have positive or negative impacts on humanitarian outcomes and on the operating space available to humanitarian agencies (Collinson et al. 2010: 14). In the best-case scenario, they have a positive impact on both. Prior to the 2010 earthquake in Haiti, for example, the UN peacekeeping mission there—the United Nations Stabilization Mission in Haiti (MINUSTAH)—was seen to have reduced armed violence, contributing both to improved protection outcomes for the population and safer access for international humanitarian agencies (Muggah 2010). More commonly, the impact of international forces is mixed. For example, during the war in Bosnia, UNPROFOR provided military protection of relief convoys, expanding operating space for the United Nations Refugee Agency (UNHCR) and other operational humanitarian agencies, but did not protect the wider civilian population.[19] Following the 1994 Rwandan genocide, the United Kingdom and the United States both sent military contingents to provide logistical support for the humanitarian response to the cholera outbreaks in refugee camps in neighbouring Zaire but not to address the insecurity in the camps.[20] In response to the Ebola outbreak

[18] See 'International military response' in Chapter 4 on the Bosnian War, 1992–1995.
[19] See 'International military response' in Chapter 4 on the Bosnian War, 1992–1995.
[20] See 'Humanitarianization of the crisis' in Chapter 5 on Genocide in Rwanda and Its Aftermath, 1994–1996.

in West Africa in 2014–2015, UK and US troops built medical facilities for the treatment of health-care professionals, which served to support the humanitarian response, but were highly risk averse such that military medics did not offer direct care to the wider community.[21] Conversely, they may provide some protection or assistance directly to the affected population even as they compromise agency operating space. There seem to be fewer examples of this in practice, but it did happen in some parts of Afghanistan, where ISAF forces were providing material assistance and medical aid to the Afghan population but their presence and activities were seen to contribute to insecurity for international humanitarian agencies and their staff.[22]

International forces also sometimes have contradictory effects on humanitarian outcomes or on operational space. NATO set up refugee camps in Macedonia for fleeing Kosovars at the same time as bombing Yugoslavia, for example, and the US military dropped both bombs and food parcels in Afghanistan (Lischer 2007). The US-led, UN-authorized Unified Task Force for Somalia (UNITAF) operation in Somalia from December 1992 to May 1993 temporarily improved security for humanitarian agencies, but working with UNITAF also reduced their autonomy, and operational constraints increased (Menkhaus 2010: S325).

Humanitarian agencies can engage with international military forces in a variety of ways, and there is significant variation in their engagement in practice, depending on the agency in question as well as the context. At one end of the spectrum are the dialogue-based activities that aid agencies carry out—some to a greater extent than others—with all armed actors. These may be focused on negotiating access to populations in need of protection or assistance or persuading the actors in question to respect international norms on the use of force. For instance, the ICRC has stepped up its engagement with UN peacekeeping forces since 2000, providing pre-departure training on relevant legal standards and maintaining dialogue in the field aimed at having a positive impact on the protection of civilians, compliance with IHL, and aid agency operating space (Carbonnier 2018a). Other humanitarian agencies are not as consistent as the ICRC in developing protection dialogue with armed actors of any kind but many will advocate for international forces to international norms. In Afghanistan from 2008, for example, international humanitarian agencies engaged in sustained lobbying and advocacy regarding violations by ISAF, which was widely credited with contributing to

[21] See 'Militarization and securitization' in Chapter 11 on Ebola in West Africa, 2014–2015.
[22] See 'International humanitarian response' in Chapter 6 on Afghanistan, 2001–2014.

a reduction in civilian facilities and in the use of force against medical facilities.[23]

There are also activities which go beyond the ways humanitarian agencies commonly engage national militaries of affected states or armed groups and imply a greater level of trust but not necessarily shared goals. For example, international forces may provide logistical support and security to international humanitarian agencies, and there may be some limited exchange of information. In Somalia in the first half of the 1990s, aid agencies relied on armed escorts first from UNITAF and then from the United Nations Operation in Somalia (UNOSOM) (Menkhaus 2010: S324–S328). The ICRC and MSF seek to avoid the use of armed escorts of any kind wherever possible, but many other agencies take a much more flexible approach (see, e.g. Labbé and Daudin 2015; Thompson 2008). UN agencies in particular often receive logistical support from UN peacekeeping forces and may also exchange information with them. In the Bosnian war, UNPROFOR provided military escorts for the UNHCR-led relief convoys.[24] More recently, in the Democratic Republic of the Congo, UNHCR not only used MONUSCO transport and escorts to access areas it would otherwise have had problems getting to but also sometimes undertook joint missions with the peacekeeping troops and exchanged information with them about military movements (Bradley 2016a: 53, 60–61). Informal engagement is probably also more common with international forces than it is with other kinds of armed actors; international aid workers sometimes socialize out-of-hours with foreign troops, especially when international forces include military personnel from the aid workers' countries of origin (Thompson 2008: 18).

Finally, there are the kinds of activities which aim to achieve coherence and complementarity between humanitarian and military approaches in pursuit of shared goals. This is exemplified by integrated missions in which the provision of relief assistance forms part of a broader package of measures aimed at 'stabilizing' a conflict zone. Purely humanitarian organizations like the ICRC and MSF generally resist working with international military forces in this way, believing that linking political, security, state-building, development, and humanitarian goals and activities threatens humanitarian principles and outcomes. Multi-mandate agencies, on the other hand, are sometimes more open to this kind of collaboration.[25] For those that embrace more transformative goals, seeking not only to save lives and reduce suffering directly in

[23] See 'Addressing violations of international humanitarian law' in Chapter 6 on Afghanistan, 2001–2014.

[24] See 'International humanitarian response' in Chapter 4 on the Bosnian War, 1992–1995.

[25] See, e.g., 'Integration and coordination' in Chapter 6 on Afghanistan, 2001–2014.

the present but also to bring about changes in society with the aim of saving lives and reducing suffering indirectly and in the future, the question is whether the goals and means of stabilization accord with those of humanitarian action (Collinson et al. 2010: 13). In practice, there is significant variation in the extent to which multi-mandate agencies collaborate with international military forces, both across different contexts and within particular contexts across time. In Afghanistan after the US-led coalition overthrew the Taliban in 2001, for example, several such agencies accepted funding to operate alongside ISAF, and some initially agreed to provide assistance explicitly aimed at garnering support for the newly-installed Afghan government or the international forces or in exchange for intelligence.[26] As time went on, however, it became clear that stabilization and post-conflict reconstruction would not be straightforward, and many agencies resisted pressure from ISAF to participate in what they saw as a battlefield clean-up operation and increasingly sought to maintain their distance from the military. In Haiti in the years prior to the 2010 earthquake, by contrast, international humanitarian agencies broadly moved in the opposite direction. Multi-mandate agencies like CARE, Concern, and Oxfam shifted from initial uncertainty about how to engage with stabilization forces to a pragmatic cooperative approach, and even within the ICRC and some MSF entities, attitudes towards MINUSTAH softened over time (Muggah 2010: S445, S448-S449, S456).

While working with international military forces or accepting armed protection from them often increases access for humanitarian agencies to some parts of the countries in which they are operating, it can also serve to reduce their access to those parts of the country controlled by opposing forces. For instance, armed protection from international military forces for humanitarian compounds and convoys in Somalia in the first half of the 1990s increased security for aid agencies at the cost of autonomy (Menkhaus 2010: S325). While the military protection facilitated humanitarian action in many respects, a perceived lack of neutrality created dangerous tensions with local militias and reduced access to some communities, while dependence on peacekeeping forces for their protection may have limited the extent to which aid agencies documented or openly criticized abuses of the population by those same forces (Menkhaus 2010: S327). In Afghanistan, many humanitarian agencies were working more closely with international military forces, and this was understood to have a negative impact on their security and

[26] See 'Instrumentalization of aid by military actors' in Chapter 6 on Afghanistan, 2001–2014.

access to populations in need in a context that already had a very high level of major security incidents involving aid workers.[27]

Very often, these more collaborative kinds of engagement with international military forces are driven by donors, political pressure, and exogenous shocks. Sometimes, donors provide funds to work in particular areas or even on condition of collecting and sharing intelligence. Where donor governments' foreign policy interests are involved, the international humanitarian agencies they are funding sometimes find themselves under pressure to contribute to support international forces in stabilization or counter-insurgency efforts (Jackson 2012a: 2). UN agencies in particular also often come under political pressure to support UN-led state-building efforts, as in Somalia (Menkhaus 2010: S328). In the face of exogenous shocks, including sudden-onset disasters, operational agencies may be more likely to turn to international forces for support and collaboration, much as they tend to be more open to working with national militaries when responding to such disasters. The major hurricanes that hit Haiti in August–September 2008 appeared to hasten integration between international military and humanitarian workers, for example (Muggah 2010: S459). While these external factors are important, humanitarian agencies always have some choice in how they work with international forces, albeit sometimes a very restricted choice. At the extreme, they can turn down funding that comes with donor pressure, for example, or even take the decision not to work in a particular context if the only way to do so involves collaboration with forces and if such collaboration is deemed too big a compromise. Some agencies are clearly more willing to work closely with international forces than others, and even if at times they would rather not do so, they may decide that collaboration is a price worth paying for the security or funding which enables them to meet needs.

Conclusions

Most aid agencies engage more with some armed actors—particularly (Western) international military forces and, sometimes, national military forces of affected states—than others—in particular, non-state armed groups. The influence of powerful states is apparent through the juxtaposition of pressure to engage with certain armed actors in integrated missions and the pressure *not* to engage with certain other armed actors as a result of counter-terror laws and regulations. However, it is not only external factors but also the

[27] See 'Insecurity and aid agency access' in Chapter 6 on Afghanistan, 2001–2014.

internal characteristics and preferences of aid agencies themselves that influence the level and kind of engagement. Avoiding all interaction with any type of actor is not often feasible, and neither is it usually desirable, but too close a relationship is equally undesirable. The reticence to engage with armed groups can feed a vicious circle of mutual distrust, and 'the perception of being allied with the "other side" in a given conflict has made it difficult for humanitarian actors to operate—as experiences in Afghanistan, Pakistan and Somalia illustrate' (Jackson 2012a: 3). If they want to maximize their access to populations in need and the chances of achieving positive humanitarian outcomes for those populations, humanitarian agencies need to engage strategically with all kinds of armed actors—the challenge is to understand each context and actor sufficiently to identify what strategic engagement looks like for any given set of agency objectives.

IV

OPERATIONS AND ACTIVITIES

Introduction to Part IV

Humanitarian Assistance and Protection

It is common to identify two main strands of international humanitarian action—assistance and protection. International humanitarian agencies work to save lives in crisis situations, primarily through the provision of food, medical services, shelter, and other forms of material assistance. However, in armed conflict and complex emergencies, saving lives requires more than material assistance, and so humanitarian agencies have increasingly incorporated the 'protection of civilians' into their missions. Definitions of protection vary but, at its core, protection is concerned with keeping people safe from violence. In addition to protecting and assisting people directly affected by humanitarian emergencies, aid agencies seek to keep their staff and their operations safe from violence through operational security management.

Instead of including a chapter on protection and a chapter on assistance, however, this part of the book is structured by activity type. Chapter 21 explores the ways humanitarian agencies collect and use evidence to support their practical work—to assess needs and capacities in any given humanitarian emergency, to understand what kinds of interventions are effective for achieving particular goals, and to inform response decisions. Chapter 22 turns to the goods and services provided by humanitarian agencies, and Chapter 23 examines the different kinds of dialogue and advocacy that humanitarian agencies engage in. Assistance tends to be equated with the provision of goods and services, and dialogue- and advocacy-based activities are often associated with protection, but the reality is more complex. In practice, the ways in which goods and services are provided can affect protection outcomes, and dialogue and advocacy are often integral to negotiating the access necessary to provide goods and services. This structure thus aims to increase conceptual clarity and to minimize gaps and overlaps.

The three chapters emphasize the importance of distinguishing concepts—needs, objectives, and activities—and interrogating the relationships between them. In humanitarian discourses, the category of 'needs' is sometimes conflated with the goods and services humanitarian agencies can offer. In general, 'assistance' refers to a set of activities, but the specific objectives of these activities—beyond 'assisting' people—are only implicit. By contrast, 'protection' is used to refer to both an objective and a set of activities, but those activities are defined in relation to the objective. For example, according to what is probably the most widely used definition in the humanitarian sector, protection encompasses 'all activities aimed at ensuring full respect for the rights of the individual in accordance with the letter and the spirit of the relevant bodies of law, i.e. human rights law, international humanitarian law and refugee law' (Giossi Caverzasio 2001: 19). This definition is problematic because it allows a wide range of humanitarian activities to be reframed as protection, and this expansiveness allows humanitarian agencies to focus on those activities in which they have experience or those that are relatively easy rather than addressing the priority issues which tend to be the more difficult tasks associated with keeping people safe (Bradley 2016a: 102).

Humanitarian agencies can help to keep people safe from violence by reducing the threat posed by violent actors or by reducing people's vulnerability or exposure to violent threats. Broadly speaking, threat reduction involves changing the behaviour of those who perpetrate violence, while vulnerability reduction accepts the level of threat as more or less given and focuses instead on facilitating the strategies people undertake to avoid those threats. Humanitarian efforts to protect civilians have mostly focused on supporting the victims to reduce their vulnerability, while human rights efforts target the perpetrators of human rights abuses with public criticism. Where humanitarian agencies have sought to change the behaviour of perpetrators, it has tended to be through indirect methods or, in the case of the International Committee of the Red Cross (ICRC) in particular, through discreet, bilateral dialogue with the (alleged) perpetrators (Bradley 2016a: 104–113). However, as discussed in Chapter 23, they also sometimes engage in public advocacy, including criticisms of states and other authorities.

Different humanitarian objectives and activities are organized into sectors in the cluster system, which the United Nations (UN) began to roll out in 2005 in an effort to improve coordination among international humanitarian actors and between those international actors and the governments of affected states. The cluster system is a coordination structure involving both UN and non-UN humanitarian organizations, and it operates at both the global level and country level. Each cluster focuses on a particular sector of emergency response: food security; camp coordination and camp management; early recovery; education; shelter; emergency telecommunications; health; logistics; nutrition; protection; and water, sanitation, and hygiene (UN OCHA 2020c). At the global level, the approach aims to strengthen preparedness and technical capacity. At country level, the aim is to improve predictability, accountability, and partnership to improve prioritization of resources and coverage of response. Each cluster has a clearly defined lead agency and a provider of last resort.

Structuring the humanitarian architecture in this way encourages thinking about needs and making response decisions around different sectors rather than around the particular constellations of needs and risks facing the population in any given humanitarian emergency. The limitations of this sector-centric approach are exacerbated by weak inter-cluster co-ordination. They reflect a more general problem, which comes through clearly in Chapters 21 and 22, that humanitarian response decisions are often made more on the basis of the convenience and experience of donors and operational agencies than on the needs and capacities of affected populations. Important needs may go without being identified, measured, or addressed if humanitarian agencies are only assessing and responding to those needs they can most easily address. Furthermore, the effectiveness, efficiency, and accountability of responses can be limited where decisions reflect organizational inertia and a desire by operational agencies to retain control of those responses (and even of the recipients of assistance). Recent moves away from refugee camps as the default response to large-scale displacement, and towards cash programming in place of in-kind assistance, suggest a shift in thinking, but this shift has been slow and has met resistance from aid agencies reluctant to relinquish control.

All three chapters in this section also highlight how the 'digital revolution' is changing humanitarian practice. Digital technologies have been widely embraced by much of the humanitarian sector and can serve to help generate important evidence regarding humanitarian needs and outcomes, as well as facilitating the delivery of cash transfers in particular, and underpinning some advocacy efforts. However, digital technology is not the panacea to humanitarian challenges that its champions claim it to be. On the one hand, many humanitarian challenges are fundamentally political in nature and are not amenable to technological solutions. On the other hand, the use of technology as part of humanitarian work has many unintended consequences and can serve to harm those the response is intended to help.

21
Needs Assessment, Evaluation, and Response Decisions

The generation and use of evidence in humanitarian policy and response matters for making humanitarian action more ethical, effective, and accountable (Knox Clarke and Darcy 2014: 5). Donors and operational agencies need good-quality evidence in order to make well-informed decisions about where to allocate resources, what goals to pursue, and about how to design programming.

The ethical distribution of humanitarian resources across the globe requires *comparable* evidence about needs and how they can be met in different contexts. Impartial humanitarian action requires assistance to be provided on the basis of (and in proportion to) need alone.[1] It therefore depends on knowing who needs what and where. At the most basic level, evidence is needed to confirm the existence of an actual or impending crisis and to assess its severity in relation to other actual or impending crises (Dijkzeul et al. 2013: S13).

Donors and operational agencies need evidence not only to determine where needs are greatest (which countries, which populations) but also where they can make the best contribution to reducing or mitigating humanitarian need; what kinds of intervention modalities work well in general; what modality would be best for the particular crisis, month, and year; who (which actors/organizations) is best placed to implement that kind of intervention,; and which populations they should target (Obrecht 2017: 10). Designing effective humanitarian programming requires knowledge of what interventions work, how they work, and the conditions under which they work. However, talking about 'what works' presupposes the goals of the programmes in question—and raises the question of whose goals are prioritized (Dijkzeul et al. 2013: S10).

Accountability requires evidence of how funding has been spent, what goals have been pursued, what activities have been carried out in pursuit of

[1] See 'Principles of humanity, impartiality, neutrality, and independence' in Chapter 13 on Politics, Principles, and Humanitarian Action.

The Politics and Everyday Practice of International Humanitarianism. Miriam Bradley, Oxford University Press.
 DOI: 10.1093/oso/9780198840695.003.0021

those goals, and what impact those activities have had. Operational humanitarian agencies need to show donors, international civil society, and the populations and governments of affected states not only that particular needs exist and merit being prioritized but also that they have made reasonable choices about the most effective and efficient ways to respond to those needs (Knox Clarke and Darcy 2014: 5). *Ex post*, they also need to be able to provide evidence regarding the effectiveness and efficiency of their chosen response.

Finding or generating sufficient evidence to inform decisions in emergency environments is inherently difficult. There are a number of practical problems involved in data collection in situations that are often characterized by insecurity, on the one hand (making access for data gathering difficult), and by urgency, on the other (meaning that the time required to gather good-quality data is simply not available). As a consequence, there is often insufficient evidence, or evidence of insufficient quality, to inform good decisions. Even where there is adequate evidence, it is not always adequately shared, with information that is important for the ethical and effective provision of humanitarian assistance unevenly distributed across donors and implementing agencies (Obrecht 2017: 18).

While access to good-quality evidence is crucial, it is not enough—ethical and effective humanitarian programming additionally depends on the will and capacity of decision-makers to design and fund programming in accordance with the evidence, even in the face of conflicting incentives. In this sense, problems associated with needs assessment and decision-making go beyond the technical issue of the (limited) availability of good-quality evidence. They also involve the issue of how evidence is used, and this is often a much more political question. Even when (albeit imperfect) evidence is available, needs assessments and programme design decisions are often driven more by personal conviction, or by political or fundraising considerations, than by that evidence.

Assessing needs, contexts, and capacities

Needs assessments are traditionally carried out by operational humanitarian agencies, who frequently face a number of obstacles to collecting the evidence necessary to build and maintain an accurate picture of humanitarian need. In theory, analysts can extrapolate estimates of the level of need across an entire population through a combination of rapid assessments or surveys and baseline data, including population figures, epidemiological risks, vaccination coverage, and modes of production (Poole and Primrose 2010: 2).

In practice, however, surveys undertaken in crisis contexts often fail to meet minimum quality criteria, and baseline data and demographic information are either not available or of poor quality (Colombo and Checchi 2018: 215, 220; Darcy and Hofmann 2003: 7). Ultimately, access, security, time, and resources limit what kind of needs assessment is possible and appropriate, and depending on the level of risk facing the population, it may be appropriate to settle for less information in some contexts than others (Darcy and Hofmann 2003: 25).

A range of factors limit the quality and usefulness of surveys in humanitarian emergencies. Armed actors sometimes prevent humanitarian agency access for data collection, counter-terror legislation can impose significant bureaucratic obstacles in areas where listed groups operate, local authorities may hinder access or expel agencies collecting sensitive information (e.g. documenting sexual violence or other violations of international humanitarian law (IHL) or international human rights law (IHRL)), and authoritarian regimes often seek to control information (Colombo and Checchi 2018: 217–218). These obstacles, as well as general insecurity, often mean that some areas are excluded from samples selected for surveying, thus limiting the representativeness of the surveys (Colombo and Checchi 2018: 221). In practice, purposive rather than probabilistic sampling is more common to select the communities, households, or individuals to be surveyed in humanitarian needs assessments, and effective assessment may require using secondary data to identify different social and economic groups or livelihoods areas and separate data collection for different sub-groups in the population (men, women, children, elderly, marginalized groups, etc.), who will have different types and levels of need (Knox Clarke and Darcy 2014: 27). Compounding the external obstacles to data collection are certain characteristics internal to the agencies carrying out the needs assessments—with skill sets and methodologies often inadequate and a culture of urgency contributing to quick but superficial data-gathering (Dijkzeul et al. 2013: S2; Obrecht 2017: 18–19). Data collection in crisis contexts additionally raises several ethical concerns, including the question of whether the resources required would be better used for relief activities, difficulties obtaining informed and voluntary consent in emergency settings, and the potential for participation in a survey to be traumatizing or expose informants to risk—and with no guarantee of benefits for participants (Colombo and Checchi 2018: 221; Knox Clarke and Darcy 2014: 20).

In many emergency contexts, basic data on population numbers and characteristics is anyway inadequate for extrapolating population-wide needs from assessments or surveys carried out on samples. Crises tend to amplify the kinds of difficulties of information management which are common in resource-poor contexts (Colombo and Checchi 2018: 217; Knox Clarke and Darcy 2014: 19). For example, disaggregated census data may not exist, and population estimates based on census data and growth projections do not take into account the ways in which crisis affects (parts of) the population through increased mortality and population movements (Colombo and Checchi 2018: 220; Knox Clarke and Darcy 2014: 20). Furthermore, basic secondary data—such as birth, death, and vaccination rates or epidemiological surveillance—may be lacking or poor quality in countries with fragile institutions, and crises often exacerbate weaknesses in data production and analysis (Colombo and Checchi 2018: 220).

While needs assessment is frequently treated as a one-off exercise, there are good reasons to repeat assessments or to monitor needs on an ongoing basis while programming is being implemented. First, where needs are urgent, aid agencies may be justified in initiating programming with only limited evidence, but over the course of an intervention, the evidence base can, and should, be reviewed and expanded (Darcy and Hofmann 2003: 25). Second, in emergency settings, needs can change very quickly, so even if an initial assessment was based on solid evidence and offered an accurate representation of needs in a given place at a given moment, it could very quickly become out of date (Knox Clarke and Darcy 2014: 26). Even in areas that see regular emergencies, however, monitoring is not always ongoing, and sequential assessments do not always use the same questionnaires and methods, making it impossible to compare results and understand changes in needs over time (Knox Clarke and Darcy 2014: 26). This seems to be a problem of organizational culture and incentives, with long-term investment in monitoring a poor fit with the shorter-term timelines of humanitarian projects and programming (Knox Clarke and Darcy 2014: 32).

A further complication arises because different actors sometimes misrepresent the level of need. Data may be politicized, with affected states wanting to avoid seeming weak and parties to conflict or other authorities facing incentives to mis-represent population and casualty numbers (Knox Clarke and Darcy 2014: 25). Several of the case studies in the first part of this book offer illustrations of the ways in which data on needs can be politicized. In the context of the Ebola outbreak in West Africa, for instance, the governments of Guinea, Liberia, and Sierra Leone sought to downplay the extent of the crisis in an effort to avoid appearing weak or deterring foreign investment.[2] In the final stages of the civil war in Sri Lanka, the government claimed

[2] See Chapter 11 on Ebola in West Africa, 2014–2015.

that far fewer people remained in the Vanni region than was the case, both to justify limiting the quantity of relief allowed into the region and to preemptively reduce casualty figures.[3] In the refugee camps in Zaire following the Rwandan genocide, by contrast, the political and military authorities of the former regime sought to inflate camp population numbers.[4] This served both to generate economic benefits, in the form of additional aid allocations, and to challenge the new Rwandan regime's claim to power. Since needs assessments are the basis for funding decisions, operational agencies themselves also have obvious incentives to inflate the number of people in need, and this has been a persistent worry (Colombo and Checchi 2018: 218; Knox Clarke and Darcy 2014: 28; Obrecht 2017: 18).

In practice, operational agencies have often been criticized for *under*- or *mis*-representing needs, in line with their capacity and skills to deliver or the level of funding they anticipate being available. Sometimes, assessment reports include information simply because it is available, even if it does little to aid understanding of needs (Knox Clarke and Darcy 2014: 28). Furthermore, humanitarian actors tend to identify needs they know they can respond to and to (mis)represent needs in accordance with the response they are best equipped to provide or can most easily provide (Keen 1998a: 322–323; Knox Clarke and Darcy 2014: 29; Vaux 2006: 77). The ambiguity of the concept of 'need' has facilitated the conflation of needs assessment with the formulation of responses, and a focus on acute risks to life, health, subsistence, and security has been proposed as a way to define the problem without presupposing the solution (Darcy and Hofmann 2003: 5, 16–17). Such an approach also helps to incorporate an assessment of, for example, threats to people's physical security—the focus of protection concerns—which do not lend themselves to the commodification implicit in talk of 'needs' or to quantitative measurement (Darcy and Hofmann 2003: 10, 17; Knox Clarke and Darcy 2014: 23–24).

A further obstacle to understanding needs within and across different settings relates to a lack of standardization and a tendency for multiple agencies to conduct independent assessments leading to gaps and overlaps. Very often, different actors—including the health authorities in the affected state and international responders—expend significant time and resources to produce situation analyses, needs assessments, and surveillance data that could more usefully inform response if it was standardized and disseminated (Colombo and Checchi 2018: 216). While duplication could improve the quality of

[3] See 'Manipulation of aid and aid agencies' in Chapter 8 on Sri Lanka, 2008–2009.

[4] See 'Diversion, manipulation, and unintended consequences of aid' in Chapter 5 on Genocide in Rwanda and Its Aftermath, 1994–1996.

assessments if results were compared and used to triangulate findings, competition between agencies reduces the incentives to standardize and share information, and in practice, each agency tends to use its own assessment methodology (Colombo and Checchi 2018: 219; Knox Clarke and Darcy 2014: 24). As a result, in some contexts, there is too much rather than too little data and analysis, with information overload and fragmentation making it difficult to extrapolate and identify humanitarian priorities (Colombo and Checchi 2018: 218–219). The lack of standardization also makes it difficult to compare humanitarian emergencies and judge their relative severity, and hence to allocate resources across the globe in line with the principle of impartiality (Darcy and Hofmann 2003: 5).

Significant progress has been made in coordinating needs assessments in recent years, an important step both in terms of improving the quality of evidence about needs and in terms of reducing unnecessary duplication. For example, a consensus seems to be emerging for early-phase assessments conducted in the first three to four weeks after a disaster to be based on a review of secondary data, supplemented with information from questionnaires put to key informants and focus groups, with communities usually selected for inclusion in assessments according to purposive rather than representative sampling (Knox Clarke and Darcy 2014: 25). The Grand Bargain agreed at the 2016 World Humanitarian Summit (WHS) included a workstream dedicated to improving joint and impartial needs assessments, and the five-year independent review of progress found that this was an area in which substantial progress had been made (Metcalfe-Hough et al. 2021: 77–84). Major donors and operational humanitarian agencies have engaged with this workstream, developing—among other tools—a joint intersectoral analysis framework, which is seen to have significant value, with the potential to improve the prioritization, articulation, and sequencing of responses (Metcalfe-Hough et al. 2021: 82).

In order to prioritize effectively and ethically, donors and operational agencies need to understand not only what affected populations need—or what risks they face—but also what capacity other local, national, or international actors and institutions have to meet those needs or to address those risks. However, capacity assessment is often even weaker than needs assessment, and international responders typically assume low national and local capacities.[5] Aid agencies tend to focus on quantifying the level of need for largely pre-determined relief items—the response they can most

[5] For discussion of the role of local and national actors in humanitarian response, see Chapter 19 on Government and Civil Society in Affected States.

easily provide—rather than analysing the broader context, including market dynamics and the capacities of local actors (Harvey 2009: 16; Knox Clarke and Darcy 2014: 28; Levine 2017: 15, 19). The governments of affected states are often assumed, rather than assessed, to lack capacity. In order to identify how best to respond in any given emergency, each international humanitarian agency needs to understand not only the capacity of local and national actors but also the capacity and programming of other international actors. The 'who is doing what and where' matrix is essential for effective coordination, but in the majority of crises, this information is not readily available in the early stages (Colombo and Checchi 2018: 216). A failure to assess the capacities of other actors can generate overlaps and duplication at the cost of gaps in coverage elsewhere, and the imposition of external assistance without due consideration of local capabilities can also have a negative impact on the quality of the international response.

Technological innovations have helped to overcome some of the obstacles to getting quality data on needs, but they cannot overcome them all, and they often bring new and different problems. For example, satellite-based imagery and mobile phone data have been used to track population movements and thus provide some indicators of certain kinds of need, even where insecurity or other obstacles limit physical access for humanitarian agencies (Colombo and Checchi 2018: 220–221). However, data on population movements may be an example of data that is collected because it is available, even if it is not necessarily a good indicator of salient needs. Mobile phone technologies have been lauded as enabling crisis-affected people to contribute their own assessments of needs and, indeed, their own formulation of responses (Bennett et al. 2016: 41). Certainly, such technology facilitates dialogue between aid agencies and affected populations, but it also has important limits. First, while mobile phone coverage and access in countries affected by humanitarian emergencies has expanded enormously since 2000, those who remain without access to such technologies are often the most marginalized (Read et al. 2016: 1322). Second, the extent to which such technology can serve emancipatory purposes, even for those who have access to them, depends on what information they are invited to contribute. Where local communities simply feed data into the machine and the power of design, funding, and analysis resides with the international agencies overseeing the assessments, the emancipatory potential is necessary limited (Read et al. 2016: 1324–1325). Furthermore, much technology is dual use with surveillance potential and may contribute not only to care but also to control. The same drone footage that can help search-and-rescue missions in the Mediterranean identify migrant boats in distress, for example, can be used by authorities to identify and prosecute those migrants who drove the boats.[6]

Even if needs assessment provided a perfect depiction of humanitarian crisis contexts, understanding needs and risks does not tell us much about how best to meet those needs or address those risks. Decisions about how to respond should ideally be based on an understanding of what kinds of interventions are effective and under what conditions.

Monitoring and evaluation

Prior to the 1990s, systematic monitoring and evaluation of humanitarian programmes was uncommon, with little pressure on humanitarian actors to demonstrate that their programming was evidence-based, partly because it was considered inappropriate, or even unethical, to question the efficiency or effectiveness of efforts to save lives and reduce suffering in extremis, efforts that were deemed good in and of themselves, almost regardless of their consequences (Darcy and Hofmann 2003: 10; Dijkzeul et al. 2013: S2). Programme monitoring refers to ongoing analysis of how a programme is being implemented, and with what impact, while programme evaluations typically capture progress and results at a particular moment in time. Many evaluations take place after completion of a project or programme, but real-time evaluations (RTEs) take place at some point while the project or programme (or even the sector-wide response) is being implemented, with the goal of feeding back findings to inform the evolution of the project, programme, or response.

Through monitoring and evaluating humanitarian responses, operational agencies can learn how needs and risks are evolving, 'what works' and what does not—and, ideally, use the lessons they learn to inform future programme design. Understanding what measures have been effective in different contexts, and how they have worked, is important to inform future response decisions. Despite the massive growth in monitoring and evaluation, however, the evidence base for much humanitarian programming remains thin (Colombo and Checchi 2018: 215). Monitoring and evaluating projects and programmes can make an important contribution to humanitarian accountability, although they arguably contribute more to upward accountability (to donors) than to downward accountability (to affected

[6] See 'Search and rescue' in Chapter 12 on the 2015–2016 European 'Migrant Crisis'.

populations). While publishing accounts of spending and programming, and analyses of their effectiveness, could potentially increase accountability to affected populations, evaluations tend to be geared towards donors and are not always easily accessible to affected populations.

Approaches to humanitarian programme monitoring and evaluation tend to focus on how far inputs have been converted into outputs rather than on how effectively these outputs contribute to addressing the needs or risks facing the affected population (Knox Clarke and Darcy 2014: 32–33). Outputs (e.g. tonnes of food aid delivered or number of households receiving food rations) are easier to measure than outcomes (e.g. reduction in hunger, malnourishment, or food insecurity). Measures of outputs can serve for upward accountability to donors as evidence that the operational agencies receiving funding did what they said they would do with those funds. Indeed, most donors have not tended to require consistent reporting or evaluations of outcomes (Obrecht 2017: 16). However, information about outputs is less useful than information about outcomes for learning what did or did not work to reduce need or risk, how it worked, and how large its effect was. Furthermore, to understand the impact of programming, it is not enough to depict the outcomes accurately; it is also necessary to show the relationship between a particular intervention or series of interventions and the outcome described (Dijkzeul et al. 2013: S9; Knox Clarke and Darcy 2014: 40). For a long time, however, evaluations did not prioritize causal analysis. Even as they sought to establish whether the intended outcomes had materialized, they largely failed to assess the contribution of the programme(s) in question to those outcomes (Hughes and Hutchings 2011: 10).

Attention to the impacts of programmes has increased, with those evaluating humanitarian action—particularly in the health sector or with a focus on cash-based programming—trying to tackle the more complex questions of causality and effectiveness (Alexander and Bonino 2015: 1; Obrecht 2017: 15). Indeed, wider concerns to improve methodologies for understanding 'what works' in humanitarian response have been driven in part by proponents of cash transfer programming seeking to demonstrate to critics that such programming is an effective and efficient way to meet a range of humanitarian needs.[7] Cash transfers are now one of the most rigorously evaluated dimensions of humanitarian response (Bailey and Harvey 2015: 2; High Level Panel on Humanitarian Cash Transfers 2015: 18).

[7] On debates about the value of cash transfer programming, see 'Cash and voucher assistance' in Chapter 22 on Material Assistance and Direct Service Provision.

Experimental approaches, including randomized controlled trials (RCTs), are sometimes considered the gold standard in assessing the impact of particular interventions but can be ethically and methodologically problematic in humanitarian contexts (Alexander and Bonino 2015: 6–7). Comparing communities that received a particular intervention with those that did not implies not providing an intervention expected to save lives or significantly reduce suffering in an emergency setting (Dijkzeul et al. 2013: S8). From a methodological perspective, experimental methods are also considered ill suited to examining complex phenomena involving multiple interacting causal factors (Colombo and Checchi 2018: 219; Dijkzeul et al. 2013: S8). However, RCTs have been used to evaluate some forms of humanitarian programming, including different therapeutic feeding options and cash transfer programming, in the latter case comparing individuals or households receiving cash transfers with those receiving in-kind assistance (Bailey and Harvey 2015: 2; Dijkzeul et al. 2013: S8). Because all individuals or households in the trials receive some assistance, the ethical problems associated with experimental approaches are minimized, and because the assistance is given at the individual or household level, it is possible to control for other factors in a way that is more difficult with community-level interventions (such as repairing a water system or operating a primary health clinic). For much humanitarian programming, however, experimental methods are not appropriate.

The most common approach to humanitarian impact assessment is to compare a set of indicators linked to anticipated outcomes before and after a programme or project has been implemented, taking into account other factors that might have an impact on the indicators being measured (Colombo and Checchi 2018: 219). However, it may be impossible to isolate the impact of the programme or project being evaluated. If, for example, a reduction in malnutrition is observed following a supplementary feeding programme, it may be difficult or impossible to say how much the programme itself contributed to the observed reduction because a range of additional factors—such as concurrent health-related interventions, seasonal factors, and higher household income—likely also had an impact (Alexander and Bonino 2015: 2).

Ultimately, a combination of experimental/quantitative, comparative/qualitative, and theory-based approaches can all contribute to understanding the factors (including, but not limited to, humanitarian agency interventions) contributing to particular outcomes on the one hand, and the range of outcomes (intended and unintended) of particular interventions on the other (Colombo and Checchi 2018: 219; Dijkzeul et al. 2013: S9). Qualitative methods are sometimes considered second-best, but in fact, they can address questions which quantitative approaches are unable to answer and vice versa (Colombo and Checchi 2018: 219; Knox Clarke and Darcy 2014: 11). For example, answering questions about the relevance of the goals pursued, and about the unintended consequences of particular interventions, tends to require qualitative methods, including narrative approaches that shed light on the perceptions and critiques of the affected population and on the broader economic, political, and social contexts in which aid is provided and on which it may have an impact (Dijkzeul et al. 2013: S10-S11). It is also important to recognize that an intervention which is effective in one humanitarian crisis may not be effective in another because social and cultural factors are often crucial for the success of humanitarian programming, and these factors can vary enormously (Knox Clarke and Darcy 2014: 21). Given they make different contributions, in many cases the most robust findings will come from 'mixed methods' approaches, which draw on both quantitative and qualitative methods (Knox Clarke and Darcy 2014: 11, 38). In practice, however, the use of mixed methods appears to be uncommon and most humanitarian evaluations rely on broadly qualitative approaches, in particular interviews and personal observations (Knox Clarke and Darcy 2014: 39).

Beyond understanding the impact of a particular intervention in a particular context, it is important to understand how far findings about the impact of that intervention are generalizable or transferable to other contexts (and to what kinds of other context) and to understand how the humanitarian sector functioned as a whole in that particular context. Assessing how far findings are generalizable requires comparison of evaluations across different contexts and/or identification of the causal mechanisms and relevant contextual factors. For the identification of causal mechanisms, comparing multiple evaluations can be productive, and the identification of relevant contextual factors will often depend on key informant interviews. Assessing the performance of the sector as a whole through inter-agency evaluations (including RTEs) has become increasingly common, and these evaluations may have more impact on future policy and practice than those focused on one or more programmes implemented by a single agency.

As with needs assessment, the quality of evidence is a key issue in monitoring and evaluation. Many of the same challenges and weaknesses apply, and it is often even more difficult to get good evidence of impact than of needs due to the requirement not only to identify the outcomes but also to establish causal relationships between programming and outcomes. For programme monitoring, which requires frequent and regular access to operations, the challenge of physical access in insecure settings is particularly acute (Knox Clarke and Darcy 2014: 33). Just as needs assessments may be politicized, with different actors seeking to inflate or downplay the size of affected populations or the numbers of people with particular needs, the quality of evaluations may suffer as a result of a range of different actors seeking to confirm their view of what has happened (Knox Clarke and Darcy 2014: 38). There are very real disincentives to generating evidence which suggests that an operational agency or a particular programme may have 'failed' (Knox Clarke and Darcy 2014: 61). Inadequate skill sets are another problem, with many humanitarian evaluations failing to meet basic quality standards, employing qualitative approaches but without using academically recognized methods (Knox Clarke and Darcy 2014: 39).

Technology—and, in particular, digital technologies—offer the potential to improve the quality of monitoring and evaluation but, once again, with some important limits and risks. Technology, including social media and customer satisfaction techniques associated with the private sector, has the ability to improve two-way communication between operational agencies and crisis-affected people, with the former providing information about their programming and the latter providing direct feedback on what they think about that programming (Bennett et al. 2016: 41, 60; Knox Clarke and Darcy 2014: 34). All of this information, in turn, can be used to improve accountability, relevance, and effectiveness. If the evidence is collected using mobile technologies, however, the source of information is limited to those with access to mobile phones, raising questions about the representativeness of the data collected and concerns about excluding the perspectives of the most marginalized (Knox Clarke and Darcy 2014: 34). Digital cash, in particular, offers the opportunity to radically improve transparency about how much aid reaches recipiences (High Level Panel on Humanitarian Cash Transfers 2015: 13). However, without the capacity to analyse, understand, and utilize the evidence gathered, the ability to gather large amounts of data quickly may be of little value, with tech solutions driven more by what is possible than what is needed or useful (Knox Clarke and Darcy 2014: 34; Read et al. 2016: 1315). Particularly in settings with scarce resources and limited capacity, low-tech options may be more effective, and in practice, face-to-face information-gathering still plays a central role in most humanitarian responses (Knox Clarke and Darcy 2014: 34; Read et al. 2016: 1320). Moreover, the use of technology for monitoring and evaluation often comes with concerns about data protection and surveillance. For instance, geo-tagging information or images can be useful but carries security and privacy risks (Knox Clarke and Darcy 2014: 34–35).

Allocating resources, choosing response modalities, and designing programmes

It is not only the quality of evidence but also the uptake of that evidence that is important. Even the highest-quality evidence from needs assessments, evaluations and other kinds of policy and academic research is of little value if it is not used by donors and operational agencies to inform a series of decisions. Which crises should they prioritize, and how should they distribute finite resources across crises? Which response modalities should they support in any given crisis? What shape should the programmes they implement take? Often, however, these decisions are driven not so much by the available evidence as by other factors, including political and fundraising considerations (Knox Clarke and Darcy 2014: 5).

Evidence from needs assessments should inform decisions on whether assistance is required, how much, for whom, and for how long, but in practice, the use of needs assessment to support these decisions is inconsistent, and several other factors also influence them (Knox Clarke and Darcy 2014: 50). Some of these decisions are taken by the donor governments that provide the vast majority of funding for international humanitarian agencies and others by those agencies themselves. Moreover, operational agencies have the power to influence donor decisions both because they are the primary sources of needs assessments and because they turn those assessments into funding proposals in which they offer a programme of work and set out a budget for it. Donors and operational agencies alike are influenced by evidence from needs assessments but also by other factors (Obrecht 2017: 12). On the one hand, political and media considerations influence donor thinking, and marketing interests influence where operational agencies want to (be seen to) be working (Darcy and Hofmann 2003: 8; Knox Clarke and Darcy 2014: 52). In the aftermath of the 2004 Indian Ocean tsunami, for example, needs assessments were mostly not coordinated or shared, and the media were the most influential source of information for donors.[8] On the other hand, contextual factors—including poor security conditions or a lack of trusted partners—may lead a donor to conclude that an intervention in a particular context would be unlikely to succeed and, hence, to focus their resources elsewhere (Knox Clarke and Darcy 2014: 52).

Donor governments are often sceptical about the quality and impartiality of needs assessments produced by the very same operational agencies

[8] See 'Funding for humanitarian response' in Chapter 7 on the 2004 Indian Ocean Tsunami.

who will then use them to justify funding requests, while the agencies doubt that objective assessment is a central factor shaping donor decisions (Darcy and Hofmann 2003: 5; Knox Clarke and Darcy 2014: 50). While both have good reason for their scepticism, this mutual distrust generates a vicious circle. Donor governments base their funding decisions on a variety of factors, not just relative levels of need. Because they do not expect donors to fund emergencies in accordance with need and need alone, operational agencies draw up budgets based on expectations about what those donors will be willing to fund. As a consequence, their budgets and funding appeals are out of synch with needs, and donors do not trust them, so they do not rely on them to make their funding decisions. Even when needs assessments are of good quality, they are often not shared in time to inform donor decisions (Knox Clarke and Darcy 2014: 51). In some humanitarian contexts—above all, rapid-onset disasters—time is of the essence, and donors might make greater use of assessments if they were more limited in scope but carried out and shared quickly in a short, succinct format (Knox Clarke and Darcy 2014: 51).

Beyond basic prioritization and allocation of resources across different humanitarian emergencies, donor governments decide which agencies and what kinds of response to fund. While they might not get too involved in detailed programme design and planning decisions, they very often earmark their funds for specific sectors and sometimes provide funding for specific projects.[9] In this way, donors can impose significant limits on what kind of response operational agencies can provide. In some cases, donor governments adopt essentially ideological decisions—for example, regarding contracting out health service provision to non-governmental organizations (NGOs) or scaling up performance-based financing—even when the available evidence is insufficient to support those decisions or contradicts them (Colombo and Checchi 2018: 218). Very often, they require the operational agencies requesting funding to provide evidence of the appropriateness and cost-effectiveness of their proposed response. For food security crises, for example, most donors expect agencies to justify the proposed response on the basis of the needs assessment and to provide a cost comparison between the proposed response and alternatives, for example between cash, vouchers, and in-kind assistance (Maxwell et al. 2013: 73). Increasingly, they have also demanded market analysis to ensure that the proposed response will not

[9] See 'Independence, earmarking, and conditionality' in Chapter 18 on Donors and the Funding of Humanitarian Action.

disrupt market systems and analyses of corruption, safety, and protection implications (Maxwell et al. 2013: 73–74). However, the focus here is on justifying a proposed response rather than comparing the expected market impacts of a range of different options (Maxwell et al. 2013: 74). Donors have also tended to demand more in terms of market analysis and monitoring for cash than for in-kind assistance (High Level Panel on Humanitarian Cash Transfers 2015: 14, 22). In general, 'assessment evidence is more likely to be used when it suggests a course of action that is possible or desirable for decision-makers, and when it fits their (often unconscious) assumptions about the situation' (Knox Clarke and Darcy 2014: 51).

Donor preferences (or perceptions of donor preferences) constrain response choices by operational agencies, but so too do a range of other factors. Perhaps the biggest determinant of the response choices of different humanitarian actors is their organizational strategy, capacity, and ethos (Knox Clarke and Darcy 2014: 51; Maxwell et al. 2013: 74). There is path dependence at play here; different operational agencies have developed different specializations and have invested in building certain capacities and competencies (Maxwell et al. 2013: 74). It would be surprising (and probably worrying) if an agency without medical expertise proposed a medical-focused response to any given crisis. On the other hand, agencies competing for funding rarely propose not responding and so might be inclined to offer a response that is not needed or should not be a priority.[10] Even where they have the capacity to respond across a wide range of response options, organizational inertia is an important factor, and decisionmakers in aid agencies often default to the ways of responding that they are most familiar with and tend to replicate intervention modalities used in previous crises without adapting them to different contexts (Colombo and Checchi 2018: 217; High Level Panel on Humanitarian Cash Transfers 2015: 22; Knox Clarke and Darcy 2014: 53; Maxwell et al. 2013: 74). In responding to the 2004 Indian Ocean tsunami, for example, operational agencies went ahead with responses based on their preconceptions about what would be needed and on what they could most easily offer, given their experience and expertise, rather than on their own assessments of what needs should be prioritized.[11] Sometimes, donors require additional monitoring, evaluation, and reporting for novel or experimental approaches, and the additional costs and obligations may further disincentivize innovation (Maxwell et al. 2013: 76).

[10] See, e.g., 'International humanitarian response' in Chapter 7 on the 2004 Indian Ocean Tsunami.
[11] See 'Emergency response' in Chapter 7 on the 2004 Indian Ocean Tsunami.

Depending on the context and the agency in question, a range of additional factors may come into play when choosing response modalities. In some emergencies, the government of the affected state may favour one response type or other, and at times, governments act to restrict programming choices, as when the Sri Lankan government refused to allow medicines on humanitarian convoys into the Vanni in 2009 or when the government in Haiti shut down food aid programming in March 2010 due to concerns that it was undermining Haitian agriculture (Maxwell et al. 2013: 76). In many cases, decision-makers in operational agencies seek to minimize risk—which could include security or reputational risks—rather than to maximize potential programme impact and may prefer high-profile, media-friendly response options to those that are less visible (Maxwell et al. 2013: 77).

None of this is to say that evidence as to what affected populations need and how aid agencies can best meet those needs plays no role in response decisions. Very often a whole series of decisions are made when choosing response options, and evidence from needs assessments, evaluations, and other kinds of research and analysis play some role in some of these decisions (Maxwell et al. 2013: 68). However, arguably, that role is not as large as it should be, and some sources of information and evidence are less likely to influence decisions than others. Evidence from professional evaluators appears more likely to influence operational decisions when the decision-makers have been engaged throughout the evaluation process, helping to ensure that the findings are relevant to operational needs, but this kind of engagement may come at the cost of objectivity (Knox Clarke and Darcy 2014: 56). The perspectives and preferences of the communities directly affected by humanitarian emergencies do not routinely have a significant impact on decision-making, and even when they have contributed information on context and needs, that information is sometimes greeted with suspicion (Knox Clarke and Darcy 2014: 34; Maxwell et al. 2013: 76). This failure to incorporate community perspectives in programme design can significantly reduce the quality of the response, as occurred in the early stages of the international response to the Ebola outbreak in West Africa in 2014.[12]

Once a response has begun to be implemented, there is still scope to adapt programme design in light of evidence emerging from ongoing monitoring of needs and of the programming being implemented. In the initial, acute phase of a humanitarian emergency, particularly large, rapid-onset emergencies, operational agencies may have to make quick and difficult decisions

[12] See 'International humanitarian response' in Chapter 11 on Ebola in West Africa, 2014–2015.

with minimal information because in the time required for quality needs assessment, the potential lifesaving impacts of any intervention decreases (Colombo and Checchi 2018: 215; Cosgrave 2009: 81). As crises continue, however, better-quality information can be collected and analysed and the response adapted accordingly. Even where good-quality evidence informs the initial response choices, without monitoring and adaptation the link between evidence and programme choice erodes over time, a particularly important issue in protracted crises.[13] RTEs are carried out while a response or programme is being implemented with the goal of producing information that can be used for 'course correction' (Knox Clarke and Darcy 2014: 56). However, it seems that recommendations are implemented in a selective manner, with operational rather than strategic changes more likely to be implemented.

Over the long term, the results of evaluations—and of research more broadly, including the outputs of academics, think tanks, and policy research institutes—can shape the direction of organizational strategy and capacity and of sector-wide thinking and approaches. However, change tends to be slow, and evidence alone may not be enough to drive it. For example, it was not only the accumulation of evidence as to the extensive benefits and limited risks of cash transfer programming but also advocacy by proponents of cash programming and more circumstantial factors—above all, the need to provide assistance in the face of severe famine in Somalia in 2011, where food aid deliveries were severely restricted—that eventually contributed to a sea change in policy and practice (Knox Clarke and Darcy 2014: 58; Maxwell et al. 2013: 77). Similarly, while evaluation evidence—specifically from the Joint Evaluation of Emergency Assistance to Rwanda—was a major factor in driving the humanitarian sector to develop minimum standards in the form of the Sphere Project, some of the most important conclusions in that evaluation were swept under the carpet (Knox Clarke and Darcy 2014: 59). The basic tenets of the localization agenda[14] have become widely accepted only after decades of evaluations—among them, sector-wide evaluations of the response to the 1994 Rwandan genocide and its aftermath (Eriksson et al. 1996), the 2004 Indian Ocean Tsunami (Cosgrave 2007), and the 2013–2016 Ebola outbreak in West Africa (DuBois et al. 2015)—reporting that the humanitarian response could have been more relevant, effective, and efficient if underpinned by better contextual understanding.[15]

[13] For a general discussion of humanitarian response in protracted crises, see 'Protracted "emergencies"' in Chapter 15 on the Shifting Parameters of Humanitarian Emergencies.

[14] See 'Localization agenda' in Chapter 19 on Government and Civil Society in Affected States.

[15] See Chapters 5, 7, and 11 respectively on the response to these three humanitarian emergencies.

Conclusions

Evidence from assessing needs and from monitoring and evaluating humanitarian response can contribute to making humanitarian work more ethical, accountable, efficient, and effective. However, there are two major obstacles which limit this contribution in practice. First, evidence is often not of sufficient quality to have maximum impact or is not available to the right people at the right time. The nature of humanitarian contexts often make collecting high-quality data difficult, but even within the bounds of externally imposed constraints, there is scope for improving the quality of evidence, not least by asking better, more open questions and enhancing capacity for analysis. Second, even to the extent that decision-makers in donor governments and operational humanitarian agencies have access to high-quality assessments, analysis, and evaluations, a range of other factors also influence their decisions—which are better described as 'evidence-informed' rather than 'evidence-based' (Knox Clarke and Darcy 2014: 52). Increasing the uptake of evidence requires a shift in will more than capacity—the will to prioritize humanitarian need, effectiveness, and accountability over political and marketing concerns, on the one hand, and to challenge organizational inertia and consider new or different approaches when compelling evidence contradicts conventional wisdom on the other.

22
Material Assistance and Direct Service Provision

When a humanitarian emergency leaves people hungry or homeless, mainstream humanitarian response has tended to focus on providing them with food or shelter—in other words, giving them the commodity they are lacking. Sometimes, the work of international humanitarian agencies extends to installing or repairing essential infrastructure, including water points or systems, clinics, and schools. Where people lack basic services, including medical and health-care services, international humanitarian agencies may provide them directly or support local actors to do so. Compared with the dialogue- and advocacy-based activities discussed in Chapter 23, the provision of material assistance and basic services is often considered less political. In general, and in contrast to most dialogue, negotiation, and advocacy, it is not aimed primarily at influencing the behaviour of states and other actors. However, the discussion in this chapter shows that programming choices may be political, and that any kind of programming can have economic and political consequences.

This chapter first discusses in general terms the provision of goods and services by international humanitarian agencies before examining in greater depth the politics of camps and the politics of food aid specifically. Camps for displaced persons and the delivery of sacks of food have a central place in the humanitarian imaginary. Material assistance goes well beyond camps and food aid, but these two archetypal dimensions of humanitarian response exemplify important issues and critiques. In many cases, programme design prioritizes donor preferences or ease of management and delivery on the part of operational agencies over and above the needs and preferences of the intended beneficiaries of the programmes. In discounting or deprioritizing their needs and preferences, these 'beneficiaries' may be dehumanized and treated as objects rather than subjects. At the same time, however, both camps and food aid can have significant political impacts.

The Politics and Everyday Practice of International Humanitarianism. Miriam Bradley, Oxford University Press.
 DOI: 10.1093/oso/9780198840695.003.0022

The chapter then turns to two significant areas of recent innovation in the delivery of humanitarian assistance: cash transfer programming and the so-called digital revolution. While cash offers advantages over in-kind assistance, many donors and international humanitarian agencies long resisted extensive cash programming, even in the face of compelling evidence as to the relative risks and benefits of cash and other kinds of programming. By contrast, digital technologies have been uncritically embraced by many humanitarian actors, despite the fact that they can increase the political complexity of humanitarian work and often carry significant risks to programming, 'beneficiaries', or both.

Providing goods and services

Some international humanitarian agencies specialize in providing one or more commodities or services but not usually to the exclusion of all else. For example, Médecins Sans Frontières (MSF) defines itself as a medical humanitarian organization, and Oxfam is known for its expertise in water and sanitation, but both also provide other kinds of assistance. In many cases, the different kinds of goods and services correspond to a particular cluster, though not all aid agencies work within the cluster system, and even those that do also sometimes work outside it—indeed, it is not activated in every humanitarian emergency. These different kinds of interventions all seek to serve an immediate purpose, related primarily to the material well-being of affected populations. However, such interventions often also have the potential to contribute to acceptance of humanitarian operations and to the (in)security of affected populations.

The provision of goods and services can, under certain conditions, help to build trust and goodwill with state and non-state authorities and, hence, to increase the acceptance of humanitarian agencies, increasing their access to affected populations. Broadly speaking, the provision of quality assistance that serves the needs of the population and directly or indirectly benefits the relevant authorities is a necessary but insufficient condition for acceptance by those authorities (Bradley 2016a: 50–51; Crombé and Hofman 2011: 56–57). Humanitarian space—which encompasses both the ability of humanitarian agencies to access those in need and the ability of those in need to access protection and assistance—must be actively constructed (Abild 2010). Thus, the more sophisticated acceptance strategies combine high-quality programming with efforts to build trust and strong relationships with all authorities and armed actors as well as specific negotiations over the scope and conditions of access.[1]

To the extent that the provision of goods and services requires or enables aid agencies to be physically present in communities in insecure contexts, their presence may also contribute to the protection of those communities by deterring potential abuses. Indeed, sometimes that is a primary reason for providing goods and services in particular places. The logic underpinning so-called protection by presence is that if aid agency staff—particularly international staff—are present in a community, armed forces and armed groups may be less likely to perpetrate abuses against the population. Of course, this logic only applies where parties to conflict or other weapons-bearers are keen to avoid (international) scrutiny. In such contexts, the possibility that humanitarian agencies could witness and report abuses may serve as a rational incentive for armed actors to change their behaviour. While this kind of passive protection by presence does seem to be effective in some contexts, there is a risk that the presence of humanitarian agencies in some communities only serves to displace violence to a different place or time (Bradley 2016a: 124).

Depending on context and programme design, providing material assistance and services can have a positive or negative impact on a range of risks. Especially in insecure contexts, poorly designed programming of almost any kind may inadvertently expose people to threats from violence. For example, poor lighting around toilet blocks in a camp may make women more vulnerable to sexual violence, and the provision of food and other commodities in large quantities may provide an incentive for attack, thus making the recipients of these commodities vulnerable to violence. By contrast, programming that is carefully designed through a protection lens may serve to reduce exposure to threats—or even to reduce threats in themselves. For example, programming that helps to reduce the need for people to walk through or work in heavily mined areas—whether by installing water points in villages, so people do not have to go far to collect water, or by developing alternative income-generating activities so that farmers do not have to travel to their fields—can reduce their vulnerability. However, the impact of different kinds of assistance and services on people's (in)security is context-specific, and these different types of programming should not be universally considered protection programming.

[1] See, respectively, 'Relationship-building interactions' and 'Access negotiations' in Chapter 23 on Dialogue, Negotiation, and Advocacy.

'Rape-stoves' exemplify the possibilities and limits of material assistance as a way to reduce people's vulnerability to violent threats. The provision of firewood to vulnerable groups in the Dadaab refugee camps in Kenya in the 1990s, and of fuel efficient stoves in camps for displaced persons in Darfur in the mid-2000s, was expected to reduce sexual violence because women and girls were at heightened risk of rape when collecting firewood outside the camps (Abdelnour and Saeed 2014: 146, 149–150, 153–155). Relegating sexual violence to one space and activity involved wilfully ignoring the sexual violence that occurred *within* the camps and facilitated a focus on the kind of technical solution humanitarian agencies often prefer (Abdelnour and Saeed 2014: 150, 154–155). The limits of such a reductionist approach to sexual violence were quickly evident. A decline in firewood-related rape around the Dadaab camps was accompanied by an increase in non-firewood-related rape (Abdelnour and Saeed 2014: 150). For camp residents in Darfur, the provision of fuel-efficient stoves did little even to reduce firewood collection, given that the low quality of some stove models limited their potential fuel savings, some recipients of the stoves used them ineffectively or disregarded them altogether, and wood was also collected for construction purposes and to sell for income (Abdelnour and Saeed 2014: 146). Furthermore, displaced people in Darfur travelled outside the camps for a variety of reasons, including a host of livelihoods activities unrelated to firewood (Abdelnour and Saeed 2014: 154).

Politics of camps

Many humanitarian crises are accompanied by large-scale displacement, prompting a need to house large numbers of people. The two clusters most centrally working on shelter and related issues are the shelter cluster, co-led by the International Federation of Red Cross and Red Crescent Societies (IFRC) and the United Nations Refugee Agency (UNHCR), and the camp coordination and camp management cluster, co-led by the International Organization for Migration (IOM) and UNHCR. For a long time, the primary international response to displacement crises was in the form of refugee and internally displaced person (IDP) camps, and in many crises, camps are still the response favoured by donors and operational agencies. An estimated 6.6 million people, or 22% of the world's refugees, live in refugee camps (UNHCR 2021b). Around 4.6 million of these live in planned or managed camps, while the other 2 million are in spontaneous or self-settled camps (UNHCR n.d.-c). Many camps house tens of thousands of people, and the very largest house hundreds of thousands, making them more populous than many cities. While camps may be an obvious response to emergency shelter needs and are intended to be temporary, where displacement becomes protracted, camps can become quasi-permanent (Turner 2016: 142).[2]

Donors and operational humanitarian agencies have historically favoured camps as a means to care for (and control) people displaced by conflict or disasters. From the perspective of operational agencies, attracting funding for their work requires not only counting the displaced for needs assessment purposes, but also representing them as helpless and dependent, and keeping people in camps facilitates both these tasks (Harrell-Bond 1998: 22). Furthermore, it is much easier to target, distribute, and coordinate assistance to people in need when they are all gathered in the same place (Crisp et al. 2012: S25–S26, S27–S28). Most donor and host governments hope that refugees will eventually repatriate and often believe that this is less likely if refugees are able to integrate (Bradley 2017: 25). In fact, those refugees who develop their skills and acquire resources during displacement may be more likely to return voluntarily when conditions in their country of origin are conducive (Harild et al. 2015; Harrell-Bond 1998: 22). Host states, who are ultimately responsible for admitting and recognizing refugees on their territory and determining the freedoms they can enjoy, nonetheless sometimes insist that refugees are housed in camps, with limited rights to movement, work, and status.[3]

While they are often the preferred response of governments and international agencies, camps are rarely the only or best way to meet the needs of the displaced and often contravene their rights, especially when displacement is protracted. It is widely held that camps offer more access to assistance from the government and aid agencies and often better legal protection, while self-settlement and urban life offer greater potential for social and economic integration (Bradley 2017: 15; Turner 2016: 141). However, recent research covering a range of contexts suggests that those in camps frequently do not have better access to assistance and services (Pape and Sharma 2020). Moreover, camps epitomize the humanitarian tendency to exercise control in tandem with offering care, serving both to keep the most vulnerable alive and to manage, contain, and police these 'undesirable' populations (Agier 2010). Camp authorities may reduce life to bare, biological, temporary survival and deprive residents of political agency, but camps also offer opportunities to create new identities, and camp residents have frequently contested

[2] On protraction, see 'Protracted "emergencies"' in Chapter 15 on the Shifting Parameters of Humanitarian Emergencies.

[3] See 'State sovereignty, responsibility, and practice' in Chapter 19 on Government and Civil Society in Affected States.

efforts to render them as bare life (Agier 2010: 40–41; Malkki 1996: 380–382; Turner 2016: 143). Camps also provide ideal breeding grounds for political and ethnic radicalization and for violence and terrorism (Harrell-Bond 1998: 22; Verdirame and Pobjoy 2013: 473). After the Rwandan genocide, for example, the *génocidaires* were able to flee across the border, to take control of the administration of camps in neighbouring countries, and to use them to regroup, restrengthen and re-arm.[4] Thus, camps can be at once depoliticizing and (re-)politicizing.

UNHCR—and other international humanitarian agencies in charge of managing refugee and IDP camps—can wield enormous power over the lives of camp residents, who have limited scope to hold UNHCR accountable for its use of that power. Camps are an archetypal example of international humanitarian agencies substituting for the state, often setting up parallel systems, attracting the best government staff with higher salaries, and undermining local institutions (Harrell-Bond 1998: 22; Verdirame and Pobjoy 2013: 473).[5] Although they come under the jurisdiction of the host state, in practice, refugee camps are often governed by different legal instruments from surrounding areas, and with varying—sometimes almost non-existent—levels of involvement from the host government (Turner 2016: 141; Verdirame and Pobjoy 2013: 473). In many cases, national laws are not enforced in refugee camps, and their residents are instead subject to written and unwritten rules laid down by the humanitarian agencies running them (Verdirame and Pobjoy 2013: 472). No wonder, then, that UNHCR has been described as a 'surrogate state, complete with its own territory (refugee camps), citizens (refugees), public services (education, health care, water, sanitation, etc.) and even ideology (community participation, gender equality)', and camp administration as 'a form of *de facto* international administration of territory' (Slaughter and Crisp 2009: 8; Verdirame and Pobjoy 2013: 473). The work of UNHCR and the agencies it subcontracts to run camps and manage refugees may be inhumane, but 'accountability is skewed in the direction of the donors who pay for the assistance, rather than the refugees' (Harrell-Bond 2002: 53).

There has been a shift in thinking such that, in principle, camps are no longer seen as the only proper response to displacement crises, even if changes in practice have not kept pace with this shift. UNHCR was slow even to adapt its thinking, with its 1997 Policy on Urban Refugees characterizing urban refugees as exceptional or deviant, and entitled to only minimal and temporary assistance (Verdirame and Pobjoy 2013: 474–477). However, a

[4] See 'Mass exodus to neighbouring countries' in Chapter 5 on Genocide in Rwanda and Its Aftermath, 1994–1996.

[5] For a more general discussion of this issue in humanitarian response, see 'Substituting for the state' in Chapter 19 on Government and Civil Society in Affected States.

number of factors in the 2000s contributed to a shift in UNHCR policy: the actions of displaced persons, who increasingly settle themselves in urban areas; the responses of host states, some of whom—including Jordan, Lebanon, and Syria as hosts to the majority of refugees fleeing the Iraq War—rejected the camp model; and campaigns by human rights and humanitarian organizations against 'warehousing' refugees in camps (Verdirame and Pobjoy 2013). More generally, there has been increased recognition of the resilience of refugees (and others affected by disasters) and of their capacity for self-reliance (Hilhorst 2018: 6). UNHCR's 2009 Urban Policy marked a significant shift in attitude, recognizing that the responsibilities of UNHCR towards refugees are the same whether those refugees are in camps or urban areas and emphasizing the primary responsibility of host states to provide refugees with protection, solutions, and assistance (Verdirame and Pobjoy 2013: 479–480). The past ten years have seen increased urbanization of humanitarian response, with increased efforts to reach those who are displaced in cities, but much remains to be done to make humanitarian assistance as accessible to displaced persons dispersed in cities as it is to those in camps.[6]

Politics of food aid

Around 2.3 billion people lacked access to adequate food in 2020, and more than 900 million of them were severely food insecure (FAO et al. 2021: xvi). The most relevant clusters are the food security cluster, led at the global level by the World Food Programme (WFP) and the Food and Agriculture Organization (FAO), and the nutrition cluster, led by the United Nations Children's Fund (UNICEF). The dominant response to food insecurity is the provision of food aid, which is often assumed to be both apolitical and benign. However, food aid is not necessarily the best way to tackle food insecurity, and it can be political in multiple ways. The economic and political self-interest of donor states sometimes drives where and how food aid is provided, and the provision of food aid sometimes has significant unintended economic and political impacts in the states receiving it.

Food aid programming has often been driven by political factors in donor countries. Critiques in the 1950s and 1960s centred on two main issues. First, food aid was motivated by a desire to dispose of surplus grains from donor countries, especially the United States, as a means to bolster agricultural

[6] On this urban turn, see 'Urbanization' in Chapter 15 on the Shifting Parameters of Humanitarian Emergencies.

incomes (Clapp 2012: 3). Second, food aid was not provided strictly in accordance with need but rather used as a tool—either carrot or stick—with which to influence policies in recipient states in the context of Cold War geopolitics (Clapp 2012: 3, 19). It was largely provided bilaterally, and recipient governments often sold the food on the open market to generate income for other activities (Barrett and Maxwell 2004: 53). In the 1970s and 1980s, donor countries reduced their surpluses and food aid came to be directed more according to the needs in recipient countries than the potential for donor influence (Clapp 2012: 25). The provision of food was also increasingly governed by multilateral institutions and norms and, by the 1980s and 1990s, had come to be seen as less controversial (Clapp 2012: 1, 25–26). Emergency food aid, in response to humanitarian crises, became the dominant form and was usually freely provided to acutely hungry people (Barrett and Maxwell 2004: 53).

By the 2000s, the focus of the debate had turned to whether or not donors imposed conditions on where food was sourced and how it was transported. As donors, the United States and Japan tend to provide funding for so-called tied food aid, which must be sourced in the donor country (Clapp 2012: 6). Since the mid-1990s, by contrast, the European Union (EU), Australia, and Canada, among others, have moved to provide funds to buy food closer to the recipients (Clapp 2012: 6, 46). Since the United States is by far the biggest food aid donor, US policies have a significant impact on the provision of food aid as a whole (Clapp 2012: 6; Murphy and McAfee 2005: 3). US agribusiness and non-governmental organizations (NGOs) have an economic interest in US food aid being tied since US law stipulates that at least 75% must be sourced, fortified, processed, and bagged in the United States, and US-based NGOs are charged with programming and delivery (Clapp 2012: 7; Murphy and McAfee 2005: 27). The US shipping industry is also a supporter of in-kind aid because US rules likewise require 75% of such aid to be transported in US vessels (Clapp 2012: 7; Murphy and McAfee 2005: 27). However, tied aid is slower and often more expensive than goods sourced more locally to recipients and sometimes consists of less appropriate foodstuffs (Barrett and Maxwell 2004: 55).

Much like any kind of external assistance, food aid can have unintended political and economic impacts in recipient countries. In Ethiopia in the 1980s, for example, food aid provided in response to famine was instrumentalized by the government to support its counter-insurgency policies.[7] More recently, in the context of the civil war in Syria, a country

[7] See 'Use of food aid for non-humanitarian ends' in Chapter 3 on Drought and Famine in Ethiopia, 1983–1985.

with a history of government subsidies for basic foodstuffs, the provision of food aid by international humanitarian agencies arguably served to bolster the power of the state (Martínez and Eng 2016). Where most people's immediate needs are met by in-kind assistance, demand in local markets suffers (Levine 2017: 6). Where used inappropriately or poorly managed, food aid thus risks lowering market prices and undermining agricultural production in recipient country economies (Barrett and Maxwell 2004: 53). Even where food aid is sourced locally, the procurement procedures of aid agencies may serve to exclude the businesses that were the main suppliers pre-crisis and to favour those from capital cities with large amounts of money, English- or French-language skills, and the ability to deal with the systems of international agencies (Levine 2017: 11). In bypassing the normal supply chain, aid contracts therefore risk exacerbating inequalities and delaying market recovery (Levine 2017: 12).

Against this background, since the late 2000s, several important actors, including WFP, have shifted the emphasis from food *aid* to food *assistance*, a wider concept that encompasses a diverse range of approaches to meeting long-term nutritional needs (WFP n.d.). This shift follows the logic of the now widespread understanding that the existence of famine, the archetypal food security crisis, does not necessarily mean that there is an aggregate shortage in the food supply but rather that not everyone has access to adequate food and nutrition.[8] This implies that increasing the food supply through the provision of food aid is not the only, nor necessarily the best, response option. In acute emergencies, the food available from local production and commercial imports is sometimes insufficient, and in such contexts, food aid can fill a crucial gap (Barrett and Maxwell 2004: 53). However, where food is available but inaccessible to many, alternative responses to increase access may be more appropriate. Even the United States, which continues to allocate the majority of its funding for food assistance to in-kind food aid, also now supports a wider range of responses, including local or regional purchase of food and the provision of cash or vouchers (Maxwell et al. 2013: 73).

Cash and voucher assistance

The idea that cash and voucher assistance (CVA) should be a standard and central component in humanitarian response is relatively new, but there are several historical examples of cash being handed out in response to particular humanitarian emergencies, including during the 1870–1871

[8] See 'Famine' in the Introduction on Humanitarian Emergencies.

Franco-Prussian war, in response to famine in nineteenth century India, and in Botswana in the 1980s (High Level Panel on Humanitarian Cash Transfers 2015: 15). The international response to the 2004 Indian Ocean tsunami marked a major turning point, with extraordinary amounts of funding for humanitarian response facilitating experimentation and a number of operational humanitarian agencies providing cash or vouchers to tsunami-affected populations at scale and, additionally, conducting research which showed that the programming was effective.[9] Then, in the context of famine in Somalia in 2011, with insecurity a major impediment to access in the worst-affected areas, a consortium of NGOs organized a large-scale cash response, providing a clear demonstration of the potential of cash transfers in highly insecure environments.[10] Since then, the role and acceptance of CVA in humanitarian response have increased dramatically, though some critics and critiques remain.

Modalities of CVA

Cash programming can employ different approaches. First, CVA can be conditional or unconditional. Depending on the context, conditions could include sending children to school, attending particular training programmes, or rebuilding part of a house (Bailey and Harvey 2017: 7). In cash-for-work programming, cash is provided as a salary in exchange for unskilled or semi-skilled labour for projects intended to contribute to recovery or community development (Food Security Cluster 2019: 3). Second, CVA can be distributed in a variety of ways, many of which do not involve literal cash. For instance, while transfers may be made as 'cash in envelopes', transfers are more often made via local financial institutions or through electronic mechanisms including banks and mobile phones (High Level Panel on Humanitarian Cash Transfers 2015: 16). Cash transfer programmes are also increasingly using the existing infrastructure of government social welfare transfer systems (Bailey and Harvey 2015: 5; High Level Panel on Humanitarian Cash Transfers 2015: 15). Third, CVA can be unrestricted, or it can be in the form of vouchers, which can only be spent on certain goods or with certain businesses. The terms 'cash-based responses' and 'cash transfer programming' have often been used to refer to the provision not only of cash but also of vouchers, which is problematic because the restrictions mean that vouchers present quite different opportunities, costs, and constraints (High Level Panel on Humanitarian Cash Transfers 2015: 16; Levine 2017: 20).

[9] See 'Emergency response' in Chapter 7 on the 2004 Indian Ocean Tsunami.
[10] See 'Cash programming' in Chapter 10 on Famine in Somalia, 2011–2012.

Pros and cons of CVA

Compared with in-kind aid, cash gives people greater choice and control over how best to meet their own needs and a greater sense of dignity. Vouchers offer greater choice than in-kind aid but less than cash itself. Unconditional and unrestricted cash transfers maximize recipient choice and agency. By empowering recipients—rather than aid agencies—to determine what they need most, cash transfers overcome the limits of aid being restricted to the kinds of goods and services operational agencies can deliver (High Level Panel on Humanitarian Cash Transfers 2015: 13). The flexibility of cash transfers means that aid can meet a greater range of needs, with recipients reporting using the money for such diverse expenditures as food, building materials, agricultural inputs, medical expenses, school fees, debt repayment, clothing, hygiene, fishing equipment, and transport (High Level Panel on Humanitarian Cash Transfers 2015: 20). In many cases, cash has been shown to be more effective than food aid at improving diet quality, with recipients reporting more varied and nutritious diets (Bailey and Harvey 2015: 3–4). However, where aid agencies give in-kind aid or vouchers with a specific objective, such as increasing the consumption of fresh food, cash may be less effective at achieving those particular objectives, which may not align with recipient priorities (Bailey and Harvey 2015: 4; High Level Panel on Humanitarian Cash Transfers 2015: 21).

Compared with providing in-kind assistance, CVA tends to be quicker and is often easier to distribute in areas which are difficult or impossible to access. Locally or regionally procured food aid (or other in-kind assistance) can often be sourced and delivered relatively quickly, but transoceanic shipments are significantly slower, especially to landlocked countries (Lentz et al. 2013: 12). Prepositioning food aid in or near areas likely to need food assistance would mean that transoceanic shipments could be delivered faster than food procured locally, but this tends to be more expensive than other forms of assistance due to higher storage and freight costs (Lentz et al. 2013: 10). Where insecurity limits access, getting assistance to people in need is hard, but greater use of cash—including through digital payments—extends the options to reach people (High Level Panel on Humanitarian Cash Transfers 2015: 14). In response to famine in Somalia in 2011, for example, cash transfers offered a way to overcome some of the security and logistical challenges which meant that large-scale food aid was not a realistic option.[11]

CVA also offers greater value for money than in-kind aid. Transporting and storing physical goods is costly and often complicated. For example, following the 2004 Indian Ocean tsunami, USD 92 million of WFP's USD 210

[11] See 'Cash programming' in Chapter 10 on Famine in Somalia, 2011–2012.

million tsunami relief fund was used for the transportation and storage of food (Oxfam 2005: 4). It usually costs less to get cash to people, although the relative cost effectiveness of cash over in-kind assistance depends on the commodity being provided and where. For instance, data from two WFP programmes in Ethiopia in 2014 suggested that cash was 25–30% cheaper to deliver than food aid (Cabot Venton et al. 2015: 21). In the context of famine in Somalia, 2.5 times more of aid budgets went directly to recipients with cash transfers rather than food aid (High Level Panel on Humanitarian Cash Transfers 2015: 19). A study comparing the two modalities in four different countries found that, overall, 18% more people could be assisted at no extra cost if everyone received cash instead of food (High Level Panel on Humanitarian Cash Transfers 2015: 19).

The greater choice afforded to recipients further increases the cost-efficiency of cash. When people receive food aid or other in-kind assistance, they often sell at least some of it in order to get cash to buy other things they need (or more appropriate food) and often for low prices. The difference between the price they receive for selling it and what it costs humanitarian agencies to provide it can represent a significant waste of limited humanitarian resources (High Level Panel on Humanitarian Cash Transfers 2015: 18). Furthermore, because cash transfers can be used to pay for a diverse range of goods and services, they offer greater value for money than in-kind assistance (High Level Panel on Humanitarian Cash Transfers 2015: 19). Each recipient likely has a distinct set of needs and priorities, and cash allows them to buy goods and services tailored to their particular circumstances in a way that aid agencies cannot easily or efficiently match with in-kind aid. Furthermore, cash can be used for things that do not have in-kind equivalents, such as debt repayment, land rental, and savings (Bailey and Harvey 2015: 4; High Level Panel on Humanitarian Cash Transfers 2015: 19). Because the element of choice is critical here, unrestricted cash transfers are generally better than vouchers, which can only be spent on particular goods or in particular businesses—and may be sold for less than their value to access cash (High Level Panel on Humanitarian Cash Transfers 2015: 19, 20). Depending on the size of the transfers, cash may also contribute to resilience, and large grants have been shown to increase future income (High Level Panel on Humanitarian Cash Transfers 2015: 20).

Cash is only useful if the markets for the goods recipients need to buy are functioning, but cash transfers can also help markets to recover. Humanitarian crises often have an impact on markets, sometimes driving price increases if stocks are lost, transport made more difficult, or risk increased, but markets rarely stop functioning altogether (Levine 2017: 5).

The absence of strong supply does not always imply an inherent weakness in the supply chain but rather may reflect depressed demand such that cash transfers could stimulate demand, and supply would increase to meet the increased demand (Levine 2017: 7). More generally, cash transfers to people in need can have positive multiplier effects on local economies as recipients spend them in local markets and so the benefits of the assistance reach beyond the immediate recipients. Voucher programmes usually have a very limited number of partner retailers, limiting their positive multiplier effects and creating monopolies, which can lead to higher prices (High Level Panel on Humanitarian Cash Transfers 2015: 19; Levine 2017: 20).

Price volatility can make the value of cash uncertain and, conversely, a large inflow of cash can sometimes cause inflation in local markets. These are real and important concerns. Where there are high levels of inflation, as for example, in Ethiopia during the food price crisis of 2008, the amounts of cash transfers could not keep up with price increases, leading to an increase in demand for in-kind food aid (Maxwell et al. 2013: 77). If currency markets are functioning well, it may be possible to mitigate such volatility by providing cash in foreign currency, but in many high-inflation contexts, in-kind aid may be preferable. The inflationary impact of cash transfers depends on how elastic the supply of goods and services people seek to purchase is. Cash can contribute to price rises by increasing the demand for goods, but where this stimulates increased supply, the price will normally fall again. However, where supply is inflexible, cash transfers can drive inflation, which not only decreases the value of the cash provided but can also make life especially difficult for those not in receipt of assistance. For example, where displaced people are provided with cash for rent in an urban setting with limited housing supply, their demand for housing may drive up rents. Depending on what the cash is intended for, programming can be designed to minimize this kind of inflationary impact, as for example, in Somalia in 2011, when the value of cash transfers was calculated in line with the price of imported rice, which was more popular than local grains and had a higher but less volatile price.[12]

Some donors, decision-makers, and staff in operational humanitarian agencies, publics in donor countries, and governments in affected states also raise concerns that recipients could spend cash on 'anti-social' goods, including weapons, drugs, and alcohol (Ali and Gelsdorf 2012: 61). Some agencies favour voucher-based programming precisely because they can restrict what

[12] See 'Cash programming' in Chapter 10 on Famine in Somalia, 2011–2012.

such assistance is used for (Levine 2017: 20). Arguably, this type of thinking reflects a prejudiced resistance to granting agency to the recipients of humanitarian assistance rather than a reasoned concern (Bailey and Harvey 2015: 3). Food (or other in-kind assistance) can also be exchanged for weapons, drugs, or alcohol. In-kind assistance is often sold so that recipients can buy other goods and services, and there is nothing to stop them engaging in 'anti-social spending' with the proceeds from selling food or other in-kind assistance. Moreover, a wealth of research shows that recipients of cash transfers do not, for the most part, use them in ways that international humanitarian actors deem unwise (Bailey and Harvey 2015: 3; High Level Panel on Humanitarian Cash Transfers 2015: 21).

Critics also argue that cash transfers are particularly prone to corruption and diversion and risk feeding war economies, but there is little evidence to support these claims. Cash transfers appear no more susceptible to theft and fraud than in-kind transfers (Bailey and Harvey 2015: 5). Risks of diversion, corruption, theft, and fraud may be different for cash transfers compared with in-kind aid, but they are not necessarily greater (High Level Panel on Humanitarian Cash Transfers 2015: 20). Even in Somalia, a highly insecure environment without a functioning government, more than USD 80 million was successfully delivered to 1.7 million people in response to famine in 2011.[13] In fact, cash transfers can be distributed more discreetly than in-kind aid—particularly when people receive payments electronically—and tend to be in transit for less time, meaning that they may be *less* vulnerable to looting or extortion (High Level Panel on Humanitarian Cash Transfers 2015: 20). Likewise, in-kind assistance may do more than cash to facilitate profiteering from war economies. When aid agencies procure in-kind assistance locally, or when they provide vouchers that can only be used in a small number of businesses, their procedures sometimes favour those with significant capital, connections, and ability to do business with international actors, enabling those with power to capture an important market for foreign currency (Levine 2017: 9).[14]

Another concern is that cash may be more difficult to target and may be more likely to be controlled by men and, hence, to disadvantage women (High Level Panel on Humanitarian Cash Transfers 2015: 18). Targeting is difficult whatever kind of support is being provided since comprehensive data on households is often lacking, and powerful actors may seek to influence the process of deciding who should receive what kinds of

[13] See 'Cash programming' in Chapter 10 on Famine in Somalia, 2011–2012.
[14] See 'Politics of food aid' in this chapter.

assistance.[15] The desirability of cash may make targeting cash marginally more difficult than other less desirable kinds of assistance, but this does not seem to have posed a fundamental problem (Bailey and Harvey 2015: 4). The question of how cash transfers affect women and gender relations lacks a clear answer. On the one hand, if in-kind assistance is more likely to be controlled by women and cash is more likely to be controlled by men, then the provision of cash could serve to reduce women's power (Bailey and Harvey 2015: 5). On the other hand, if cash is provided to women and they retain control of it, it could serve to empower them and elevate their status in the household (Bailey and Harvey 2015: 5). Both these scenarios seem equally plausible, and the evidence to date is inconclusive—most likely, the impacts depend on the context. However, there is evidence to suggest that cash often helps to reduce stress and arguments linked to difficulties meeting household needs (Bailey and Harvey 2015: 5).

While cash transfers clearly offer considerable advantages over in-kind assistance, and the critiques and concerns discussed here may be overstated, some of the concerns should not be completely dismissed. Donors and operational agencies seem to be especially demanding of evidence that cash transfers will not have negative unintended consequences, whether in terms of corruption, market dynamics, gender relations, or something else. For instance, most donors demand market studies to justify CVA, and the same is not required for transoceanic food aid or other forms of in-kind assistance, despite the fact that there is no evidence to suggest they represent lower (as opposed to different) risks to the functioning of markets (Levine 2017: 20; Maxwell et al. 2013: 73). Thus, the feared negative consequences may not have materialized in practice precisely because careful analysis has gone into deciding when to provided cash transfers.

Recent developments in CVA

In 2015, the central recommendation of the High Level Panel on Cash Transfers was to use cash programming more widely and to shift the default humanitarian assistance from in-kind aid to cash transfers, urging donors and operational agencies to consider cash the 'first-best' response to crises and to ask the questions 'Why not cash?' and 'If not now, when?' in every crisis (High Level Panel on Humanitarian Cash Transfers 2015: 23). The Panel

[15] For a broader discussion of the challenges inherent in needs assessment, which is related—albeit analytically prior—to targeting decisions, see 'Assessing needs, contexts, and capacities' in Chapter 21 on Needs assessment, Evaluation, and Response Decisions.

was very clear that cash cannot meet all humanitarian needs and called for cash transfers to be complemented by work to provide additional goods and services that the market will not provide effectively, including security and public health, or that require specialist inputs, such as treating acute child nutrition (High Level Panel on Humanitarian Cash Transfers 2015: 21). The Panel further argued that by increasing the volume of cash assistance, aid delivery could be streamlined, thus freeing up time for aid workers to provide these and other complementary services, including help to secure land rights or technical assistance to build disaster-resistant housing and to be present in affected communities affected by crisis, bearing witness to their suffering (High Level Panel on Humanitarian Cash Transfers 2015: 21).

The 2016 World Humanitarian Summit (WHS) also embraced cash programming, recognizing its importance for increasing efficiency, supporting people's agency and dignity, and stimulating local economies, with several donors and operational agencies committing to increase the proportion of total humanitarian expenditure going to cash programming (UN Secretary-General 2016: 14). Workstream 3 of the Grand Bargain focused on increasing the use and coordination of cash and, in keeping with the idea of making cash a default part of any response, signatories committing to increasing the *routine* use of cash (Metcalfe-Hough et al. 2021: 63).

Since then, there have been major changes—both in the quantity of cash programming and in the way much of the humanitarian sector thinks about it. The importance of CVA has grown significantly relative to in-kind aid, from an estimated 10.6% of international humanitarian assistance in 2016 to 17.9% in 2019 (Cash Learning Partnership 2020: 9). The resetting of thinking about the default response also seems to have resonated, with 'Why not cash?' seized as a commonsense question (Bailey and Harvey 2017: 5). By 2020, several operational agencies had adopted 'cash-first' policies for a range of programming objectives, such as basic needs and food security (Metcalfe-Hough et al. 2021: 60). The myth that cash is more risky than in-kind assistance has been effectively debunked and cash assistance is now a norm in humanitarian practice, with an increasing range of programming objectives pursued with cash (Metcalfe-Hough et al. 2021: 60). While these developments have been almost universally welcomed, it is clear that there is still some way to go to maximize the scope and benefits of cash programming—with an urgent need to find ways to reach the most vulnerable and marginalized, who are often (unintentionally) excluded from CVA and to improve inter-agency coordination of multi-purpose cash (Doornbos 2022; Metcalfe-Hough et al. 2021: 61).

Technology and the digital revolution

Recent decades have seen a so-called tech or digital 'revolution' in the delivery of humanitarian goods and services, with proponents promising that technology can help overcome many of the challenges facing humanitarian agencies. However, it is doubtful that technology can overcome the kinds of challenges that derive primarily from political—as opposed to logistical—complexity. Furthermore, in many cases, the technological solutions employed by humanitarian actors actually serve to exacerbate political complexity, and even successful applications of technology often come with unintended and unwanted consequences (Jacobsen 2015).

Certainly, technology has facilitated some significant developments in the delivery of humanitarian services and material assistance, and tech solutions have been widely—although not universally—welcomed in the humanitarian sector. In many settings, for example, digital technologies have played an important role in both delivering and evaluating cash transfer programming. As stated by the High Level Panel, the use of 'digital payments makes disbursing and receiving transfers cheaper, improves their transparency and traceability, increases security for recipients and can give people an entry point into other formal financial systems' (High Level Panel on Humanitarian Cash Transfers 2015: 13). Biometric data is often used to identify and verify recipients of digital cash transfers and sometimes also of other kinds of assistance, potentially making the provision of assistance more efficient, increasing accountability to donors, and reducing the scope for duplication and fraud (Holloway et al. 2021: 14, 17–18). Biometric identification is also lauded as a means to provide a legal identity to people who might otherwise lack one, including refugees and stateless persons, among other things helping to safeguard their rights to education and health care (Holloway et al. 2021: 12). The use of biometrics illustrates the diversity of attitudes to digital technology among international humanitarian agencies—with UNHCR in particular embracing biometric use, while some other organizations are much more cautious, with the ICRC, MSF, and Oxfam opting to use biometric technologies in only quite limited ways (Açıkyıldız 2021; Oxfam 2021).

Tech solutions also have some important limitations, and dependence on digital technologies can carry significant risks. Where biometrics are used to verify the recipients of humanitarian goods and services, the 'problems' the technology purports to solve are arguably not the most fundamental or urgent problems. For example, most fraud in the humanitarian supply chain happens at the level of procurement rather than through double registration at the point of delivery (Holloway et al. 2021: 17). Whether fuel-efficient stoves or digital cash, the tendency to embrace new technologies and techniques may come at the expense of tackling more tenacious problems in the architecture and ethos of the formal humanitarian system (Bennett et al. 2016: 42). Indeed, the addition of technology has often served to reinforce rather than challenge unequal power dynamics between providers and recipients of aid (Holloway et al. 2021: 7). Furthermore, dependence on technology for the delivery of assistance risks (further) marginalizing those without access to the necessary technology. Thus, for example, the increase in mobile phone penetration means that cash transfers made using mobile phone technology can reach an ever growing number of people, but those without mobile phones may be excluded (Capgemini Consulting 2019: 30–31; Cash Learning Partnership 2020: 115; High Level Panel on Humanitarian Cash Transfers 2015: 21). Where the technology in question fails, the delivery of assistance may also fail. For instance, where iris recognition technology is used to prevent refugees claiming the same package of assistance twice, false matches can lead to refugees being denied the assistance to which they are entitled (Jacobsen 2015: 61–64).

The technologies employed by humanitarian agencies often also have the potential to do harm. In particular, many of these technologies, and much of the data collected, facilitate not only the provision of goods and services to individuals in need but also the surveillance of those individuals. The issue of humanitarian action serving both to care and to control is not unique to digital technology but may be exacerbated by it as such technology facilitates the policing of populations. Where data collected or stored by humanitarian agencies is shared or stolen, it can also be used by non-humanitarian actors for surveillance, and in some cases, the data may then be used against those who humanitarian agencies are supposed to be helping.

Humanitarian agencies sometimes share data voluntarily with other actors, including states, but they can also be forced to share data. UNHCR sometimes shares biometric technology or refugee data with host states, who may use it to track and control people's movements and may even share it with the regimes from which refugees have fled (Jacobsen 2015: 76–79). There have been cases of biometric data from Syrian refugees being leaked to the Syrian government (Jacobsen 2015: 76). Search-and-rescue (SAR) NGOs operating in the Mediterranean can use drones to help spot migrant boats in distress, and on at least one occasion, authorities have used SAR drone footage—which NGOs could be legally bound to hand over in the event of a criminal investigation—to identify and arrest the individual who had allegedly driven a migrant boat.[16] A hack of the ICRC servers for the Red Cross and Red Crescent Movement's Restoring Family Links services in November 2021 compromised the personal data of more than half a million people worldwide (ICRC 2022). The ICRC has said that the attack was 'state-like', raising concerns that a government in a country of transit or destination could use that data to track down undocumented migrants or a government in a country of origin could use it to identify members of a minority group seeking to maintain contact with people who have fled. Concerns about the (mis-)use of data for surveillance purposes are not unique to humanitarian settings, but the ethics are further complicated if those whose data is collected and processed are not in a position to give truly voluntary and informed consent—and where humanitarian assistance is conditional on biometric identity checks, for example, it is difficult to see how consent can be truly voluntary.

Conclusions

Increasing recognition of the agency and self-reliance of people affected by humanitarian agencies has contributed to shifts in thinking about how best to assist them. The move away from refugee camps, and the move towards cash programming, reflect these shifts in thinking, but the resistance those moves met with is indicative of a resistance among aid agencies to relinquish control. Such resistance stands in contrast to the uncritical acceptance of much technology by some humanitarian agencies—technology which often serves to increase, rather than decrease, their control over humanitarian response and over affected populations. The widespread acceptance of cash transfers as a central and default component of international responses to humanitarian crises is arguably one of the most important developments in the humanitarian sector this century, but neither cash programming nor the digital revolution is a panacea for the more structural problems of international humanitarianism.

[16] See 'Search and rescue' in Chapter 12 on the 2015–2016 European 'Migrant Crisis'.

23
Dialogue, Negotiation, and Advocacy

Beyond the more concrete tasks associated with the delivery of material assistance and the provision of services such as water and health care, humanitarian agencies must communicate and liaise with a variety of actors in a variety of ways. Dialogue and advocacy can potentially impact on how different actors treat populations in need and on the ability of aid agencies to access those populations and fulfil their institutional mandates.

International humanitarian agencies can engage in private dialogue or negotiation with de jure and de facto authorities, strategically share information with select third parties, and make public statements which criticize authorities. This chapter examines each of these three broad approaches in turn, explaining how different humanitarian agencies utilize these different approaches, what and whose behaviour they seek to influence, how effective they have been in achieving their goals, and what unintended consequences their efforts have had in different contexts. International humanitarian agencies also can, and do, engage in dialogue with affected communities as part of needs assessment, implementation, and evaluation of programming and with donors to discuss funding and programming. Their publicity and marketing campaigns can also be understood as a kind of public advocacy. The focus of this chapter, however, is on the use of dialogue, negotiation, and different kinds of advocacy with the goal of changing the behaviour of authorities who contribute to suffering and/or have control over humanitarian access.

While direct dialogue with authorities may be more often associated with negotiating access and public criticism more associated with efforts to improve the safety of the civilian population, all three approaches can be directed at increasing either access or protection. However, the use of these different approaches varies depending on the humanitarian agency in question and on the nature of the authorities whose behaviour they seek to influence. In particular, many agencies seem to engage in private dialogue and direct negotiation more with state rather than non-state actors, whereas public criticism in some cases is focused more on armed groups, with government forces sometimes getting off lightly due to aid agency fears of being expelled.

The Politics and Everyday Practice of International Humanitarianism. Miriam Bradley, Oxford University Press.
 DOI: 10.1093/oso/9780198840695.003.0023

Direct engagement with authorities

Dialogue and negotiation with de jure and de facto authorities can help humanitarian agencies to gain acceptance for their work and security guarantees for their staff and to change the way their interlocutors conduct themselves towards the wider population. Many such interactions take place in private or even in secret, so it is impossible to know about all of them. However, it seems that while many aid agencies engage in significant dialogue with the government of the affected state, and also with civil society, regarding acceptance, access, and aid agency security, direct structured dialogue with armed non-state actors is often more limited. Furthermore, there appears to be much less negotiation (with either state or non-state interlocutors) focused on the conduct of these different actors towards the wider civilian population.

The International Committee of the Red Cross (ICRC) and Médecins Sans Frontières (MSF) seem to be the main exception to these two general patterns, or at least, they are more open about the kinds of dialogue they engage in with different interlocutors than some other humanitarian agencies. The ICRC is committed to confidential dialogue and discretion, so it does not tend to share details of its interactions with particular actors but is quite open about its general commitment to engaging with both state and non-state parties to armed conflict in a dialogue on protection, principally aimed at increasing compliance with international humanitarian law (IHL) (Bradley 2013). MSF is more open about the specifics of its engagement with different actors, at least once some time has passed,publishing several accounts of different challenges and dilemmas the agency has faced, including a book focused on humanitarian negotiations (Magone et al. 2011).

There are several reasons why humanitarian agencies engage more with state actors than with armed groups, including restrictions imposed by donor governments or the governments of affected states and concern and uncertainty over the legal implications of engaging with armed groups designated as terrorist.[1] In addition, however, there seems to be a general fear of armed groups in some humanitarian agencies and a perception that they are inherently dangerous and illegitimate. Even the ICRC seems to share this perception when it comes to criminal groups. While it has a clear mandate and commitment to engaging with all armed parties to both international and non-international armed conflicts, the ICRC's mandate and commitment

[1] For a more detailed discussion of the reasons engagement with armed groups is so often limited, see 'Non-state armed groups' in Chapter 20 on Armed Actors. On counter-terror policy specifically, see 'Counter-terror legislation' in Chapter 14 on Law and International Humanitarianism.

is less clear in what it refers to as 'other situations of violence', and institutional policy suggests much greater caution in engaging with criminal actors (Bradley 2020).

Relationship-building interactions

Good-quality programming may be necessary for building trust with affected communities and state and non-state authorities, but the humanitarian agencies with the greatest commitment to maximizing their access to populations in need additionally place significant emphasis on building relationships with a range of interlocutors.

For the ICRC, access is understood to be determined by acceptance of the organization by the government, armed actors, and local communities themselves (Bradley 2016a: 50). Acceptance, in turn, is perceived to result primarily from demonstrating the benefits of ICRC work—both for affected communities and authorities—and by maintaining neutrality[2] and confidentiality (Geremia 2009: 5). The ICRC thus sees quality programming as key but also emphasizes the importance of communication activities with military forces, armed groups, and affected communities (Bradley 2016a: 51, 62–63). These communications activities encompass explaining the work of the ICRC and its principled approach and are aimed at gaining acceptance for its work. There are also multiple examples of MSF staff building relationships with a range of de jure and de facto authorities, highlighting the importance of such activities as drinking tea with community leaders in Afghanistan or golfing with the generals in Myanmar (De Gryse 2013; Terry 2011a). It is difficult to gauge how much most other aid agencies invest in interacting with state and non-state actors with the explicit goal of building relationships and trust with those actors.

Efforts to build relationships and trust with both de jure and de facto authorities may be especially important—and especially difficult—where those authorities have antagonistic relationships with Western powers. For instance, the military government in Myanmar prior to the transition to nominally civilian government in 2011 (and, to some degree, again since the military coup in February 2021) was an isolationist regime with pariah status in the West that was suspicious of the Western-headquartered and largely Western-funded aid agencies that dominate the international humanitarian system, imposing severe restrictions on where they could work and what they

[2] On neutrality, see 'Principles of humanity, impartiality, neutrality, and independence' in Chapter 13 on Politics, Principles, and Humanitarian Action.

could do. More commonly, it is armed groups, and, in recent years, especially those associated with Islamic terrorism, that are ostracized by states in the global North and often distrustful of the large international humanitarian agencies. Since the US-led coalition overthrew the Taliban government at the end of 2001, for example, Afghanistan has been an incredibly difficult operating environment for aid agencies. During Operation Enduring Freedom, the ICRC credited its—admittedly limited—ability to operate there on its intense efforts to build trust with the Taliban and demonstrate its worth to all sides (Terry 2011b).[3] Conversely, it was precisely the lack of trust and established relationships with local communities and Al-Shabaab that prevented agencies from rapidly scaling up their activities in South-Central Somalia after funding was increased following the declaration of famine in 2011.[4]

Remote control or remote management operations—underpinned by digital technology—reduce the need for the kinds of day-to-day contact and meetings on which trust and relationships are built. One of the reasons 'digital humanitarianism' has been so enthusiastically embraced is the increasing risk aversion among aid agencies, who use a combination of technology and sub-contracting of local actors to orchestrate operations in challenging or politically difficult environments without the need for expatriate staff to be physically present (Duffield 2016: 148–149). In this way, tech solutions offer a means to assess needs and deliver assistance from a distance, facilitating humanitarian response in insecure contexts where relationships are weak or non-existent, but they also increase the epistemological and social distance between international humanitarian actors and affected communities and authorities. Thus, they may serve to undermine this important, face-to-face aspect of humanitarian work and can feed a vicious cycle according to which aid agencies are ever more distant—physically, socially, and epistemologically—from those they seek to assist.

Building trust is important because it may be a prerequisite for access negotiations and dialogue on more concrete or more sensitive issues, but close relationships (or relationships that are perceived to be close) with authorities carry some risks. Drinking tea with the warlords or golfing with the generals can affect how humanitarian actors are perceived by competing authorities. In contexts where trust is so important, humanitarian actors have to tread a fine line between building sufficient relationships with one authority and losing the trust of another.

[3] See 'Aid agency responses to these challenges' in Chapter 6 on Afghanistan, 2001–2014.
[4] See 'International humanitarian response' in Chapter 10 on Famine in Somalia, 2011–2012.

Access negotiations

In all humanitarian emergencies, aid agencies need the consent of the state in order to operate and, hence, must negotiate with the government in question.[5] In conflict contexts, humanitarian agencies generally also need to negotiate with other parties to conflict, in particular where they control access to territory in which the humanitarian agencies seek to work. Negotiations with state and non-state authorities concern not only consent for humanitarian operations but also the terms of those operations, which may include stipulations regarding the kinds of activities humanitarian agencies can undertake, where, and when.

The United Nations (UN) Emergency Relief Coordinator is mandated by the General Assembly to facilitate access to emergency contexts for operational humanitarian agencies and is expected to be assisted in this role at the country level by the Office for the Coordination of Humanitarian Affairs (OCHA) and the Humanitarian Coordinator (UN General Assembly 1991; UN OCHA 2020b: 2). Throughout the 1990s, the UN took the lead in negotiating with armed groups on behalf of much of the wider humanitarian community and has continued to play a central role in some negotiations with both de jure and de facto authorities. For example, in Darfur between 2004 and 2006, OCHA and the UN Department of Safety and Security (UNDSS) coordinated near-daily negotiations with rebels, which enabled largely unfettered access to rebel-held territories (Jackson and Davey 2014: 9). OCHA and UNDSS have also negotiated access to Gaza with Hamas on behalf of the wider humanitarian community, a role that is particularly important since the designation of Hamas as a terrorist group has left many humanitarian agencies concerned that they might be liable to prosecution if they were to negotiate directly (Jackson and Davey 2014: 15).[6] In both these examples, 'the relatively large degree of separation of UN humanitarian entities from political concerns, combined with resources dedicated to engagement' facilitated effective coordination by the UN (Jackson and Davey 2014: 28).

For a variety of reasons, operational agencies nonetheless often negotiate access for themselves. OCHA's leadership of humanitarian negotiations with armed groups in particular has diminished significantly since 9/11.[7] Where the UN supports the government, many non-governmental organizations (NGOs) have sought separation from UN-led negotiations, fearful

[5] See 'State sovereignty, responsibility, and practice' in Chapter 19 on Government and Civil Society in Affected States.

[6] See 'Counter-terror legislation' in Chapter 14 on Law and International Humanitarianism.

[7] See 'Non-state armed groups' in Chapter 20 on Armed Actors.

that association with the UN taints how they are perceived by opposition groups (Jackson and Davey 2014: 27).[8] Even where OCHA leads negotiations, individual agencies may also need to negotiate directly, not least because, in many contexts, access depends on consistent and comprehensive engagement, often involving repeated negotiations at multiple levels (Svoboda et al. 2018: 7). Humanitarian agencies operating in the 'messy' post-1995 political environment in Somalia, for example, had to negotiate access continuously and often on a village-by-village basis (Menkhaus 2010: S329). While OCHA may be well placed to negotiate with higher-level state (and sometimes non-state) authorities, operational agencies may still need to undertake their own negotiations on the ground when passing through checkpoints, etc. Indeed, most aid agencies seem to focus their efforts at the local level, although the ICRC and MSF also invest at national and regional levels, and most access agreements are made bilaterally, either through direct dialogue or through each agency's own intermediaries (Harmer et al. 2018: 10–11).

Access negotiations can be grounded in legal arguments, to a greater or lesser degree, depending on the nature of the humanitarian emergency. Armed conflict is regulated by IHL, according to which state and non-state parties to conflict must not arbitrarily withhold their consent for humanitarian relief operations, and when they consent, they must allow and facilitate rapid and unimpeded passage for relief.[9] In other kinds of humanitarian emergencies, the legal obligation of states not to arbitrarily withhold consent is less clear. Acceptance for humanitarian operations is often most contested in armed conflict contexts, but there are also examples of states denying or limiting access to international humanitarian agencies in the context of other crises.[10] Furthermore, states sometimes deny the existence of conflict on their territory to argue that IHL does not apply and thus that they do not have any obligation to consent to humanitarian access and assistance (Jackson and Davey 2014: 8–9). Thus, the government of Pakistan has defined its military campaigns against the Pakistani Taliban as law enforcement campaigns, denied the applicability of IHL, and restricted or prevented relief and protection work (Jackson and Davey 2014: 9). Even where IHL is applicable and accepted by parties to conflict, legal arguments alone are often insufficient to convince those authorities to grant humanitarian agencies the access they seek (Harmer et al. 2018: 7).

[8] See, e.g. 'Integration and coordination' in Chapter 6 on Afghanistan, 2001–2014 and 'Background' in Chapter 10 on Famine in Somalia, 2011–2012.

[9] See 'International humanitarian law' in Chapter 14 on Law and International Humanitarianism.

[10] See 'State sovereignty, responsibility, and practice' in Chapter 19 on Government and Civil Society in Affected States.

In practice, the shape and outcome of negotiations is, in large part, determined by the conditions authorities seek to impose on humanitarian operations, the leverage the humanitarian actors can bring to bear, and the compromises they are prepared to make. While aid agencies often consider the 'taxation' or diversion of a proportion of aid to be the price of doing business in contested environments, sometimes authorities set conditions which go far beyond this. For instance, in the late 1990s and early 2000s, the Myanmar government restricted where international aid agencies worked and what kind of programming they implemented, insisting on projects that matched the government's own priorities (Bradley 2016a: 17–18; Terry 2011a: 114–120). During the war in Afghanistan, the armed opposition at various times demanded that aid agencies pay a tax, hire local staff selected by the Taliban, employ only male staff, or refrain from pursuing projects directed at women or girls (Jackson and Giustozzi 2012: 5, 23–24, 27). Access depends not simply on insisting on adherence to humanitarian principles and IHL but on taking a pragmatic and transactional approach, which appeals to the interests of the relevant authorities without necessarily giving way to their demands (Allié 2011: 3; Harmer et al. 2018: 10).

The leverage that aid agencies have is context-specific. For example, where an authority is keen for the assistance a humanitarian agency is providing (or proposing to provide), then the threat of withdrawal may be effective. Where an authority seeks international legitimacy, the threat—explicit or implicit—of going public on obstructions or denial of access may be effective. While Hamas has sought to regulate aid agency activities, for example, its desire for international recognition and for support from the population, many of whom depend on international assistance, have enabled humanitarians to negotiate more favourable terms of access (Jackson and Davey 2014: 15).

In addition to exercising leverage, gaining access tends to require compromises on the part of humanitarian agencies, and different agencies are prepared to make different kinds of compromise. In the early 2000s in Myanmar, for instance, the ICRC and the United Nations Refugee Agency (UNHCR) took quite different approaches. Seeking to establish a presence in the south of Myanmar, UNHCR acquiesced to doing only community assistance projects with the non-displaced population in exclusively government-controlled areas, which the ICRC saw as collusion with the government and contrary to humanitarian principles (Bradley 2016a: 17). Even within a single organization, the Dutch, French, and Swiss sections of MSF each had different levels of tolerance for the government conditions (Terry 2011a).

A lack of clear and shared policies on acceptable compromises can weaken the negotiating hand of humanitarian actors. International agencies do not always have clear red lines or well-developed internal policies and are reluctant to articulate what conditions they are willing or unwilling to accept (Harmer et al. 2018: 11). For instance, many of the agencies operating in Afghanistan during the 2001–2014 war appeared to lack clear guidelines or official policies regarding the kinds of compromises that could be made in order to gain acceptance from the Taliban or local power-holders (Jackson and Giustozzi 2012: 5, 15, 23, 24, 27, 29). Moreover, where agencies have different red lines, authorities may be able to play them off against one another (Harmer et al. 2018: 12; Svoboda et al. 2018: 7). In Myanmar in the early 2000s, for example, it became increasingly difficult for the ICRC to do its classic protection work as the government pushed it to be more like UNHCR and the International Organization for Migration (IOM) and to stop raising concerns about the protection of internally displaced persons (IDPs), whose existence the government preferred to deny (Bradley 2016a: 17). The competition and lack of unity among aid agencies in Sri Lanka in 2008–2009 similarly reduced the potential for collective pressure on the government, first vis-à-vis aid convoys into the Vanni and later the camps in which Tamil civilians were interned.[11]

Effective negotiation thus requires skills on the part of the negotiators, clarity on each agency's position and willingness to compromise on relevant issues, and coordination across different agencies. Coordination does not necessarily mean joint negotiations and sometimes takes the form of joint operating principles, guidelines, or ground rules at the country level, which can provide a framework for (joint or bilateral) negotiations (Harmer et al. 2018: 12). There is some suggestion that humanitarian negotiations were more coordinated in the past, with agreed sets of ground rules, and that they have become increasingly fragmented (Jackson and Davey 2014: 23, 26–28). There are, nonetheless, many recent examples of more ad hoc, local-level coordination, which can help agencies avoid manipulation (Harmer et al. 2018: 12). However, agreeing red lines is not enough—they also need to be implemented, and while there are examples of this happening, there are many more examples of lines constantly being redrawn to accommodate demands and threats on the ground (Svoboda et al. 2018: 7).

[11] See 'Manipulation of aid and aid agencies' and 'Advocacy vs access and assistance?' in Chapter 8 on Sri Lanka, 2008–2009.

Protection dialogue

International humanitarian agencies can also maintain a protection-focused dialogue with the parties to conflict and with other kinds of weapons-bearers—which, in other contexts characterized by large-scale violence, may include the police and criminal groups. In essence, this means engaging with the perpetrators of violence in an effort to persuade them to reduce the threat they pose to the population. Humanitarian actors can seek to convince weapons-bearers that it is in their own self-interest to change their behaviour or that it is simply the right and moral thing to do, and in either case, they can use legal or non-legal argumentation, and engagement can range from more official written interventions to unofficial verbal interactions (Bradley 2016a: 110–111).

Some humanitarian agencies have a particular mandate for protection in international law, albeit each with a distinct focus. The ICRC has a mandate for the protection of civilians and others hors de combat, which is grounded in IHL. UNHCR has a mandate for the protection of refugees, linked to its supervisory role for international refugee law (IRL). UNICEF is mandated to promote and protect child rights (as enshrined in the 1989 UN Convention on the Rights of the Child) by supporting the work of the Committee on the Rights of the Child, but its mandate is for child protection broadly conceived rather than for the protection of children during humanitarian emergencies per se. The OHCHR is mandated by the international community to promote and protect the human rights that are guaranteed under international law and stipulated in the Universal Declaration of Human Rights of 1948, which is also not primarily concerned with humanitarian emergencies. The different foci of each of these mandates thus imply different understandings of protection and different kinds of protection dialogue.

For the ICRC, protection dialogue is directed at improving the conduct of combatants towards the rest of the population. While the formal policy includes a range of other activities, the understanding of many within the ICRC is that bilateral, confidential dialogue with the perpetrators of violence is the main protection activity at the field level (Bradley 2016a: 162). Drawing on the testimony of victims or their families, ICRC staff document alleged violations of IHL and present those allegations to state- and non-state armed actors with the aim that the violations will be investigated by the organization in question and individual violators punished (Bradley 2016a: 162). Where presenting specific allegations might put someone at risk, or where they want to bring attention to patterns of abuse, ICRC staff gather information on multiple incidents of the same kind of abuse and present a thematic intervention in which they discuss a trend with alleged perpetrators (Bradley 2016a: 163). In line with its role as the 'guardian of IHL' and the institutional expectation that reference to rules and norms is more persuasive than general moral arguments, the ICRC relies heavily—though not exclusively—on legal argumentation, and particularly on IHL (Bradley 2013).

UNHCR engages in dialogue with states on refugee protection, and while this shares some characteristics with the classic humanitarian protection dialogue of the ICRC, it is different in important ways. Like the ICRC, UNHCR has a preference for legal argumentation but based on IRL rather than IHL. Its main interlocutors are states, and UNHCR seeks to get them to respect the principle of non-refoulement and to grant asylum to those who meet the criteria for refugee status.[12] Beyond its legal mandate, the institutional mandate of UNHCR also encompasses IDP protection, particularly in armed conflict contexts, and here UNHCR also engages with the state, seeking compliance with the Guiding Principles on Internal Displacement and promoting the development of public policies that would benefit the displaced. Unlike the ICRC, UNHCR's protection dialogue is principally about vulnerability reduction rather than threat reduction.[13] Furthermore, whereas the ICRC seeks to engage with all parties to conflict, including national militaries and non-state armed groups, UNHCR engages primarily with the civilian agencies of the state. Where UNHCR dialogue does raise concrete protection concerns with the aim of threat reduction, it does not often extend beyond advocacy with the state, although there are some exceptions, as in eastern Democratic Republic of the Congo in the mid-2010s (Bradley 2016a: 163).

Other operational agencies have no mandate in international law that requires them to protect particular categories of person, but many nonetheless receive funding for, and implement, programming under the rubric of 'protection'. Most of this programming, however, is focused on providing goods and services and tends to emphasize vulnerability reduction or mitigating the consequences of violence after the fact rather than threat reduction. No other agency engages in protection dialogue with such a wide range of weapons-bearers or so consistently as the ICRC, but some do so on an occasional basis. There is broad consensus among humanitarian policy analysts that direct engagement with armed groups on protection issues is a necessary (if insufficient) condition for improving protection outcomes and

[12] On *non-refoulement* and asylum, see 'International displacement and migration law' in Chapter 14 on Law and International Humanitarianism.

[13] On threat reduction and vulnerability reduction as protection strategies, see the Introduction to Part IV on Humanitarian Assistance and Protection.

is currently inadequate (ADH 2011; Jackson 2012b; Zeender 2005). The same can probably be said of engagement with states. While humanitarian agencies engage much more with states on other issues (including access, coordinating responses, and collaborating in some sectors, e.g. health care), systematic and robust engagement with the governments of affected states regarding protection issues seems to be lacking.

The extent and effectiveness of protection dialogue depend not only on the strength of the relationship between the humanitarian agency in question and the authorities it is engaging with but also on the substantive protection concerns under discussion. With any given authority, humanitarian actors may choose to discuss some issues and avoid more sensitive concerns or to address them in a more limited way (Bradley 2013: 125–128, 2016a: 165–166). Where an armed group makes use of landmines and kidnapping, and the ICRC does not expect to be able to stop this behaviour altogether, for example, rather than pushing for full compliance with IHL, ICRC delegates might instead negotiate with the armed group to encourage it to inform civilians of the locations of mines or to allow the ICRC to bring medicine to hostages and transmit news of them to their families (Bradley 2016a: 166). In general, with any given party to conflict or other authority, humanitarian agencies report more success addressing some protection concerns than others and may have a more positive impact on protection outcomes if they focus on the areas where they can make most difference rather than pushing for full compliance with IHL or other international norms.

Sharing information with third parties

Where direct engagement with authorities is not yielding adequate results, operational agencies may choose to engage in private advocacy, either instead of or as well as their more direct efforts. Private advocacy or 'mobilization' entails sharing information in a discreet way with selected third parties who have the capacity to influence those actors whose behaviour a humanitarian agency seeks to change (Bradley 2016a: 168–169; ICRC 2005: 396–7; Slim and Bonwick 2005: 81). In particular, this may involve sharing information with the diplomatic community, who may be able to put pressure on governments and other authorities either publicly or behind closed doors, and international human rights organizations, who may be able to go public with criticisms of authorities more easily than operational humanitarian agencies, who need acceptance and access for much of their work. Mobilization is unlikely to be aimed at building relationships but can be about gaining, maintaining, or expanding access to populations in need or about convincing de jure or de facto authorities to treat those populations better.

Given that this kind of approach is discreet, it is difficult to know the extent to which different humanitarian agencies work to mobilize third parties this way. However, there are many anecdotal examples of aid agencies using this approach both to secure access for their operations and in an effort to reduce the threats posed to the civilian population. In terms of access, for instance, in 2007–2008, MSF sought assistance from Western diplomats and representatives of UN agencies to help pressurize the Ethiopian government to allow access to Ogaden and stop misappropriating aid (Binet 2011: 39). With respect to protection aims, in the years immediately following the 2011 transition to a civilian government, UNHCR in Myanmar and the Protection Working Group in Rakhine State specifically were producing advocacy notes for the diplomatic community and sharing information with key donors (Bradley 2016a: 169). This was done in the hope that these third parties could pressurize the Myanmar government to reduce the threat posed by the armed forces to the civilian population, and the embassies and donors reported using the information and analysis they received in their own bilateral dialogue with the government (Bradley 2016a: 170).

Since the logic of private advocacy or mobilization involves third parties pressurizing or influencing particular authorities, it depends on the humanitarian agencies in question having good enough relationships with third-party actors who have the will and capacity to exert some leverage over the authorities in question. Where authorities are not much concerned with how they are perceived in the global North, for example, Western diplomats may not be able to bring effective pressure to bear on those authorities. Thus, when the ICRC was having difficulties with the Myanmar government in 2006 and 2007, it shared its concerns with US diplomats but asked them not to intervene, anticipating that Asia-Pacific countries could better influence the Myanmar government (Bradley 2016a: 169). On the other hand, third parties that are too close to the authority in question may lack the will to exert what leverage they could. In Ethiopia in 2007–2008, MSF thus sought to work with Western diplomats and UN representatives instead of the African Union because Ethiopia itself occupied such a prominent role in the organization (Binet 2011: 39). Given that the largest international humanitarian agencies are Western-headquartered, largely Western-funded, and tend to have stronger relationships with other Western actors, they may find it more difficult to mobilize non-Western third parties, and in some contexts, this is likely to limit the scope of private advocacy.

The logic of this approach additionally depends on the humanitarian agencies sharing information that will either motivate or enable the third parties in question to put pressure on the target actors. The examples of MSF in Ethiopia and the Protection Working Group in Myanmar show that humanitarian agencies sometimes do this in a deliberate way. However, there have also been important missed opportunities, and in some cases, third parties have assumed that the absence of information coming from humanitarian agencies implied an absence of grave access restrictions or abuses of the population. At the end of the war in Sri Lanka, when most international agencies were forced to evacuate from the Vanni, for example, the UN did not explain the full circumstances of this evacuation (notably, that the main threat to aid agencies came from the government side, so when the same government claimed it could not guarantee their safety, this was disingenuous) to member states, with the consequence that the Sri Lankan government did not face significant criticism for expelling from the main theatre of war almost all the international actors who could have witnessed abuses by the army.[14]

Finally, the logic of mobilization depends on the third-party actors engaging with humanitarian agencies, responding to the information they provide, and using it to put pressure on the relevant authorities. They are sometimes willing to prioritize humanitarian concerns, but in some contexts, they have other priorities. In Sri Lanka, for example, the weakness of a competitive and uncoordinated humanitarian sector, which failed to share pertinent information with the diplomatic community and donor states, was exacerbated due to a lack of political will in the UN and a lack of unity among donors.[15] Following a renewed outbreak of conflict in the South Kordofan and Blue Nile states of Sudan in 2011, donor governments were largely preoccupied with the political negotiations between Sudan and South Sudan, and high-level diplomatic engagement on government restrictions on aid agency access to rebel-held areas within Sudan was lacking (Jackson and Davey 2014: 9).

Even when humanitarian agencies have good relationships with appropriate third parties and share pertinent information with those third parties, who use that information to call out—either publicly or privately—the behaviour of the target authorities, this kind of mobilization does not always achieve the desired effect. Thus, humanitarian agencies may decide to publicize information about access restrictions or abuses of the population either as well as or instead of their efforts to change behaviour through direct dialogue or private advocacy.

[14] See 'Manipulation of aid and aid agencies' in Chapter 8 on Sri Lanka, 2008–2009.
[15] See 'Manipulation of aid and aid agencies' in Chapter 8 on Sri Lanka, 2008–2009.

Public advocacy and criticism of authorities

Many international humanitarian agencies engage in some kind of public criticism of de jure or de facto authorities, sometimes referred to simply as humanitarian advocacy, a modus operandi more often employed by human rights organizations like Amnesty International and Human Rights Watch. Public advocacy can take various forms, including issuing press releases, media interviews, and speeches in high-level political forums like UN General Assembly or Security Council meetings. This kind of activity is most often associated with protection goals but can also be used in pursuit of access and humanitarian space—and sometimes, humanitarian agencies make public statements less out of an expectation that they will persuade the targets of their criticism to change their conduct and more as a means to avoid being (seen to be) complicit in that conduct.

Neutrality, access, and humanitarian space

Unlike international human rights organizations such as Amnesty International and Human Rights Watch, the core work of humanitarian agencies—providing material assistance and services—generally requires them to be physically present, and many are committed to the humanitarian principle of neutrality. They face dilemmas because public criticism of authorities is often assumed to be non-neutral and may have repercussions that threaten access to the victims they seek to support. Thus, the debate on humanitarian advocacy is dominated by discussion of apparent or assumed tensions between public advocacy and neutrality or access.

The relationship between public advocacy and neutrality in humanitarian response elicits significant concern but often without concomitant analysis. According to one perspective, neutrality requires silence (Fox 2001). However, neither the ICRC nor MSF—the two agencies most committed to neutral humanitarian action—see it this way.[16] Advocacy can be consistent with neutral intentions, but this depends on the goals and the content of the advocacy. Where humanitarian agencies advocate for structural change to address the causes of crises, for example, or where they publicly criticize one side in a conflict in an effort to delegitimize it, their advocacy cannot be

[16] See 'Interpretation and implementation of these principles in practice' in Chapter 13 on Politics, Principles, and Humanitarian Action.

said to be neutral. However, much public criticism has limited aims, such as increasing access to relief or reducing the targeting of civilians, which are consistent with neutral intentions (Bradley 2022: 1038–1039). Speaking out is not inherently less neutral than more practical activities undertaken by humanitarian agencies—both can have either limited or more transformative goals, and even with limited goals, both very often have unintended consequences (Bradley 2022: 1039).

Even if advocacy does not necessarily pose a problem in terms of neutrality, it can often have negative repercussions in terms of access and humanitarian space. In some cases, public statements by humanitarian agencies—in particular, those criticizing the authorities—appear to have led directly to the expulsion of the agencies in question (and to violence against them and their staff), and governments are increasingly demanding confidentiality as a condition for their consent to operations (Weissman 2011a). During the Sri Lankan Civil War, for example, the government made visas for aid workers, authorizations to travel within the country, and import licences explicitly or implicitly conditional on the aid agency requesting them not making public statements on the conduct of the government or the armed forces.[17] Where both direct negotiations and public advocacy are focused on expanding access, there can be a tension between the two approaches, and advocacy in intergovernmental forums like the Security Council, in particular, can quickly politicize the dialogue in the eyes of relevant authorities, undermining direct negotiations with them (Harmer et al. 2018: 10).

In practice, the relationship between humanitarian advocacy and access is more complex, and advocacy does not always have a negative impact on staff security and access. Indeed, it can be explicitly focused on access issues; access for humanitarian agencies, and acceptance of their work by authorities, are very often the product of both direct negotiation and public advocacy (DuBois 2007; HPG 2007). Public advocacy can also be undertaken in conjunction with private advocacy, and this combination of approaches was adopted by MSF in Sri Lanka in 2006 (Weissman 2011b: 20). Of course, just as direct negotiations and private advocacy are not always successful in improving access, neither is public advocacy, but the fact that it sometimes plays a role in extending humanitarian access shows that the relationship between advocacy and access is not straightforward.

[17] See 'Manipulation of aid and aid agencies' in Chapter 8 on Sri Lanka, 2008–2009.

Threat reduction

Naming and shaming is most often associated with effort to reduce the threats faced by the population. Arguments that humanitarian actors have a responsibility to speak out against abuses are frequently made in the context of criticizing one or more international humanitarian agencies for failing to speak out in a specific setting. In many cases, such criticisms are based on the explicit or implicit assumption that public criticism would have led to improved conduct by the target of the criticism and, hence, to better protection outcomes for the affected population (Bradley 2014; Niland 2014). The expectation is that in being called out for its abuses, the criticized party would be concerned for its reputation and potential retribution and so would adapt its behaviour in the desired direction, hence reducing the threat. The evidence base on humanitarian advocacy is thin, but the extensive literature on naming and shaming by human rights organizations suggests that it is only effective under certain conditions and can sometimes be counter-productive (see, e.g. Franklin 2008; Hafner-Burton 2008; Hendrix and Wong 2013; Krain 2012; Murdie and Davis 2012). On the other hand, where humanitarian agencies fail to call out abuses they have witnessed, or have compelling evidence of, the perpetrators of those abuses may see this as a signal that the international community is not concerned with their conduct and take it as a green light to continue with more or worse abuses.[18]

Of the largest international humanitarian agencies, MSF is the most associated with 'speaking out' or *témoignage*, while the ICRC is most associated with discretion, but the reality is more nuanced. According to the origin myth surrounding MSF, *témoignage* has always been a central institutional principle, but in fact, this principle emerged over time, and in practice, MSF usually undertakes its medical work without making public statements about abuses in its zones of operations.[19] That said, in some contexts (as, for instance, in response to the treatment of migrants in Europe), advocacy is considered an essential component of MSF's work and even a motivating factor.[20] On the other hand, the ICRC commitment to confidentiality and discretion is conditional, not absolute. The ICRC does, on occasion, publicly criticize authorities—particularly states, as the primary

[18] This is what seemed to happen in the final stages of the Sri Lankan Civil War. See 'Advocacy vs access and assistance?' in Chapter 8 on Sri Lanka, 2008–2009.

[19] See 'Longer-term significance of the humanitarian response to the war in Biafra' in Chapter 2 on the Nigerian Civil War, 1967–1970.

[20] See 'Search and rescue' in Chapter 12 on the 2015–2016 European 'Migrant Crisis' and 'Europeanization' in Chapter 15 on the Shifting Parameters of Humanitarian Emergencies.

duty-bearers in IHL—although it has a strong preference for seeking to persuade authorities to change their behaviour through bilateral and confidential dialogue.

Public advocacy aimed at threat reduction does not have to name the perpetrators of abuses, and often it involves publicizing the suffering of the population in a specific context, without attributing blame to any given actor (Bradley 2016a: 170). Thematic advocacy campaigns that relate to a particular kind of threat—such as sexual violence in war—but not to a specific perpetrator or context can also be conducted at the global level. Where the perpetrators of abuses are identified, this is not always done in an even-handed way. In the last months of the Sri Lankan Civil War, for example, most of the humanitarian community—with the exception of MSF—avoided public criticism of the government even as they spoke out about abuses by the Liberation Tigers of Tamil Eelam (LTTE).[21] There is some evidence to suggest that there is a wider tendency for UN humanitarian agencies in particular to call out armed non-state actors more often than they do the governments of affected states (Bradley 2016a: 171). While this might reflect a tendency among UN agencies to support the state, and calls into question their neutrality, it might also reflect fears about the negative impact of public statements on humanitarian space and operations, given that states tend to have greater control over access.

Sometimes public advocacy is limited to identifying problems, but sometimes, humanitarian agencies also advocate for particular solutions. For instance, in the mid-2000s, some humanitarian agencies emphasized demands for a stronger peacekeeping force in Darfur in their public advocacy (HPG 2007: 2). In 2014, MSF called for immediate deployment of civilian and military assets with expertise in biohazard containment to address the Ebola outbreak in West Africa.[22] As they increasingly move to address migration, several humanitarian agencies have also gone beyond calling on authorities to fulfil their duties, with the ICRC advocating (at the global level) for states to limit their use of detention for migrants and MSF calling on European states to reform their border policies and increase safe and legal migration routes.[23] In prescribing solutions, such advocacy will often be non-neutral, particularly in conflict contexts. For those agencies without a strong commitment to neutrality, the question then becomes one of weighing advocacy

[21] See 'A failure of protection' in Chapter 8 on Sri Lanka, 2008–2009.
[22] See 'A public health emergency of international concern' in Chapter 11 on Ebola in West Africa, 2014–2015.
[23] See 'People on the move' in Chapter 15 on the Shifting Parameters of Humanitarian Emergencies.

against its expected impact on access and programming. In responding to the crisis in Darfur in the mid-2000s, for example, many agencies developed a 'more pragmatic' form of neutrality, which was 'sufficiently non-partisan to facilitate access to affected communities, while also sufficiently flexible to allow advocacy' (HPG 2007).

In light of fears that publicly criticizing an authority—especially if that authority is a state—may compromise the terms of access or lead to expulsion from the country, in many contexts, this kind of activity is seen as a last resort. Several examples from MSF's own accounts of negotiations, private and public advocacy, suggest a preference for the first two, and recognition that in many cases, MSF (or the specific national section doing so) would have to leave the country in question if and when it went public with criticisms of authorities (Magone et al. 2011). Only where bilateral, confidential dialogue and mobilization have been tried and deemed ineffective will the ICRC make a public statement (ICRC 2005: 397). ICRC policy distinguishes two kinds of public statement: public declarations and public condemnations. A public declaration comments on the quality of dialogue with a particular party to conflict, or on the responses of that party to ICRC recommendations, but it does not detail the violations perpetrated by that party or the content of the recommendations made to the party by the ICRC (ICRC 2005: 397). By contrast, a public condemnation enumerates IHL violations perpetrated by a named party to conflict and is seen as the last resort (ICRC 2005: 398).

Complicity

Ostensibly, public criticisms of authorities are aimed at changing the behaviour of those authorities—either to improve access of humanitarian agencies to people in need or to improve the way the authorities treat those people—but, in many cases, public criticism is also a strategy through which humanitarian agencies seek to avoid being complicit in abuses. Complicity can be said to occur when humanitarian actors inadvertently facilitate, or contribute to, abuses, something they may do in a variety of ways, including by paying a tax to the perpetrator of those abuses or by absolving the perpetrator of some of its responsibilities and so freeing up time and resources with which to perpetrate abuses. Examples of such complicity can be found in most of the case study chapters in the first part of this book, indicating how widespread the phenomenon is.

The accusations of moral complicity sometimes levelled at humanitarian agencies working alongside those responsible for serious abuses 'can carry a strong if unfocused moral charge and are frequently the source of significant moral unease' (Buth et al. 2018: 299). The sense of complicity and associated unease may be felt more intensely in the 'home' regions of humanitarian agencies, where 'their' governments or regional institutions are directly involved in wrongdoing.[24] Where they find themselves complicit in abuses, agencies can choose to accept complicity and the associated unease as a price worth paying for the support they are able to provide to people in need, they can suspend their operations and withdraw, or they can publicly criticize the abuses their work is inadvertently contributing to. Of course, sometimes, that kind of criticism will lead to restrictions on their work or expulsion from the place in question.

Speaking out about wrongdoing is seen as a way to avoid—or compensate for—complicity in that wrongdoing, but public criticisms do not always serve the best interests of the populations whose safety and well-being is at stake. As Fiona Terry has put it, 'sometimes public denunciation feels good [. . .] but it doesn't necessarily improve the situation for those we seek to help' (Terry 2022). Sometimes, humanitarian actors, when faced with their complicity in abuses, 'may focus more on their image as self-consciously good actors than on the interests of potential beneficiaries' (Buth et al. 2018: 299). However, there are at least three reasons—beyond vanity and feeling good—why a consequentialist logic might lead humanitarian agencies to denounce abuses in which they are complicit, even at risk of expulsion. First, sometimes a humanitarian agency might assess that the indirect harm they are doing by contributing to wrongdoing outweighs the good they are doing there. Second, complicity in wrongdoing can cause reputational damage, and publicly calling out those abuses may help to minimize that damage. This obviously serves agencies' self-interest but may also be essential to generating the funds necessary to support those in need in other humanitarian emergencies. Third, to keep working alongside the perpetrators of abuses without calling them out can send a message to would-be abusers elsewhere in the world that their wrongdoing will be tolerated and so contribute to harm on a wider scale. Under some conditions, then, complicity might be a good motive for naming and shaming, but without careful analysis, there is significant risk of moral narcissism (Buth et al. 2018).

[24] See, e.g., 'Humanitarian response' in Chapter 12 on the 2015–2016 European 'Migrant Crisis' and 'Europeanization' in Chapter 15 on the Shifting Parameters of Humanitarian Emergencies.

Conclusions

The effectiveness of these three approaches—direct engagement, mobilization, and public advocacy—is unclear and likely highly goal- and context-dependent. Most humanitarian agencies seem more comfortable engaging directly with state, rather than non-state, actors, and counter-terror legislation and domestic laws often pose an obstacle to direct engagement with armed groups. NGOs in particular may fear prosecution, which suggests an important role for OCHA, but in practice, its role has been limited in recent decades, especially in terms of engaging with armed opposition in contexts where the UN is supporting the state. Anecdotally, most examples of private advocacy seek to mobilize third parties to bring pressure to bear on governments, and it is not clear that the same strategy would be expected to have much effect with most armed groups. By contrast, in many settings, public criticisms of authorities focus on the armed opposition, with humanitarian agencies fearful of the repercussions of speaking out about abuses by the state. The different approaches are not mutually exclusive, and in some cases, all three can be effectively combined to bring about desired changes. However, they can be in tension with one another, as when speaking out over protection concerns damages relationships and reduces the access necessary for productive bilateral dialogue. There is scope for different agencies to adopt different approaches in ways that are complementary, but the activities of any given humanitarian agency can limit the scope of action of others—either because they are collectively punished by authorities or because they are played off against one another.

24
Conclusion
International Humanitarianism after COVID-19

Since I began writing this book, the COVID-19 pandemic struck. The pandemic has had obvious and significant impacts on humanitarian needs, responses, and outcomes. However, as I write this concluding chapter from the vantage point of late 2022, the experiences of the pandemic do not alter the book's central claims. In this short conclusion, I draw out some of the main arguments of the book and reflect briefly on how the pandemic has—and has not—affected different dimensions of international humanitarianism. For all the talk of critical turning points in the early months of the pandemic, the impacts of COVID-19 and the ways that humanitarian donors and agencies have responded seem to confirm much of what we already knew about humanitarian needs and practices and to crystallize or accelerate some of the pre-existing trends in the sector.

The COVID-19 pandemic compounded existing humanitarian crises. In the sense that it exceeded the capacity of every country to respond, the pandemic could itself be conceptualized as a global humanitarian emergency. At the same time, in much of the world, the pandemic did not require the kind of external institutionalized response we associate with international humanitarianism. It is also quite different from the case studies in the first part of the book in that each of those was delimited in space and time, while the pandemic has affected the whole globe, and its temporal extension is not yet clear. In many countries around the world, however, it did increase the kinds of needs we normally classify as humanitarian needs. As a result of economic contractions worldwide, extreme poverty has increased for the first time since 1998, and food insecurity has increased due to movement restrictions and disruptions to supply chains (Allouche and te Lintelo 2022: 3). The Global Humanitarian Assistance Report attributed higher than ever needs in 2020 to the compounding effects of COVID-19, with an additional 19 million people and 10 more countries in need of humanitarian assistance compared with 2019 (Development Initiatives 2021: 33).

The Politics and Everyday Practice of International Humanitarianism. Miriam Bradley, Oxford University Press.
 DOI: 10.1093/oso/9780198840695.003.0024

The pandemic offers a clear illustration of the argument introduced in Chapter 1 regarding the importance of economic, political, and other structural factors in determining the vulnerability of people and societies to crises. Whether we compare across or within countries, vulnerability to COVID and the wider impacts of the pandemic was not distributed equally, and often those who were already marginalized have suffered the most. In the United Kingdom, for example, black, Asian, and minority ethnic communities are disproportionately represented on the frontline of health-care services and more likely to live and work in conditions that increase vulnerability to the virus (Allouche and te Lintelo 2022: 3). In the context of lockdowns and other restrictions, those working in the informal sector, including many displaced persons, have seen their livelihoods severely disrupted (Allouche and te Lintelo 2022: 4). Police enforcement of quarantine policies poses a particular threat to undocumented migrants. The pandemic and the measures adopted to limit its spread have thus generated a socio-economic crisis as well as a public health crisis and have further deepened structural inequalities (Allouche and te Lintelo 2022: 3).

Funding for humanitarian response through the COVID-19 pandemic has followed familiar patterns. Total humanitarian funding invariably falls well short of needs, and the pandemic has exacerbated the shortfall as it has multiplied needs without a concomitant increase in resources (Development Initiatives 2021: 31, 2022: 15). In many emergencies, donors have prioritized COVID aid over other issues such that the overall volume of humanitarian funding for requirements not related to the pandemic fell in 2020 for the first time since 2015 (Development Initiatives 2021: 35). It is not only the quantity but also the quality of humanitarian funding that matters, and the urgency of the COVID-19 response reinforced the need for more flexible funding (Development Initiatives 2021: 76). Donors initially responded positively to the call for greater flexibility, with the volume and proportion of unearmarked funding received by UN agencies increasing in 2020, reversing the previous year's drop (Development Initiatives 2021: 76). However, the proportion of unearmarked funding remained below the targets set in the Grand Bargain and fell again in 2021 (Development Initiatives 2021: 76, 2022: 18).

In recent decades, the concept of the humanitarian emergency has been stretched, and COVID has intersected with its shifting parameters in a number of ways. For example, the number of countries experiencing protracted crisis—defined here as countries with five or more consecutive years of UN-coordinated appeals, as of the year of analysis—doubled to thirty-four in the six years to 2020, and increased again to thirty-six in 2021

(Development Initiatives 2021: 12, 89, 2022: 10). Almost three-quarters of all people in humanitarian need live in these thirty-six countries, where the average COVID vaccination rate is considerably lower than elsewhere, including other developing countries (Development Initiatives 2021: 12, 2022: 10). Over the past ten to fifteen years, international humanitarian agencies have also increasingly responded to migration, going beyond their traditional work supporting refugees in the global South to assist those arriving in the global North and those who do not qualify for refugee status. Much of the suffering endured by migrants is the direct consequence of the state border policies, which serve to generate humanitarian need at the same time as they pose significant challenges and dilemmas to those humanitarian actors who seek to respond. There was some hope that the pandemic could drive lasting migration reform as it offered a compelling insight into the extent to which many countries in the global North rely on migrant workers in health care and other essential sectors (Kumar 2021). In practice, however, the COVID-19 pandemic has served as a motivation or pretext for further tightening border and migration policies and weakening protection regimes (Allouche and te Lintelo 2022: 4–6).

The historical case studies, as well as the chapters in the second part of this book, show that the boundaries between humanitarianism and development, on the one hand, and between humanitarianism and human rights, on the other, have long been contested. In protracted crises and urban responses, humanitarian agencies very often undertake activities akin to development programming, and when working in their home regions and/or to support migrants, they often adopt some of the goals and methods of human rights agencies. Grounded in the assumption that reducing needs is more efficient than simply meeting them, the New Way of Working (NWoW) endorsed at the World Humanitarian Summit (WHS) in 2016 seeks to bring together humanitarian, development, human rights, and peace and security efforts to address the causes and not just the symptoms of humanitarian emergencies. As COVID-19 has widened the gap between humanitarian needs and resources, the efficiency motivation may be stronger than ever.

The localization agenda, through which humanitarian responses are expected to be more efficient, more developmental, and more human rights-focused, has—in some respects—been given a boost by the COVID-19 pandemic. As in other emergencies, local and national civil society actors played a crucial role in the pandemic response around the world, with people and communities stepping up to provide care and assistance (Aneja and DuBois 2020). While their role is often underacknowledged and sidelined by international responders, the scale of needs resulting from the pandemic, combined with travel restrictions and a reduced international presence in many humanitarian settings, reinforced the centrality of local and national actors (Allouche and te Lintelo 2022: 7; Development Initiatives 2021: 68). This underscores the need for more funding for local and national actors, yet the Grand Bargain target of least 25% of humanitarian funding going to local and national responders as directly as possible by 2020 has not been met. In fact, following an increase in 2020, direct funding to local and national actors dropped by almost two-thirds in 2021 to only 1.2% of total humanitarian assistance (Development Initiatives 2022: 17).

While international funding and recognition for local and national efforts might be inadequate, local actors are nonetheless driving what appear to be broader shifts in thinking about humanitarianism, solidarity, and neutrality. Even before COVID, very few mainstream humanitarian policy analysts maintained the view that neutrality was a necessary condition for action to be humanitarian, and most had long seen a role for differentiated approaches to encompass the classical humanitarianism of the International Committee of the Red Cross (ICRC) as well as more expansive, flexible, and integrated forms of relief (Bennett et al. 2016: 53). However, local responses to the pandemic, and concurrent events, have breathed new life into age-old debates on the value of neutrality and on the relationship between neutrality and solidarity. An important characteristic of many community-level responses to COVID-19 has been an expanded remit that goes beyond material aspects to relieve the immaterial and emotional impacts of the pandemic (Allouche and te Lintelo 2022: 7). The citizen volunteer role many have adopted in response to the pandemic intersected with the Black Lives Matter movement, and widespread action against social injustice, and in the Ukraine war and post-coup Myanmar, local humanitarian responders have taken a leading role with a self-consciously solidaristic approach (Aneja and DuBois 2020; O'Callaghan et al. 2022).

Finally, two major trends in international humanitarianism in the twenty-first century—the growth in the use of cash transfer programming and the so-called digital revolution—have been catalysed by pandemic-related needs and restrictions. If the 2004 Indian Ocean tsunami and the 2011–2012 famine in Somalia were major milestones in the process of convincing much of the humanitarian sector of the value of cash programming, the pandemic served as an accelerator. Many operational agencies favoured the use of cash transfers to provide humanitarian assistance, given the ability to scale up quickly and deliver remotely (Development Initiatives 2021: 65). The pandemic contributed not only to a significant expansion in the quantity of cash programming but also to shifts in how cash is provided, with emergency cash transfers increasingly incorporated into existing social safety net programmes in such a way as to support local actors (Development Initiatives 2021: 80; Metcalfe-Hough et al. 2021: 61).

Technology already served to facilitate remote management approaches and, hence, to support a reduced international presence. With pandemic-related restrictions on international travel, the combined trend towards reliance on remote management and digital technologies was further entrenched. On the surface, by supporting a reduced international presence, digital technologies can be seen as facilitating localization, but the way they are used in practice tends to pull in the opposite direction, with tech allowing international actors to retain power, even from a distance (Aneja and DuBois 2020). More generally, reliance on technology to assess needs and deliver humanitarian assistance carries some obvious benefits and some less obvious risks. The pandemic has served as a catalyst not only to the uptake of new technologies but also to recognition of the need for better risk management procedures, increasing the urgency and focus on inter-agency dialogue on data protection requirements, for example Metcalfe-Hough et al. 2021: 61). However, progress on addressing the risks around safe programming and data management has not been as rapid as the take-up of the use of biometrics and digital technologies, developments which are all the more concerning given that tech tends to be 'sticky' so that new tech adopted during the pandemic is likely to persist for years to come (Holloway et al. 2021: 16).

It is too soon to be certain about the long-term impacts of the COVID-19 pandemic on international humanitarian action, but the evidence to date suggests that it is mostly business as usual. Some changes that were already in motion have been accelerated, but opportunities for more fundamental rethinking of humanitarianism have not been seized, and in many cases, previously emerging patterns have become entrenched.

Bibliography

Abdelnour, Samer, and Akbar M. Saeed. 2014. 'Technologizing humanitarian space: Darfur advocacy and the rape-stove panacea'. *International Political Sociology* 8 (2): 145–163. https://doi.org/10.1111/ips.12049.

Abild, Erik. 2010. 'Creating humanitarian space: A case study of Somalia'. *Refugee Survey Quarterly* 29 (3): 67–102. Accessed 9 January 2023. http://rsq.oxfordjournals.org/content/29/3/67.abstract.

Açıkyıldız, Çağlar. 2021. 'Digitalization in humanitarian response: Biometric practices of international humanitarian organizations'. International Humanitarian Studies Conference, 3–5 November 2021, Sciences-Po, Paris.

Adams, Lesley. 2007. *Learning from cash responses to the tsunami*. London: HPG/ODI.

Addario, Lynsey, and Lydia Polgreen. 2009. 'Aid groups' expulsion, fears of more misery.' New York Times 22 March 2009.

ADH. 2011. *Rules of engagement: Protecting civilians through dialogue with armed non-state actors*. Geneva: Geneva Academy of International Humanitarian Law and Human Rights.

African Rights. 1994. *Humanitarianism unbound? Current dilemmas facing multi-mandate relief operations in political emergencies*. London: African Rights.

African Union. 2009. African Union Convention for the Protection and Assistance of Internally Displaced Persons in Africa ('Kampala Convention'). 22 October 2009.

Agamben, Giorgio. 1998. *Homo Sacer: Sovereign Power and Bare Life*, trans. Daniel Heller-Roazen. Stanford, CA: Stanford University Press.

Agier, Michel. 2010. 'Humanity as an identity and its political effects (a note on camps and humanitarian government)'. *Humanity: An International Journal of Human Rights, Humanitarianism, and Development* 1 (1): 29–45.

Akande, Dapo, and Emanuela-Chiara Gillard. 2016. 'Oxford guidance on the law relating to humanitarian relief operations in situations of armed conflict'. Commissioned by the United Nations Office for the Coordination of Humanitarian Affairs.

Alexander, David. 1997. 'The study of natural disasters, 1977–97: Some reflections on a changing field of knowledge'. *Disasters* 21 (4): 284–304. https://doi.org/10.1111/1467-7717.00064.

Alexander, Jessica, and Francesca Bonino. 2015. 'Addressing causation in humanitarian evaluation: A discussion on designs, approaches and examples'. Active Learning Network for Accountability and Performance (ALNAP).

Ali, Degan, and Kirsten Gelsdorf. 2012. 'Risk-averse to risk-willing: Learning from the 2011 Somalia cash response'. *Global Food Security* 1 (1): 57–63. https://doi.org/http://dx.doi.org/10.1016/j.gfs.2012.07.008.

Allain, Jean. 2001. 'The jus cogens nature of non-refoulement'. *International Journal of Refugee Law* 13 (4): 533–558. https://doi.org/10.1093/ijrl/13.4.533.

Allié, Marie-Pierre. 2011. 'Introduction: Acting at any price?' In *Humanitarian negotiations revealed: The MSF experience*, ed. Claire Magone, Michaël Neuman, and Fabrice Weissman. London: Hurst, 1–11.

Allouche, Jeremy, and Dolf J. H. te Lintelo. 2022. 'Editorial: Covid-19 responses: Insights into contemporary humanitarianism'. *IDS Bulletin* 53 (2): 1–10. https://doi.org/10.19088/1968-2022.113.

ALNAP (Active Learning Network for Accountability and Performance in Humanitarian Action). 2012. *Meeting the urban challenge: Adapting humanitarian efforts to an urban world*. London: ALNAP/Overseas Development Institute.

AMISOM (African Union Mission in Somalia). n.d. 'AMISOM background.' Accessed 10 February 2020. https://amisom-au.org/amisom-background.

Amnesty International. 2019. *Targeting solidarity: Criminalization and harassment of people defending refugee and migrant rights in northern France*. London: Amnesty International.

Anderson, Emma-Louise, and Alexander Beresford. 2016. 'Infectious injustice: The political foundations of the Ebola crisis in Sierra Leone'. *Third World Quarterly* 37 (3): 468–486. https://doi.org/10.1080/01436597.2015.1103175.

Anderson, Mary B. 1999. *Do no harm: How aid can support peace—or war*. London: Lynne Rienner.

Andersson, Ruben, and David Keen. 2019. *Partners in crime? The impacts of Europe's outsourced migration controls on peace, stability and rights*. London: Saferworld.

Aneja, Urvashi, and Marc DuBois. 2020. 'Covid-19 futures in humanitarian action'. Humanitarian Practice Network. 29 September 2020. Accessed 9 January 2023. https://odihpn.org/publication/covid-19-futures-in-humanitarian-action.

Anthony, Douglas. 2014. '"Ours is a war of survival": Biafra, Nigeria and arguments about genocide, 1966–70'. *Journal of Genocide Research* 16 (2–3): 205–225. https://doi.org/10.1080/14623528.2014.936701.

Austin, Lois, and Sebastien Chessex. 2018. *The case for complementarity: Working together within the International Red Cross and Red Crescent Movement in armed conflict and other situations of violence*. London: British Red Cross.

Bailey, Sarah, and Veronique Barbelet. 2014. *Towards a resilience-based approach to the Syrian refugee crisis: A critical review of vulnerability criteria and frameworks*. London: Humanitarian Policy Group/Overseas Development Institute (HPG/ODI).

Bailey, Sarah, and Paul Harvey. 2015. *State of evidence on humanitarian cash transfers. Background note for the high level panel on humanitarian cash transfers*. London: ODI.

Bailey, Sarah, and Paul Harvey. 2017. *Time for change: Harnessing the potential of humanitarian cash transfers*. London: ODI.

Balaji, Murali. 2011. 'Racializing pity: The Haiti earthquake and the plight of "others"'. *Critical Studies in Media Communication* 28 (1): 50–67. https://doi.org/10.1080/15295036.2010.545703.

Balzacq, Thierry, Sarah Léonard, and Jan Ruzicka. 2016. '"Securitization" revisited: Theory and cases'. *International Relations* 30 (4): 494–531. https://doi.org/10.1177/0047117815596590.

Barbelet, Veronica, and Caitlin Wake. 2017. *Livelihoods in displacement: From refugee perspectives to aid agency response*. London: HPG.

Barbelet, Veronique, Gemma Davies, Josie Flint, and Eleanor Davey. 2021. *Interrogating the evidence base on humanitarian localisation: A literature study*. London: HPG/ODI.

Barnes, Cedric, and Harun Hassan. 2007. 'The rise and fall of Mogadishu's Islamic Courts'. *Journal of Eastern African Studies* 1 (2): 151–160. https://doi.org/10.1080/17531050701452382.

Barnett, Michael. 2002. *Eyewitness to a genocide: The United Nations and Rwanda*. Ithaca, NY: Cornell University Press.

Barnett, Michael. 2005. 'Humanitarianism transformed'. *Perspectives on Politics* 3 (4): 723–740.

Barnett, Michael. 2011. *Empire of humanity: A history of humanitarianism*. Ithaca, NY; London: Cornell University Press.

Barnett, Michael. 2012. 'International paternalism and humanitarian governance'. *Global Constitutionalism* 1 (3): 485–521.

Barnett, Michael. 2018. 'Human rights, humanitarianism, and the practices of humanity'. *International Theory* 10 (3): 314–349.

Barnett, Michael, and Raymond Duvall, eds. 2005. *Power in global governance*. Cambridge: Cambridge University Press.

Barnett, Michael, and Martha Finnemore. 2004. *Rules for the world: International organizations in global politics*. Ithaca, NY: Cornell University Press.

Barnett, Michael, and Thomas G. Weiss. 2008. 'Humanitarianism: A brief history of the present'. In *Humanitarianism in question: Politics, power, ethics*, ed. Michael Barnett and Thomas G. Weiss. Ithaca, NY: Cornell University Press, 1–48.

Barnett, Michael, and Thomas G. Weiss. 2011. *Humanitarianism contested: Where angels fear to tread*. London: Routledge.
Barrett, Christopher B., and Daniel G. Maxwell. 2004. 'PL480 food aid: We can do better'. *Choices* 19 (3): 53–57.
Barry, Brian. 1990. 'How not to defend liberal institutions'. *British Journal of Political Science* 20 (1): 1–14. http://www.jstor.org/stable/193676.
Barutciski, Michael. 2002. 'A Critical View on UNHCR's Mandate Dilemmas'. *International Journal of Refugee Law* 14 (2–3): 365–381.
Basaran, Tugba. 2015. 'The saved and the drowned: Governing indifference in the name of security'. *Security Dialogue* 46 (3): 205–220. https://doi.org/10.1177/0967010614557512.
Baudendistel, Rainer. 2006. *Between bombs and good intentions: The Red Cross and the Italo-Ethiopian War, 1935-1936*. Oxford: Berghahn Books.
Beardsley, Kyle, and Brian McQuinn. 2009. 'Rebel groups as predatory organizations: The political effects of the 2004 tsunami in Indonesia and Sri Lanka'. *Journal of Conflict Resolution* 53 (4): 624–645. https://doi.org/10.1177/0022002709336460.
Beerli, Monique J. 2018. 'Saving the saviors: Security practices and professional struggles in the humanitarian space'. *International Political Sociology* 12 (1): 70–87. https://doi.org/10.1093/ips/olx023.
Benelli, Prisca, Antonio Donini, and Norah Niland. 2012. *Afghanistan: Humanitarianism in uncertain times*. Somerville, MA: Feinstein International Center.
Bennett, Christina, Matthew Foley, and Sara Pantuliano. 2016. *Time to let go: Remaking humanitarian action for the modern era*. London: HPG/ODI.
Best, Geoffrey. 1997. *War and law since 1945*. Oxford: Clarendon Press.
Binder, Martin. 2009. 'Humanitarian crises and the international politics of selectivity'. *Human Rights Review* 10 (3): 327–348. http://dx.doi.org/10.1007/s12142-009-0121-7.
Binet, Laurence. 2011. 'Ethiopia: A fool's game in Ogaden'. In *Humanitarian negotiations revealed: The MSF experience*, ed. Claire Magone, Michaël Neuman, and Fabrice Weissman. London: Hurst, 35–39.
Binet, Laurence. 2013. *Famine and forced relocations in Ethiopia 1984–1986*. Médecins Sans Frontières.
Biquet, Jean-Marc. 2013. 'Haiti: Between emergency and reconstruction'. *International Development Policy* 4 (3): 129–135. https://doi.org/10.4000/poldev.1600.
Black, Maggie. 1992. *A cause for our times: Oxfam, the first 50 years*. Oxford: Oxford University Press.
Borton, John. 1996. 'The joint evaluation of emergency assistance to Rwanda: Study III principal findings and recommendations'. Relief and Rehabilitation Network (RRN) Network Paper 16. London: Relief and Rehabilitation Network/ODI.
Borton, John, and Sarah Collinson. 2017. 'Responses to mixed migration in Europe: Implications for the humanitarian sector'. Humanitarian Practice Network (HPN) Network Paper 81.
Bothe, Michael. 2005. 'Customary international humanitarian law: Some reflections on the ICRC study'. *Yearbook of International Humanitarian Law* 8: 143–178.
Bradbury, Mark, and Robert Maletta. 2012. 'When state-building fails: Famine, counterterrorism, and the politicization of humanitarian action in Somalia'. In *The golden fleece: Manipulation and independence in humanitarian action*, ed. Antonio Donini. Boulder & London: Kumarian, 109–136.
Bradley, Miriam. 2013. 'International humanitarian law, non-state armed groups and the International Committee of the Red Cross in Colombia'. *Journal of International Humanitarian Legal Studies* 4 (1): 108–134.
Bradley, Miriam. 2014. 'Sri Lanka: Limited Humanitarian Action—or a Lesson in the Limits of Humanitarian Action?' *International Development Policy* 6 (1): 149–154.
Bradley, Miriam. 2016a. *Protecting civilians in war: The ICRC, UNHCR, and their limitations in internal armed conflicts*. Oxford: Oxford University Press.

Bradley, Miriam. 2016b. 'UNHCR and accountability for IDP protection in Colombia'. In *UNHCR and the struggle for accountability*, ed. Kristin Bergtora Sandvik and Katja Lindskov Jacobsen. Abingdon: Routledge, 81–100.

Bradley, Miriam. 2017. 'The impact of armed conflict on displacement'. Report commissioned by Concern Worldwide. https://doi.org/10.13140/RG.2.2.33905.94562/1.

Bradley, Miriam. 2019a. 'All lives are equal but some lives are more equal than others: Staff security and civilian protection in the humanitarian sector'. *Journal of Humanitarian Affairs* 1 (2): 13–22.

Bradley, Miriam. 2019b. 'Unintended consequences of adjacency claims: The function and dysfunction of analogies between refugee protection and IDP protection in the work of UNHCR'. *Global Governance* 25 (3): 620–644.

Bradley, Miriam. 2020. 'From armed conflict to urban violence: Transformations in the International Committee of the Red Cross, international humanitarianism, and the laws of war'. *European Journal of International Relations* 26 (4): 1061–1083. https://doi.org/10.1177/1354066120908637.

Bradley, Miriam. 2022. 'Five shades of grey: Variants of "political" humanitarianism'. *Disasters* 46 (4): 1027–1048. https://doi.org/10.1111/disa.12512.

Bradley, Miriam. 2023. 'A humanitarian agency in global migration governance: The International Committee of the Red Cross's migration policy and practice'. In *Research handbook on the institutions of global migration governance*, ed. Antoine Pécoud and Hélène Thiollet. Cheltenham: Edward Elgar.

Brauman, Rony. 2009. 'Global media and the myths of humanitarian relief: The case of the 2004 tsunami'. In *Humanitarianism and suffering: The mobilization of empathy*, ed. Richard Ashby Wilson and Richard D Brown. New York: Cambridge University Press, 108–117.

Brauman, Rony. 2012. 'Médecins sans Frontières and the ICRC: Matters of principle'. *International Review of the Red Cross* 94 (888): 1–13.

Bridges, K. M. 2010. 'Between aid and politics: Diagnosing the challenge of humanitarian advocacy in politically complex environments—the case of Darfur, Sudan'. *Third World Quarterly* 31 (8): 1251–1269. https://doi.org/10.1080/01436597.2010.541084.

Brun, Cathrine. 2016. 'There is no future in humanitarianism: Emergency, temporality and protracted displacement'. *History and Anthropology* 27 (4): 393–410. https://doi.org/10.1080/02757206.2016.1207637.

Bryant, John. 2019. 'Remittances in humanitarian crises'. HPG Working Paper. London: HPG/ODI.

Burniske, Jessica, Naz K. Modirzadeh, and Dustin A. Lewis. 2014. *Counter-terrorism laws and regulations: What aid agencies need to know*. HPN Network Paper 79. London: HPN/ODI.

Buth, Pete, Benoit de Gryse, Sean Healy, Vincent Hoedt, Tara Newell, Giovanni Pintaldi, Hernan Del Valle, Julian C Sheather, and Sidney Wong. 2018. '"He who helps the guilty, shares the crime"? INGOs, moral narcissism and complicity in wrongdoing'. *Journal of Medical Ethics* 44 (5): 299–304. https://doi.org/10.1136/medethics-2017-104399.

Buzan, Barry, Ole Wæver, and Jaap de Wilde, eds. 1998. *Security: A new framework for analysis*. Boulder, CO: Lynne Rienner Publishers.

Cabot Venton, Courtenay, Sarah Bailey, and Sophie Pongracz. 2015. 'Value for money of cash transfers in emergencies'. Department for International Development (DFID).

Calhoun, Craig. 2008. 'The imperative to reduce suffering: Charity, progress, and emergencies in the field of humanitarian action'. In *Humanitarianism in question: Politics, power, ethics*, ed. Michael Barnett and Thomas G Weiss. Ithaca, NY: Cornell University Press, 73–97.

Calhoun, Craig. 2009. 'The idea of emergency: Humanitarian action and global (dis)order'. IILJ International Legal Theory Colloquium Spring 2009: Virtues, Vices, Human Behavior and Democracy in International Law, NYU Law School.

Capgemini Consulting. 2019. *Technological innovation for humanitarian aid and assistance*. Brussels: European Parliamentary Research Service.

Carbonnier, Gilles. 2018a. 'ICRC and peacekeeping operations: Speech of ICRC Vice President Gilles Carbonnier at the 8th Xiangshan Forum'. 29 October. Accessed 26 September 2022. https://www.icrc.org/en/document/china-xiangshan-forum-icrc-and-peacekeeping-operations.

Carbonnier, Gilles. 2018b. 'Revisiting the nexus: Numbers, principles and the issue of social change'. Humanitarian Law & Policy (blog), ICRC. 4 October. Accessed 9 January 2023. https://blogs.icrc.org/law-and-policy/2018/10/04/revisiting-nexus-numbers-principles-issue-social-change.

Caremel, Jean-Francois, Sylvain Landry B. Faye, and Ramatou Ouedraogo. 2017. 'The "humanitarian" response to the Ebola epidemic in Guinea: Between routines and exceptions'. In *The politics of fear: Médecins sans Frontières and the West African Ebola epidemic*, ed. Michiel Hofman and Sokhieng Au. New York: Oxford University Press, 63–83.

Carrera, Sergio, Valsamis Mitsilegas, Jennifer Allsopp, and Lina Vosyliūtė. 2019. *Policing humanitarianism: EU policies against human smuggling and their impact on civil society*. Oxford: Hart.

Cash Learning Partnership. 2020. *The state of the world's cash 2020: Cash and voucher assistance in humanitarian aid*. Cash Learning Partnership.

CDC (Centers for Disease Control and Prevention). 2021. 'Cholera in Haiti'. Centers for Disease Control and Prevention. Last modified 14 April 2021. Accessed 14 August 2021. https://www.cdc.gov/cholera/haiti/index.html.

Chandler, David. 2001. 'The road to military humanitarianism: How the human rights NGOs shaped a new humanitarian agenda'. *Human Rights Quarterly* 23 (3): 678–700.

Chari, Tendai. 2010. 'Representation or misrepresentation? The New York Times's framing of the 1994 Rwanda genocide'. *African Identities* 8 (4): 333–349. https://doi.org/10.1080/14725843.2010.513242.

Checchi, Francesco, and W Courtland Robinson. 2013. *Mortality among populations of southern and central Somalia affected by severe food insecurity and famine during 2010–2012*. Rome; Washington, DC: Food and Agriculture Organization (FAO)/Food Security and Nutrition Analysis Unit (FSNAU) and Famine Early Warning Systems Network (FEWSNET).

Chimni, B. S. 2000. 'Globalization, humanitarianism and the erosion of refugee protection'. *Journal of Refugee Studies* 13 (3): 243–263. http://jrs.oxfordjournals.org/content/13/3/243.abstract.

Chouliaraki, Lilie. 2012. 'The theatricality of humanitarianism: A critique of celebrity advocacy'. *Communication and Critical/Cultural Studies* 9 (1): 1–21. https://doi.org/10.1080/14791420.2011.637055.

Clapham, Christopher. 1991. 'The structure of regional conflict in Northern Ethiopia'. *Disasters* 15 (3): 244–253. https://doi.org/10.1111/j.1467-7717.1991.tb00458.x.

Clapp, Jennifer. 2012. *Hunger in the balance: The new politics of international food aid*. Ithaca, NY: Cornell University Press.

Cloud, Dana L. 2014. 'Shock therapy: Oprah Winfrey, celebrity philanthropy, and disaster "relief" in Haiti'. *Critical Studies in Media Communication* 31 (1): 42–56. https://doi.org/10.1080/15295036.2013.864047.

Cohen, Cynthia Price. 1990. 'The role of nongovernmental organizations in the drafting of the Convention on the Rights of the Child'. *Human Rights Quarterly* 12 (1): 137–147. https://doi.org/10.2307/762172.

Collier, Paul, V. L. Elliott, Håvard Hegre, Anke Hoeffler, Marta Reynal-Querol, and Nicholas Sambanis. 2003. *Breaking the conflict trap: Civil war and development policy: A World Bank policy research report*. Washington, DC: World Bank and Oxford University Press.

Collinson, Sarah, Samir Elhawary, and Robert Muggah. 2010. 'States of fragility: stabilisation and its implications for humanitarian action'. *HPG Working Paper*. London: HPG.

Colombo, Sandro, and Francesco Checchi. 2018. 'Decision-making in humanitarian crises: Politics, and not only evidence, is the problem'. *Epidemiologia e prevenzione* 42 (3–4): 214–225.

Cook, Maria Lorena. 2011. '"Humanitarian aid is never a crime": Humanitarianism and illegality in migrant advocacy'. *Law & Society Review* 45 (3): 561–591. https://doi.org/https://doi.org/10.1111/j.1540-5893.2011.00444.x.

Cooley, Alexander, and James Ron. 2002. 'The NGO scramble: Organizational insecurity and the political economy of transnational action'. *International Security* 27 (1): 5–39.

Cooper, Glenda. 2019. '#AidToo? The 2018 humanitarian scandals in Oxfam GB and save the children UK'. In *The Routledge companion to media and scandal*, ed. Howard Tumber and Silvio Waisbord. Abingdon: Routledge, 342–353.

Cosgrave, John. 2007. *Joint evaluation of the international response to the Indian Ocean tsunami (synthesis report: expanded summary).* London: Tsunami Evaluation Coalition.

Cosgrave, John. 2009. 'Chapter 4: Humanitarian funding and needs assessment'. In *The humanitarian response index 2008: Donor accountability in humanitarian action*, ed. Silvia Hidalgo, Augusto López-Claros, and Laura Altinger. Madrid: DARA, 75–88.

Cotroneo, Angela. 2017. 'Specificities and challenges of responding to internal displacement in urban settings'. *International Review of the Red Cross* 99 (904): 283–318.

Crack, Angela. 2018. 'The Oxfam scandal has taught us there is no reward for honest charities'. The Guardian, 16 March. Accessed 6 December 2021. https://www.theguardian.com/voluntary-sector-network/2018/mar/16/government-donors-reward-honest-charities-oxfam.

Crawford, Nicholas, John Cosgrave, Simone Haysom, and Nadine Walicki. 2015. *Protracted displacement: Uncertain paths to self-reliance in exile.* London: HPG.

Crisp, Jeff, Tim Morris, and Hilde Refstie. 2012. 'Displacement in urban areas: New challenges, new partnerships'. *Disasters* 36: S23–S42. http://dx.doi.org/10.1111/j.1467-7717.2012.01284.x.

Crombé, Xavier, and Michiel Hofman. 2011. 'Afghanistan: Regaining leverage'. In *Humanitarian negotiations revealed: The MSF experience*, ed. Claire Magone, Michaël Neuman, and Fabrice Weissman. London: Hurst, 48–66.

Cryer, Robert. 2015. 'The role of international criminal prosecutions in increasing compliance with international humanitarian law in contemporary African conflicts'. In *Inducing compliance with international humanitarian law: Lessons from the African Great Lakes region*, ed. Heike Krieger. Cambridge: Cambridge University Press, 188–216.

Cryer, Robert. 2017. 'The International Committee of the Red Cross' "Interpretative guidance on the notion of direct participation in hostilities": See a little light'. In *Humanizing the laws of war: The Red Cross and the development of international humanitarian law*, ed. Robin Geiss, Andreas Zimmermann, and Stefanie Haumer. Cambridge: Cambridge University Press, 113–138.

Cubie, Dug. 2017. *The international legal protection of persons in humanitarian crises: Exploring the acquis humanitaire.* Studies in international law vol. 67. Oxford; Portland, OR: Hart Publishing.

Cusumano, Eugenio. 2018a. 'Migrant rescue as organized hypocrisy: EU maritime missions offshore Libya between humanitarianism and border control'. *Cooperation and Conflict* 54 (1): 3–24. https://doi.org/10.1177/0010836718780175.

Cusumano, Eugenio. 2018b. 'The sea as humanitarian space: Non-governmental Search and Rescue dilemmas on the Central Mediterranean migratory route'. *Mediterranean Politics* 23 (3): 387–394. https://doi.org/10.1080/13629395.2017.1302223.

Cusumano, Eugenio. 2019. 'Straightjacketing migrant rescuers? The code of conduct on maritime NGOs'. *Mediterranean Politics* 24 (1): 106–114. https://doi.org/10.1080/13629395.2017.1381400.

Cusumano, Eugenio, and Kristof Gombeer. 2020. 'In deep waters: The legal, humanitarian and political implications of closing Italian ports to migrant rescuers'. *Mediterranean Politics* 25 (2): 245–253. https://doi.org/10.1080/13629395.2018.1532145.

Cusumano, Eugenio, and Matteo Villa. 2021. 'From "angels" to "vice smugglers": The criminalization of sea rescue NGOs in Italy'. *European Journal on Criminal Policy and Research* 27: 23–40. https://doi.org/10.1007/s10610-020-09464-1.

Cuttitta, Paolo. 2018. 'Repoliticization through search and rescue? Humanitarian NGOs and migration management in the central Mediterranean'. *Geopolitics* 23 (3): 632–660. https://doi.org/10.1080/14650045.2017.1344834.

Cutts, Mark. 1999. 'The humanitarian operation in Bosnia, 1992–95: Dilemmas of negotiating humanitarian access'. New Issues in Refugee Research (Working Paper No. 8).

Dallaire, Roméo. 2003. *Shake hands with the devil: The failure of humanity in Rwanda.* London: Arrow Books.

DARA (Development Assistance Research Associates). 2011. *Focus on Haiti: Building back better?* Madrid: DARA.

Darcy, James. 2004. 'Locating responsibility: The Sphere humanitarian charter and its rationale'. *Disasters* 28 (2): 112–123. https://doi.org/10.1111/j.0361-3666.2004.00247.x.

Darcy, James, and Charles-Antoine Hofmann. 2003. *According to need? Needs assessment and decision-making in the humanitarian sector.* London: HPG/ODI.

Davey, Eleanor. 2015. *Idealism beyond borders: The French revolutionary Left and the rise of humanitarianism, 1954–1988*. Cambridge: Cambridge University Press.

Davies, Sara E., and Simon Rushton. 2016. 'Public health emergencies: A new peacekeeping mission? Insights from UNMIL's role in the Liberia Ebola outbreak'. *Third World Quarterly* 37 (3): 419–435. https://doi.org/10.1080/01436597.2015.1110015.

Davis, Morris. 1975. 'Audits of international relief in the Nigerian Civil War: Some political perspectives'. *International Organization* 29 (2): 501–512. https://doi.org/10.1017/S0020818300005002.

Davitti, Daria. 2019. 'The rise of private military and security companies in European Union migration policies: Implications under the UNGPs'. *Business and Human Rights Journal* 4 (1): 33–53. https://doi.org/10.1017/bhj.2018.21.

De Gryse, Benoit. 2013. 'Hospitality lessons during a mobile clinic in Afghanistan'. MSF. 18 April. Accessed 29 November 2022. https://www.doctorswithoutborders.org/latest/hospitality-lessons-during-mobile-clinic-afghanistan.

de Silva, M. W. Amarasiri. 2009. 'Ethnicity, politics and inequality: Post-tsunami humanitarian aid delivery in Ampara District, Sri Lanka'. *Disasters* 33 (2): 253–273. https://doi.org/10.1111/j.1467-7717.2008.01073.x.

de Torrenté, Nicolas. 2004. 'Humanitarianism sacrificed: Integration's false promise'. *Ethics & International Affairs* 18 (2): 3–12. http://dx.doi.org/10.1111/j.1747-7093.2004.tb00461.x.

de Ville de Goyet, Claude, and Lezlie C Morinière. 2006. *The role of needs assessment in the tsunami response: Executive summary*. London: Tsunami Evaluation Coalition.

de Waal, Alex. 1989. *Famine that kills: Darfur, Sudan, 1984–1985*. Oxford: Clarendon.

de Waal, Alex. 1990. 'A re-assessment of entitlement theory in the light of the recent famines in Africa'. *Development and Change* 21 (3): 469–490. https://doi.org/10.1111/j.1467-7660.1990.tb00384.x.

de Waal, Alex. 1997. *Famine crimes: Politics and the disaster relief industry in Africa. African issues*. Oxford: Currey.

de Waal, Alex. 2008. 'The humanitarian carnival: A celebrity vogue'. *World Affairs* 171 (2): 43–55. www.jstor.org/stable/20672853.

de Waal, Alex. 2014. 'Militarizing global health'. Boston Review. Accessed 9 January 2023. http://bostonreview.net/world/alex-de-waal-militarizing-global-health-ebola.

del Valle, Hernan. 2016. 'Search and rescue in the Mediterranean Sea: Negotiating political differences'. *Refugee Survey Quarterly* 35 (2): 22–40. https://doi.org/10.1093/rsq/hdw002. https://doi.org/10.1093/rsq/hdw002.

Des Forges, Alison. 1999. *'Leave none to tell the story': Genocide in Rwanda*. New York: Human Rights Watch.

Desgrandchamps, Marie-Luce. 2012. '"Organising the unpredictable": The Nigeria–Biafra war and its impact on the ICRC'. *International Review of the Red Cross* 94 (888): 1409–1432. https://doi.org/10.1017/S1816383113000428.

Desgrandchamps, Marie-Luce. 2014. 'Dealing with "genocide": The ICRC and the UN during the Nigeria–Biafra war, 1967–70'. *Journal of Genocide Research* 16 (2–3): 281–297. http://dx.doi.org/10.1080/14623528.2014.936705.

Desmond, Alan. 2015. 'The triangle that could square the circle: The UN International Convention on the Protection of the Rights of All Migrant Workers and Members of Their Families, the EU and the universal periodic review'. *European Journal of Migration and Law* 17 (1): 39–69. https://heinonline.org/HOL/P?h=hein.journals/ejml17&i=43.

Development Initiatives. 2019. *Global humanitarian assistance report 2019*. Bristol: Development Initiatives.

Development Initiatives. 2021. *Global humanitarian assistance report 2021*. Bristol: Development Initiatives.

Development Initiatives. 2022. *Global humanitarian assistance report 2022*. Bristol: Development Initiatives.

Dieter, Heribert, and Rajiv Kumar. 2008. 'The downside of celebrity diplomacy: The neglected complexity of development'. *Global Governance* 14 (3): 259–264. http://www.jstor.org/stable/27800708.

Dijkzeul, Dennis, Dorothea Hilhorst, and Peter Walker. 2013. 'Introduction: Evidence-based action in humanitarian crises'. *Disasters* 37 (s1): S1–S19. https://doi.org/10.1111/disa.12009.

Donini, Antonio. 2011. 'Between a rock and a hard place: Integration or independence of humanitarian action?' *International Review of the Red Cross* 93 (881): 141–157. https://doi.org/10.1017/S1816383110000639.

Doornbos, Amos. 2022. 'Cash policy commitments and getting over humps'. Choice and Dignity for People in Crisis (CALP). 17 October. Accessed 9 January 2023. https://www.calpnetwork.org/blog/cash-policy-commitments-getting-over-humps.

Doron, Roy. 2014. 'Marketing genocide: Biafran propaganda strategies during the Nigerian civil war, 1967–70'. *Journal of Genocide Research* 16 (2-3): 227–246. https://doi.org/10.1080/14623528.2014.936702.

Douzinas, Costas. 2007. 'The many faces of humanitarianism'. *Parrhesia* 2: 1–28.

Dreher, Axel, Andreas Fuchs, and Peter Nunnenkamp. 2013. 'New donors'. *International Interactions* 39 (3): 402–415. https://doi.org/10.1080/03050629.2013.784076.

Driessens, Olivier, Stijn Joye, and Daniel Biltereyst. 2012. 'The X-factor of charity: A critical analysis of celebrities' involvement in the 2010 Flemish and Dutch Haiti relief shows'. *Media, Culture & Society* 34 (6): 709–725. https://doi.org/10.1177/0163443712449498.

Drury, A. Cooper, Richard Stuart Olson, and Douglas A. Van Belle. 2005. 'The politics of humanitarian aid: U.S. foreign disaster assistance, 1964–1995'. *Journal of Politics* 67 (2): 454–473. https://doi.org/10.1111/j.1468-2508.2005.00324.x.

DuBois, Marc. 2007. 'Civilian protection and humanitarian advocacy: Strategies and (false?) dilemmas'. *Humanitarian Exchange* 39: 12–15.

DuBois, Marc, Caitlin Wake, Scarlett Sturridge, and Christina Bennett. 2015. *The Ebola response in West Africa: Exposing the politics and culture of international aid*. London: HPG/ODI.

Duffield, Mark. 1994. *An account of relief operations in Bosnia*. London: RRN/ODI.

Duffield, Mark. 1997. 'NGO relief in war zones: Towards an analysis of the new aid paradigm'. *Third World Quarterly* 18 (3): 527–542. https://doi.org/10.1080/01436599714858.

Duffield, Mark. 2001. *Global governance and the new wars: The merging of development and security*. London: Zed Books.

Duffield, Mark. 2007. *Development, security and unending war: Governing the world of peoples*. Cambridge: Polity Press.

Duffield, Mark. 2016. 'The resilience of the ruins: Towards a critique of digital humanitarianism'. *Resilience* 4 (3): 147–165. https://doi.org/10.1080/21693293.2016.1153772.

Duffield, Mark, and John Prendergast. 1994. *Without troops & tanks: The emergency relief desk and the cross border operation into Eritrea and Tigray*. Lawrenceville, NJ: Red Sea Press.

Dupuy, Alex. 2010. 'Disaster capitalism to the rescue: The international community and Haiti after the earthquake'. *NACLA Report on the Americas* 43 (4): 14–19. https://doi.org/10.1080/10714839.2010.11722192.

Duvall, Spring-Serenity. 2015. 'Masculine interventions and transnational celebrity activism in the aftermath of the 2010 earthquake in Haiti'. *Celebrity Studies* 6 (4): 583–600. https://doi.org/10.1080/19392397.2015.1037779.

Eck, Kristine, and Lisa Hultman. 2007. 'One-sided violence against civilians in war: Insights from new fatality data'. *Journal of Peace Research* 44 (2): 233–246.

Eilstrup-Sangiovanni, Mette, and J. C. Sharman. 2021. 'Enforcers beyond borders: Transnational NGOs and the enforcement of international law'. *Perspectives on Politics* 19 (1): 131–147. https://doi.org/10.1017/S153759271900344X.

Eriksson, John, Howard Adelman, John Borton, Hanne Christensen, Krishna Kumar, Astri Suhrke, David Tardif-Douglin, Stein Villumstad, and Lennart Wohlgemuth. 1996. The international response to conflict and genocide: Lessons from the Rwanda experience. Synthesis Report. Copenhagen: Danish International Development Agency (DANIDA).

Esperti, Marta. 2020. 'Rescuing migrants in the Central Mediterranean: The emergence of a new civil humanitarianism at the maritime border'. *American Behavioral Scientist* 64 (4): 436–455. https://doi.org/10.1177/0002764219882976.

Fallah, Mosoka P., Laura A. Skrip, and Jonathan Enders. 2018. 'Preventing rural to urban spread of Ebola: Lessons from Liberia'. *The Lancet* 392 (10144): 279–280. https://doi.org/10.1016/S0140-6736(18)31435-1.

Fan, Lilianne. 2013. *Disaster as opportunity? Building back better in Aceh, Myanmar and Haiti.* London: HPG/ODI.

FAO (Food and Agriculture Organization), IFAD (International Fund for Agricultural Development), UNICEF (United Nations Children's Fund), WFP (World Food Programme), and WHO (World Health Organization). 2021. *The state of food security and nutrition in the world 2021: Transforming food systems for food security, improved nutrition and affordable healthy diets for all.* Rome: FAO.

Fassin, Didier. 2007. 'Humanitarianism as a politics of life'. *Public Culture* 19 (3): 499–520. https://doi.org/10.1215/08992363-2007-007.

Fassin, Didier, and Paula Vasquez. 2005. 'Humanitarian exception as the rule: The political theology of the 1999 Tragedia in Venezuela'. *American Ethnologist* 32 (3): 389–405. https://doi.org/10.1525/ae.2005.32.3.389.

Fast, Larissa. 2017. *Upending humanitarianism: Questions emerging 'from the ground up'*. London: HPG/ODI.

Featherstone, Andy. 2012. *Strengthening principled humanitarian response capacities: Afghanistan case study*. Oslo; London: Norwegian Refugee Council/HPG.

Feldman, Ilana. 2009. 'Gaza's humanitarianism problem'. *Journal of Palestine Studies* 38 (3): 22–37. Accessed 9 January 2023. http://www.jstor.org/stable/10.1525/jps.2009.xxxviii.3.22.

Fink, Guenther, and Silvia Redaelli. 2011. 'Determinants of international emergency aid—humanitarian need only?' *World Development* 39 (5): 741–757. https://doi.org/10.1016/j.worlddev.2010.09.004.

Fink-Hooijer, Florika. 2015. 'Civil protection and humanitarian aid in the Ebola response: Lessons for the humanitarian system from the EU experience'. *Humanitarian Exchange* 64: 3–5.

Fishstein, Paul, and Andrew Wilder. 2012. *Winning hearts and minds? Examining the relationship between aid and security in Afghanistan*. Medford, MA: Feinstein International Center.

Fisman, David, Edwin Khoo, and Ashleigh Tuite. 2014. 'Early epidemic dynamics of the West African 2014 Ebola outbreak: Estimates derived with a simple two-parameter model'. *PLoS Currents* 6. https://doi.org/10.1371/currents.outbreaks.89c0d3783f36958d96ebbae97348d571.

Food Security Cluster. 2019. 'Cash for Work (CFW): Guidance note'. Food Security Cluster.

Forsythe, David P. 1997. 'International humanitarian assistance: The role of the Red Cross'. *Buffalo Journal of International Law* 3 (2): 235–260.

Forsythe, David P. 2005. *The humanitarians: The International Committee of the Red Cross.* Cambridge: Cambridge University Press.

Fortune. 2021a. 'Fortune 500'. Accessed 4 December 2021. https://fortune.com/fortune500/2020.

Fortune. 2021b. 'Fortune 500: Coca-Cola'. Last modified 2 August 2021. Accessed 4 December 2021. https://fortune.com/company/coca-cola/fortune500.

Fox, Fiona. 2001. 'New humanitarianism: Does it provide a moral banner for the 21st century?' *Disasters* 25 (4): 275–289.

FRA (EU Agency for Fundamental Rights). 2021. 'June 2021 update—search and rescue (SAR) operations in the Mediterranean and fundamental rights'. FRA. Last modified 18 June 2021. Accessed 29 December 2021. https://fra.europa.eu/en/publication/2021/june-2021-update-ngo-ships-sar-activities.

Franklin, James C. 2008. 'Shame on you: The impact of human rights criticism on political repression in Latin America'. *International Studies Quarterly* 52 (1): 187–211. http://dx.doi.org/10.1111/j.1468-2478.2007.00496.x.

Franks, Suzanne. 2006a. 'How famine captured the headlines'. *Media History* 12 (3): 291–312. https://doi.org/10.1080/13688800601014025.

Franks, Suzanne. 2006b. 'The CARMA report: Western media coverage of humanitarian disasters'. *Political Quarterly* 77 (2): 281–284. https://doi.org/10.1111/j.1467-923X.2006.00771.x.

Franks, Suzanne. 2010. 'The neglect of Africa and the power of aid'. *International Communication Gazette* 72 (1): 71–84. https://doi.org/10.1177/1748048509350339.

Fredriksen, Aurora. 2016. 'Crisis in "a normal bad year": Spaces of humanitarian emergency, the Integrated Food Security Phase Classification scale and the Somali famine of 2011'. *Environment and Planning A: Economy and Space* 48 (1): 40–57. https://doi.org/10.1177/0308518X15597446.

Frelick, Bill. 1992. '"Preventive protection" and the right to seek asylum: A preliminary look at Bosnia and Croatia'. *International Journal of Refugee Law* 4 (4): 439–454. Accessed 9 January 2023. http://ijrl.oxfordjournals.org/content/4/4/439.abstract.

Fritz Institute. 2005. *Lessons from the tsunami: Top line findings*. Fritz Institute.

Gagnon, V. P. Jr. 1990. 'Yugoslavia: prospects for stability'. *Foreign Affairs* (3): 17–35. Accessed 9 January 2023. https://heinonline.org/HOL/P?h=hein.journals/fora70&i=489.

Gaillard, Philippe. 2002. 'Rwanda 1994: ". . . kill as many people as you want, you cannot kill their memory.' Speech by Philippe Gaillard, head of the International Committee of the Red Cross's (ICRC's) delegation in Rwanda, 1993–1994; given at the Genocide Prevention Conference, London, January 2002, organized by the Aegis Trust and the UK Foreign Office. ICRC. Accessed 19 September 2020. https://www.icrc.org/en/doc/resources/documents/misc/5xfncq.htm.

Garcés Mascareñas, Blanca. 2020. 'Four years of the EU–Turkey deal'. *CIDOB Opinion* 617.

Gayle, Damien. 2018. 'Timeline: Oxfam sexual exploitation scandal in Haiti'. The Guardian, 15 June. Accessed 6 December 2021. https://www.theguardian.com/world/2018/jun/15/timeline-oxfam-sexual-exploitation-scandal-in-haiti.

Geremia, Maurizio. 2009. 'Neutrality, impartiality and independence in Colombia: An ICRC perspective'. *Humanitarian Exchange* 45: 4–6.

GHD (Good Humanitarian Donorship). n.d. '24 principles and good practice of humanitarian donorship'. Accessed 16 April 2022. https://www.ghdinitiative.org/ghd/gns/principles-good-practice-of-ghd/principles-good-practice-ghd.html.

Ghezelbash, Daniel, Violeta Moreno-Lax, Natalie Klein, and Brian Opeskin. 2018. 'Securitization of search and rescue at sea: The response to boat migration in the Mediterranean and offshore Australia'. *International and Comparative Law Quarterly* 67 (2): 315–351. https://doi.org/10.1017/S0020589317000562.

Giladi, Rotem, and Steven R. Ratner. 2015. 'The role of the International Committee of the Red Cross'. In *The 1949 Geneva Conventions: A commentary*, ed. Andrew Clapham, Paola Gaeta and Marco Sassòli. Oxford: Oxford University Press, 525–548.

Gillard, Emanuela-Chiara. 2013. 'The law regulating cross-border relief operations'. *International Review of the Red Cross* 95 (890): 351–382. https://doi.org/10.1017/S1816383114000307.

Gingerich, Tara R., and Marc J. Cohen. 2015. 'Turning the humanitarian system on its head: Saving lives and livelihoods by strengthening local capacity and shifting leadership to local actors'. *Oxfam Research Reports*. Oxford: Oxfam GB.

Giossi Caverzasio, Sylvie, ed. 2001. *Strengthening protection in war: A search for professional standards*. Geneva: ICRC.

Goodwin-Gill, Guy S. 2020. 'The Office of the United Nations High Commissioner for Refugees and the sources of international refugee law'. *International and Comparative Law Quarterly* 69 (1): 1–41. https://doi.org/10.1017/S002058931900054X.

Goodwin-Gill, Guy S, and Jane McAdam. 2007. *The refugee in international law*. Oxford: Oxford University Press.

Gordon, Eleanor, and Henrik Kjellmo Larsen. 2021. 'The violent inaction of the state and the camp as site of struggle: The perspectives of humanitarian actors in Moria Camp, Lesvos'. *European Journal of International Security* 6(4): 418–438. https://doi.org/10.1017/eis.2021.9.

Gordon, Stuart. 2010. 'The United Kingdom's stabilisation model and Afghanistan: The impact on humanitarian actors'. *Disasters* 34: S368–S387. https://doi.org/10.1111/j.1467-7717.2010.01208.x.

Gordon, Stuart, and Antonio Donini. 2015. 'Romancing principles and human rights: Are humanitarian principles salvageable?' *International Review of the Red Cross* 97 (897–898): 77–109. https://doi.org/10.1017/S1816383115000727.

Griffin, Christopher. 2015. 'French military policy in the Nigerian Civil War, 1967–1970'. *Small Wars & Insurgencies* 26 (1): 114–135. https://doi.org/10.1080/09592318.2014.959766.

Grépin, Karen A. 2015. 'International donations to the Ebola virus outbreak: Too little, too late?' *British Medical Journal* 350. Accessed 9 January 2023. http://www.bmj.com/content/350/bmj.h376.abstract.

Guha-Sapir, Debarati, and Willem Gijsbert van Panhuis. 2009. 'Health impact of the 2004 Andaman Nicobar earthquake and tsunami in Indonesia'. *Prehospital and Disaster Medicine* 24 (6): 493–499. https://doi.org/10.1017/S1049023X00007391.

Guichaoua, Andre. 2017. *From war to genocide: Criminal politics in Rwanda, 1990–1994*. Transl. Don E. Webster. Madison, WI: University of Wisconsin Press.

Hafner-Burton, Emilie M. 2008. 'Sticks and stones: Naming and shaming the human rights enforcement problem'. *International Organization* 62 (4): 689–716.

Hammerstad, Anne. 2014. *The rise and decline of a global security actor: UNHCR, refugee protection, and security*. First edn. Oxford: Oxford University Press.

Hammond, Laura, and Hannah Vaughan-Lee. 2012. *Humanitarian space in Somalia: A scarce commodity*. London: ODI/HPG.

Hanrieder, Tine, and Claire Galesne. 2021. 'Domestic humanitarianism: The Mission France of Médecins Sans Frontières and Médecins du Monde'. *Third World Quarterly* 42 (8): 1715–1732. https://doi.org/10.1080/01436597.2021.1916393.

Hansen, Lene. 2006. *Security as practice: discourse analysis and the Bosnian war*. London: Routledge.

Hansen, Lene. 2011. "Theorizing the image for Security Studies'. *European Journal of International Relations* 17 (1): 51–74. https://doi.org/10.1177/1354066110388593.

Harild, Niels, Asger Christensen, and Roger Zetter. 2015. *Sustainable refugee return: Trigger, constraints, and lessons on addressing the development challenges of forced displacement*. Washington, DC: World Bank.

Harmer, Adele, and Lin Cotterrell. 2005. *Diversity in donorship: The changing landscape of official humanitarian aid*. London: HPG/ODI.

Harmer, Adele, and Ellen Martin. 2010. *Diversity in donorship: Field lessons*. London: HPG/ODI.

Harmer, Adele, Abby Stoddard, and Alexandra Sarazen. 2018. *Humanitarian access in armed conflict: A need for new principles?* Humanitarian Outcomes.

Harrell-Bond, Barbara. 1998. 'Camps: literature review'. *Forced Migration Review* 2: 22–23.

Harrell-Bond, Barbara. 2002. 'Can humanitarian work with refugees be humane?' *Human Rights Quarterly* 24 (1): 51–85.

Harvey, Paul. 2009. *Towards good humanitarian government: The role of the affected state in disaster response*. London: HPG/ODI.

Haysom, Simone. 2013. *Sanctuary in the city? Urban displacement and vulnerability*. London: HPG/ODI.

Haysom, Simone, and Ashley Jackson. 2013. '"You don't need to love us": Civil-military relations in Afghanistan, 2002–13'. *Stability: International Journal of Security and Development* 2 (2): 1–16. http://dx.doi.org/10.5334/sta.by.

Hedlund, Kerren, Nisar Majid, Dan Maxwell, and Nigel Nicholson. n.d. *Final evaluation of the unconditional cash and voucher response to the 2011–12 crisis in southern and central Somalia: Executive summary*. Nairobi: UNICEF Somalia.

Heerten, Lasse. 2016. '"A" as in Auschwitz, "B" as in Biafra: The Nigerian civil war, visual narratives of genocide, and the fragmented universalization of the Holocaust'. In *Humanitarian photography: A history*, ed. Heide Fehrenbach and Davide Rodogno. New York: Cambridge University Press, 249–274.

Heerten, Lasse. 2017. *The Biafran war and postcolonial humanitarianism: Spectacles of suffering*. Cambridge: Cambridge University Press.

Heller Pérache, André. 2015. '"To put out this fire, we must run into the burning building": A review of MSF's call for biological containment teams in West Africa'. *Humanitarian Exchange* 64: 5–7.

Hendrickson, Ryan C. 2005. 'Crossing the Rubicon'. NATO Review 3.

Hendrie, Barbara. 1989. 'Cross-border relief operations in Eritrea and Tigray'. *Disasters* 13 (4): 351–360. https://doi.org/10.1111/j.1467-7717.1989.tb00729.x.

Hendrix, Cullen S., and Wendy H. Wong. 2013. 'When is the pen truly mighty? Regime type and the efficacy of naming and shaming in curbing human rights abuses'. *British Journal of Political Science* 43 (3): 651–672. href="http://dx.doi.org/10.1017/S0007123412000488.

High Level Panel on Humanitarian Cash Transfers. 2015. *Doing cash differently: How cash transfers can transform humanitarian aid.* London: ODI.

Hilhorst, Dorothea. 2018. 'Classical humanitarianism and resilience humanitarianism: Making sense of two brands of humanitarian action'. *Journal of International Humanitarian Action* 3 (15): 1–12. https://doi.org/10.1186/s41018-018-0043-6.

Hilhorst, Dorothea, Maria Hagan, and Olivia Quinn. 2021. 'Reconsidering humanitarian advocacy through pressure points of the European "migration crisis"'. *International Migration* 59 (3): 125–144. https://doi.org/https://doi.org/10.1111/imig.12802.

Hillbruner, Chris, and Grainne Moloney. 2012. 'When early warning is not enough—lessons learned from the 2011 Somalia Famine'. *Global Food Security* 1 (1): 20–28. https://doi.org/http://dx.doi.org/10.1016/j.gfs.2012.08.001.

Hobbs, Christina, Mark Gordon, and Brian Bogart. 2012. 'When business is not as usual: Decision-making and the humanitarian response to the famine in South Central Somalia'. *Global Food Security* 1 (1): 50–56. https://doi.org/http://dx.doi.org/10.1016/j.gfs.2012.07.005.

Holloway, Kerrie, Reem Al Masri, and Afnan Abu Yahia. 2021. *Digital identity, biometrics and inclusion in humanitarian responses to refugee crises.* London: HPG/ODI.

Holmes, John. 2014. 'Humanitarian action and protection: A comment on Norah Niland's "Sri Lanka: Unrestricted warfare and limited protective humanitarian action"'. *International Development Policy* 6 (1): 143–148.

Hopgood, Stephen. 2019. 'When the music stops: Humanitarianism in a post-liberal world order'. *Journal of Humanitarian Affairs* 1 (1): 4–14. https://doi.org/10.7227/JHA.002.

Hopkins, Nick. 2012. 'Afghan schools and clinics built by British military forced to close'. The Guardian, 27 September. Accessed 3 December 2020. https://www.theguardian.com/world/2012/sep/27/afghan-schools-clinics-built-british-close.

Hughes, Karl, and Claire Hutchings. 2011. *Can we obtain the required rigour without randomisation? Oxfam GB's non-experimental Global Performance Framework.* New Delhi: International Initiative for Impact Evaluation (3ie).

HPG (Humanitarian Policy Group). 2007. 'Humanitarian advocacy in Darfur: The challenge of neutrality'. HPG Policy Brief 28.

HPG. 2010. 'Humanitarian space in Sri Lanka: What lessons can be learned?' Round-table at ODI (HPG), London, 18-19 November 2010, meeting summary. London: HPG/ODI.

Hutchinson, John F. 1996. *Champions of charity: War and the rise of the Red Cross.* Oxford: Westview Press.

Hutchison, Emma. 2014. 'A global politics of pity? Disaster imagery and the emotional construction of solidarity after the 2004 Asian tsunami'. *International Political Sociology* 8 (1): 1–19. https://doi.org/10.1111/ips.12037.

ICG (International Crisis Group). 2010. 'War crimes in Sri Lanka'. Asia Report No. 191, 17 May. ICG.

ICISS (International Commission on Intervention and State Sovereignty). 2001. *The responsibility to protect.* Ottawa: International Dispute Resolution Centre (IDRC).

ICRC (International Committee of the Red Cross). 1969. *Annual report 1968.* Geneva: ICRC.

ICRC. 1995. *Annual report 1994.* Geneva: ICRC.

ICRC. 2005. 'Action by the International Committee of the Red Cross in the event of violations of international humanitarian law or of other fundamental rules protecting persons in situations of violence'. *International Review of the Red Cross* 87 (858): 393–400.

ICRC. 2007a, 'Myanmar: ICRC denounces major and repeated violations of international humanitarian law'. 29 June. Accessed 9 January 2023. http://www.icrc.org/eng/resources/documents/news-release/2009-and-earlier/myanmar-news-290607.htm.

ICRC. 2007b, 'Myanmar: No progress in talks. ICRC closes offices'. 15 March. Accessed 9 January 2023. http://www.icrc.org/eng/resources/documents/news-release/2009-and-earlier/myanmar-news-150307.htm.

ICRC. 2015. *Activities for migrants*. Geneva: ICRC.

ICRC. 2016. *Protracted conflict and humanitarian action: Some recent ICRC experiences*. Geneva: ICRC.

ICRC. 2017. *Migration: Our work in Europe & Central Asia*. Geneva: ICRC.

ICRC. 2018. *The roots of restraint in war*. Geneva: ICRC.

ICRC. 2022. 'Cyber-attack on ICRC: What we know'. Accessed 25 November 2022. https://www.icrc.org/en/document/cyber-attack-icrc-what-we-know.

ICVA (International Council of Voluntary Agencies). 2017a. *The Grand Bargain explained: An ICVA briefing paper*. Geneva: ICVA.

ICVA. 2017b. *The 'New Way of Working' examined: An ICVA briefing paper*. Geneva: ICVA.

IDMC (Internal Displacement Monitoring Centre). 2015. *Global overview 2015: People internally displaced by conflict and violence*. Geneva: IDMC.

IDNDR (International Decade for Natural Disaster Reduction) 1994. *Yokohama strategy and plan of action for a safer world: Guidelines for natural disaster prevention, preparedness and mitigation*. Geneva: IDNDR/UN Department of Humanitarian Affairs.

IFRC (International Federation of Red Cross and Red Crescent Societies). 2005. *World disasters report 2005: Focus on information in disasters*. Bloomfield, CT; London: Kumarian Press and Eurospan.

IFRC. 2007. *Law and legal issues in international disaster response: A desk study*. Geneva: IFRC.

IFRC. 2011. *Introduction to the Guidelines for the domestic facilitation and regulation of international disaster relief and initial recovery assistance*. Geneva: IFRC.

IFRC. 2013. 'Chile: Earthquake'. In Emergency appeal final report: emergency appeal No. MDRCL006 GLIDE No. EQ-2010-000034-CHL, 29 July.

IFRC. 2016. *World disasters report 2016:* Resilience: Saving lives today, investing for tomorrow. Geneva: IFRC.

Internal Review Panel. 2012. 'Report of the Secretary-General's Internal Review Panel on United Nations action in Sri Lanka'. United Nations.

IRC (International Rescue Committee). 2016. 'The Ebola lessons reader: What's being said, what's missing and why it matters'. IRC.

IRIN (Integrated Regional Information Networks) News. 2009. 'SUDAN: NGO expulsion to hit Darfur's displaced'. 09 March 2009.

IRRC (International Review of the Red Cross). 2019. 'The ICRC and the "humanitarian–development–peace nexus" discussion: In conversation with Filipa Schmitz Guinote, ICRC Policy Adviser'. *International Review of the Red Cross* 101 (912): 1051–1066. https://doi.org/10.1017/S1816383120000284.

Jackson, Ashley. 2012a. 'Talking to the other side: Humanitarian engagement with armed non-state actors'. HPG Policy Brief 47. London: HPG/ODI.

Jackson, Ashley. 2012b. *Talking to the other side: Humanitarian engagement with armed non-state actors*. London: Humanitarian Policy Group.

Jackson, Ashley, and Abdi Aynte. 2013. *Talking to the other side: Humanitarian negotiations with Al-Shabaab in Somalia*. London: ODI/HPG.

Jackson, Ashley, and Eleanor Davey. 2014. *From the Spanish civil war to Afghanistan: Historical and contemporary reflections on humanitarian engagement with non-state armed groups*. London: ODI/HPG.

Jackson, Ashley, and Antonio Giustozzi. 2012. 'Talking to the other side: humanitarian engagement with the Taliban in Afghanistan'. HPG Working Paper. London: HPG/ODI.

Jackson, Ashley, and Simone Haysom. 2013. *The search for common ground: Civil–military relations in Afghanistan, 2002–13*. London: HPG/ODI.

Jacobsen, Katja Lindskov. 2015. *The politics of humanitarian technology: Good intentions, unintended consequences and insecurity*. London; New York: Routledge.

Jansson, Kurt, Michael Harris, and Angela Penrose. 1987. *The Ethiopian famine: The story of the emergency relief operation*. London: Zed.

Jaspars, Susanne. 2021. 'Protracted crisis, food security and the fantasy of resilience in Sudan'. *Security Dialogue* 52 (3): 195–212. https://doi.org/10.1177/0967010620927279.

Jütersonke, Oliver, Robert Muggah, and Dennis Rodgers. 2009. 'Gangs, urban violence, and security interventions in Central America'. *Security Dialogue* 40 (4–5): 373–397. Accessed 9 January 2023. http://sdi.sagepub.com/content/40/4-5/373.abstract.

Kahn, Clea. 2015. 'Ebola and humanitarian protection'. *Humanitarian Exchange* 64: 10–12.

Kahn, Clea, and Andrew Cunningham. 2013. 'Introduction to the issue of state sovereignty and humanitarian action'. *Disasters* 37 (s2): S139–S150. https://doi.org/10.1111/disa.12018.

Kalmanovitz, Pablo, and Miriam Bradley. forthcoming. 'Individualization and collectivization in contexts of organized criminal violence: The case of Mexico's "war on drugs"'. In *The individualization of war: Rights, liability, and accountability in contemporary armed conflict*, ed. Jennifer Welsh, Dapo Akande, and David Rodin. Oxford: Oxford University Press.

Kamradt-Scott, Adam. 2016. 'WHO's to blame? The World Health Organization and the 2014 Ebola outbreak in West Africa'. *Third World Quarterly* 37 (3): 401–418. https://doi.org/10.1080/01436597.2015.1112232.

Kapoor, Ilan. 2013. *Celebrity humanitarianism: The ideology of global charity*. Abingdon: Routledge.

Keen, David. 1994. *The benefits of famine: A political economy of famine and relief in southwestern Sudan, 1983–1989*. Princeton, NJ; Chichester: Princeton University Press.

Keen, David. 1998a. 'Aid and violence, with special reference to Sierra Leone'. *Disasters* 22 (4): 318–328.

Keen, David. 1998b. *The economic functions of violence in civil wars*. Oxford: Oxford University Press.

Keen, David. 2009. 'Compromise or capitulation: Report on WFP and the humanitarian crisis in Sri Lanka'. In *Humanitarian assistance in conflict and complex emergencies, 23–25 June 2009*, ed. WFP. Rome: WFP, 49–102.

Keen, David. 2014. '"The camp" and "the lesser evil": Humanitarianism in Sri Lanka'. *Conflict, Security & Development* 14 (1): 1–31. https://doi.org/10.1080/14678802.2013.856176.

Keen, David. 2021. 'The functions and legitimization of suffering in Calais, France'. *International Migration* 59 (3): 9–28. https://doi.org/https://doi.org/10.1111/imig.12800.

Keller, Edmond J. 1992. 'Drought, war, and the politics of famine in Ethiopia and Eritrea'. *Journal of Modern African Studies* 30 (4): 609–624. Accessed 9 January 2023. http://www.jstor.org/stable/161267.

Kellow, Christine L., and H. Leslie Steeves. 1998. 'The role of radio in the Rwandan genocide'. *Journal of Communication* 48 (3): 107–128. https://doi.org/10.1111/j.1460-2466.1998.tb02762.x.

Kennedy, Jim, Joseph Ashmore, Elizabeth Babister, and Ilan Kelman. 2008. 'The meaning of "build back better": Evidence from post-tsunami Aceh and Sri Lanka'. *Journal of Contingencies and Crisis Management* 16 (1): 24–36. https://doi.org/10.1111/j.1468-5973.2008.00529.x.

Kentikelenis, Alexander, Lawrence King, Martin McKee, and David Stuckler. 2015. 'The International Monetary Fund and the Ebola outbreak'. *Lancet Global Health* 3 (2): e69–e70. https://doi.org/10.1016/S2214-109X(14)70377-8.

Kevlihan, Rob, Karl DeRouen, and Glen Biglaiser. 2014. 'Is US humanitarian aid based primarily on need or self-interest?' *International Studies Quarterly* 58 (4): 839–854. http://dx.doi.org/10.1111/isqu.12121.

Khasalamwa, Sarah. 2009. 'Is "build back better" a response to vulnerability? Analysis of the post-tsunami humanitarian interventions in Sri Lanka'. *Norsk Geografisk Tidsskrift—Norwegian Journal of Geography* 63 (1): 73–88. https://doi.org/10.1080/00291950802712152.

Knox Clarke, Paul, and James Darcy. 2014. *Insufficient evidence? The quality and use of evidence in humanitarian action*. London: ALNAP.

Knox Clarke, Paul, Abby Stoddard, and Luminita Tuchel. 2018. *The state of the humanitarian system*. London: ALNAP.

Kotsioni, Ioanna. 2016. 'Detention of migrants and asylum-seekers: The challenge for humanitarian actors'. *Refugee Survey Quarterly* 35 (2): 41–55. https://doi.org/10.1093/rsq/hdw004.

Krain, Matthew. 2012. 'J'accuse! Does naming and shaming perpetrators reduce the severity of genocides or politicides?' *International Studies Quarterly* 56 (3): 574–589. http://dx.doi.org/10.1111/j.1468-2478.2012.00732.x.

Krause, Keith. 2016. 'From armed conflict to political violence: Mapping & explaining conflict trends'. *Daedalus* 145 (4): 113–126. https://doi.org/10.1162/DAED_a_00416.

Krever, Tor. 2011. '"Mopping-up": UNHCR, Neutrality and Non-refoulement since the Cold War'. *Chinese Journal of International Law* 10 (3): 587–608.

Krill, Francoise. 2001. 'The ICRC's policy on refugees and internally displaced civilians'. *International Review of the Red Cross* 83 (843): 607–628.

Krähenbühl, Pierre. 2013. 'There are no "good" or "bad" civilians in Syria—we must help all who need aid'. The Guardian, 3 March. Accessed 21 February 2021. https://www.theguardian.com/commentisfree/2013/mar/03/red-cross-aid-inside-syria.

Kumar, Claire. 2021. 'A year on, migrants' vital contributions to the Covid-19 response must drive lasting reforms'. ODI. Accessed 10 December 2022. https://odi.org/en/insights/a-year-on-migrants-vital-contributions-to-the-covid-19-response-must-drive-lasting-reforms.

Kumar, Gopalakrishna. 1987. *Ethiopian famines 1973–1985: A case-study*. Helsinki: World Institute for Development Economics Research.

Kumar, Rajeesh. 2019. *The International Committee of the Red Cross in internal armed conflicts: Is neutrality possible?* Singapore: Palgrave Pivot.

Kurasawa, Fuyuki. 2013. 'The sentimentalist paradox: On the normative and visual foundations of humanitarianism'. *Journal of Global Ethics* 9 (2): 201–214. http://dx.doi.org/10.1080/17449626.2013.818461.

Labbé, Jérémie, and Pascal Daudin. 2015. 'Applying the humanitarian principles: Reflecting on the experience of the International Committee of the Red Cross'. *International Review of the Red Cross* 97 (897–898): 183–210. https://doi.org/10.1017/S1816383115000715.

Landgren, Karin. 1995. 'Safety zones and international protection: A dark grey area'. *International Journal of Refugee Law* 7 (3): 436–458.

Lanz, David. 2009. 'Save Darfur: A Movement and its Discontents'. *African Affairs* 108 (433): 669–677. https://doi.org/10.1093/afraf/adp061. https://dx.doi.org/10.1093/afraf/adp061.

Lautze, Sue, Winnie Bell, Luca Alinovi, and Luca Russo. 2012. 'Early warning, late response (again): The 2011 famine in Somalia'. *Global Food Security* 1 (1): 43–49. https://doi.org/http://dx.doi.org/10.1016/j.gfs.2012.07.006.

Leader, Nicholas. 1998. 'Proliferating principles; or how to sup with the devil without getting eaten'. *Disasters* 22 (4): 288–309.

Lentz, Erin C., Simone Passarelli, and Christopher B. Barrett. 2013. 'The timeliness and cost-effectiveness of the local and regional procurement of food aid'. *World Development* 49: 9–18. https://doi.org/https://doi.org/10.1016/j.worlddev.2013.01.017.

Lessing, Benjamin. 2021. 'Conceptualizing criminal governance'. *Perspectives on Politics* 19 (3): 854–873. https://doi.org/10.1017/S1537592720001243.

Levine, Simon. 2017. *Markets in crises: The implications for humanitarian action*. London: HPG/ODI.

Levine, Simon, Sarah Bailey, Béatrice Boyer, and Cassandra Mehu. 2012a. *Avoiding reality: Land, institutions and humanitarian action in post-earthquake Haiti*. London: ODI/HPG.

Levine, Simon, Adam Pain, Sarah Bailey, and Lilianne Fan. 2012b. 'The relevance of "resilience"?' In HPG Policy Brief 49. London: HPG.

Lie, Jon Harald Sande. 2020. 'The humanitarian–development nexus: Humanitarian principles, practice, and pragmatics'. *Journal of International Humanitarian Action* 5 (18). https://doi.org/10.1186/s41018-020-00086-0.

Lindley, Anna. 2011. 'Between a protracted and a crisis situation: Policy responses to Somali refugees in Kenya'. *Refugee Survey Quarterly* 30 (4): 14–49. https://doi.org/10.1093/rsq/hdr013.

Lischer, Sarah Kenyon. 2003. 'Collateral damage: Humanitarian assistance as a cause of conflict.' *International Security* 28 (1): 79–109.

Lischer, Sarah Kenyon. 2007. "Military intervention and the humanitarian force multiplier.' *Global Governance* 13 (1): 99–118. Accessed 10 January 2023. https://heinonline.org/HOL/P?h=hein.journals/glogo13&i=107.

Little, Peter D. 2012. 'On the Somalia dilemma: Adding layers of complexity to an already complex emergency'. *African Studies Review* 55 (1): 191–195. https://doi.org/10.1353/arw.2012.0018.

Loescher, Gil. 2001. *The UNHCR and world politics: A perilous path*. Oxford: Oxford University Press.

Louis, Marieke, and Lucile Maertens 2021. 'Why international organizations hate politics: The case of the ILO and UNEP'. The Global (blog). 11 October 2021. Accessed 10 January 2023. https://theglobal.blog/2021/10/11/why-international-organizations-hate-politics-the-case-of-the-ilo-and-unep.

Mackintosh, Kate, and Patrick Duplat. 2013. *Study of the impact of donor counter-terrorism measures on principled humanitarian action*. Geneva: Norwegian Refugee Council.

Macrae, Joanna, and Adele Harmer. 2004. 'Beyond the continuum: An overview of the changing role of aid policy in protracted crises'. In *Beyond the continuum: An overview of the changing role of aid policy in protracted crises*, ed. Adele Harmer and Joanna Macrae. London: HPG, 1–13.

Macrae, Joanna, and Anthony B Zwi. 1992. 'Food as an instrument of war in contemporary African famines: A review of the evidence'. *Disasters* 16 (4): 299–321. http://dx.doi.org/10.1111/j.1467-7717.1992.tb00412.x.

Magone, Claire, Michaël Neuman, and Fabrice Weissman, eds. 2011. *Humanitarian negotiations revealed: The MSF experience*. London: Hurst.

Majid, Nisar, and Stephen McDowell. 2012. 'Hidden dimensions of the Somalia famine'. *Global Food Security* 1 (1): 36–42. https://doi.org/http://dx.doi.org/10.1016/j.gfs.2012.07.003.

Malkki, Liisa H. 1996. 'Speechless emissaries: Refugees, humanitarianism, and dehistoricization'. *Cultural Anthropology* 11 (3): 377–404. http://dx.doi.org/10.1525/can.1996.11.3.02a00050.

Mamiya, Ralph. 2016. 'A history and conceptual development of the protection of civilians'. In *Protection of civilians*, ed. Haidi Willmot, Ralph Mamiya, Scott Sheeran, and Marc Weller. Oxford: Oxford University Press, 63–87.

Maritime Executive. 2017. 'Three NGOs suspend migrant rescue operations off Libya'. Accessed 29 December 2021. https://maritime-executive.com/article/three-ngos-suspend-migrant-rescue-operations-off-libya.

Martínez, José Ciro, and Brent Eng. 2016. 'The unintended consequences of emergency food aid: neutrality, sovereignty and politics in the Syrian civil war, 2012–15'. *International Affairs* 92 (1): 153–173.

Maxwell, Daniel, and Merry Fitzpatrick. 2012. 'The 2011 Somalia famine: Context, causes, and complications'. *Global Food Security* 1 (1): 5–12. https://doi.org/http://dx.doi.org/10.1016/j.gfs.2012.07.002.

Maxwell, Daniel, Nicholas Haan, Kirsten Gelsdorf, and David Dawe. 2012. 'The 2011–12 famine in Somalia: Introduction to the special edition'. *Global Food Security* 1 (1): 1–4. https://doi.org/http://dx.doi.org/10.1016/j.gfs.2012.07.007.

Maxwell, Daniel G., John W. Parker, and Heather C. Stobaugh. 2013. 'What drives program choice in food security crises? Examining the "response analysis" question'. *World Development* 49 (0): 68–79. Accessed 10 January 2023. http://www.sciencedirect.com/science/article/pii/S0305750X13000284.

Maxwell, Daniel, Nisar Majid, Guhad Adan, Khalif Abdirahman, and Jeeyon Janet Kim. 2016. 'Facing famine: Somali experiences in the famine of 2011'. *Food Policy* 65: 63–73. https://doi.org/10.1016/j.foodpol.2016.11.001.

Mbaku, John Mukum. 2013. 'International justice: The International Criminal Court and Africa'. In *Foresight Africa: Top Priorities for the Continent in 2014*, ed. Brookings Institution. Washington, DC: Brookings Institution, 9–12.

Mc Evoy, Claire, and Gergely Hideg. 2017. *Global violent deaths 2017: Time to decide*. Geneva: Small Arms Survey.

Mc Hugh, Gerard, and Manuel Bessler. 2006. *Humanitarian negotiations with armed groups: A manual for practitioners*. New York: United Nations Office for the Coordination of Humanitarian Affairs (UN OCHA).

Mc Hugh, Gerard, and Simar Singh. 2013. 'Preserving the integrity of humanitarian negotiations'. *Humanitarian Exchange* 58: 3–6.

McGovern, Mike. 2014. 'Bushmeat and the politics of disgust'. *Fieldsights*. 7 October. Accessed 10 January 2023. https://culanth.org/fieldsights/bushmeat-and-the-politics-of-disgust.

McGuire, Bill. 2012. *Waking the giant: How a changing climate triggers earthquakes, tsunamis, and volcanoes*. Oxford: Oxford University Press.

McInnes, Colin. 2016. 'Crisis! What crisis? Global health and the 2014–15 West African Ebola outbreak'. *Third World Quarterly* 37 (3): 380–400. https://doi.org/10.1080/01436597.2015.1113868.

Melvern, Linda. 2001. 'Missing the story: The media and the Rwandan genocide'. *Contemporary Security Policy* 22 (3): 91–106. https://doi.org/10.1080/13523260512331391248.

Mena, Rodrigo, and Dorothea Hilhorst. 2021. 'The (im)possibilities of disaster risk reduction in the context of high-intensity conflict: The case of Afghanistan'. *Environmental Hazards* 20 (2): 188–208. https://doi.org/10.1080/17477891.2020.1771250.

Menkhaus, Ken. 2009. 'Somalia: "They created a desert and called it peace(building)"'. *Review of African Political Economy* 36 (120): 223–233. https://doi.org/10.1080/03056240903083136.

Menkhaus, Ken. 2010. 'Stabilisation and humanitarian access in a collapsed state: The Somali case'. *Disasters* 34: S320–S341. https://doi.org/10.1111/j.1467-7717.2010.01204.x.

Menkhaus, Ken. 2012. 'No access: critical bottlenecks in the 2011 Somali famine'. *Global Food Security* 1 (1): 29–35. https://doi.org/http://dx.doi.org/10.1016/j.gfs.2012.07.004.

Mercier, Michèle. 1995. *Crimes without punishment: Humanitarian action in former Yugoslavia*. London: Pluto Press.

Meredith, Catherine. 2015. 'A bottom-up approach to the Ebola response'. *Humanitarian Exchange* 64: 15–17.

Metcalfe-Hough, Victoria, Wendy Fenton, Barnaby Willitts-King, and Alexandra Spencer. 2021. *The Grand Bargain at five years: An independent review*. London: HPG/ODI.

Michaels-Strasser, Susan, Miriam Rabkin, Maria Lahuerta, Katherine Harripersaud, Roberta Sutton, Laurence Natacha Ahoua, Bibole Ngalamulume, Julie Franks, and Wafaa M. El-Sadr. 2015. 'Innovation to confront Ebola in Sierra Leone: The community-care-centre model'. *The Lancet Global Health* 3 (7): e361–e362. https://doi.org/10.1016/S2214-109X(15)00045-5.

Miliband, David, 2016, 'From sector to system: Reform and renewal in humanitarian aid'. 27 April. Accessed 10 January 2023. https://www.rescue.org/press-release/sector-system-reform-and-renewal-humanitarian-aid.

Mills, Kurt. 2013. 'Constructing humanitarian space in Darfur'. *International Journal of Human Rights* 17 (5–6): 605–618. http://dx.doi.org/10.1080/13642987.2013.805125.

Moeller, Susan D. 2002. 'A hierarchy of innocence: The media's use of children in the telling of international news'. *Harvard International Journal of Press/Politics* 7 (1): 36–56. Accessed 10 January 2023. http://hij.sagepub.com/content/7/1/36.abstract.

Moeller, Susan D. 2006. '"Regarding the pain of others": Media, bias and the coverage of international disasters'. *Journal of International Affairs* 59 (2): 173–196. Accessed 10 January 2023. http://www.jstor.org/stable/24358432.

Monson, Sarah. 2017. 'Ebola as African: American media discourses of panic and otherization'. *Africa Today* 63 (3): 3–27. https://doi.org/10.2979/africatoday.63.3.02.

Mooney, Erin D. 1995. 'Presence, ergo protection? UNPROFOR, UNHCR and the ICRC in Croatia and Bosnia and Herzegovina'. *International Journal of Refugee Law* 7 (3): 407–435. Accessed 10 January 2023. http://ijrl.oxfordjournals.org/content/7/3/407.abstract.

Moreno-Lax, Violeta, Daniel Ghezelbash, and Natalie Klein. 2019. 'Between life, security and rights: Framing the interdiction of "boat migrants" in the Central Mediterranean and Australia'. *Leiden Journal of International Law* 32 (4): 715–740. https://doi.org/10.1017/S0922156519000451.

Mosel, Irina, and Simon Levine. 2014. *Remaking the case for linking relief, rehabilitation and development: How LRRD can become a practically useful concept for assistance in difficult places*. London: HPG/ODI.

Moser, Caroline O N, and Cathy McIlwaine. 2006. 'Latin American urban violence as a development concern: towards a framework for violence reduction'. *World Development* 34 (1):89–112.

Mowjee, Tasneem, and Lydia Poole. 2014. 'Study of the Central Emergency Response Fund (CERF) and humanitarian donor decision-making'. Policy2Practice Team.

MSF (Médecins sans Frontiéres). 2014a. 'Ebola in West Africa: Epidemic requires massive deployment of resources'. Last modified 21 June 2014. Accessed 2 January 2021. https://www.msf.org/ebola-west-africa-epidemic-requires-massive-deployment-resources.

MSF. 2014b. 'Official MSF response to the WHO declaring Ebola an international public health emergency'. Last modified 8 August 2014. Accessed 29 April 2020. https://www.msf.org/ebola-official-msf-response-who-declaring-ebola-international-public-health-emergency.

MSF. 2015. 'Pushed to the limit and beyond: A year into the largest ever Ebola outbreak'. MSF.

MSF. 2016, 'EU States' dangerous approach to migration places asylum in jeopardy worldwide: MSF to no longer take funds from EU Member States and institutions'. 17 June. Accessed 10 January 2023. https://www.msf.org/eu-states%E2%80%99-dangerous-approach-migration-places-asylum-jeopardy-worldwide.

MSF. 2021. 'Financial independence and accountability'. Accessed 17 January 2022. https://www.msf.org/reports-and-finances.

MSF. n.d. 'The MSF Charter'. Accessed 24 February 2023. https://www.msf.org/msf-charter.

Muggah, Robert. 2010. 'The effects of stabilisation on humanitarian action in Haiti'. *Disasters* 34: S444–S463. https://doi.org/10.1111/j.1467-7717.2010.01205.x.

Muggah, Robert, and Katherine, Aguirre Tobón. 2018. *Citizen Security in Latin America: Facts and Figures*. Rio de Janeiro: Igarapé Institute.

Müller, Tanja R. 2013a. '"The Ethiopian famine" revisited: Band Aid and the antipolitics of celebrity humanitarian action'. *Disasters* 37 (1): 61–79. http://dx.doi.org/10.1111/j.1467-7717.2012.01293.x.

Müller, Tanja R. 2013b. 'The long shadow of Band Aid humanitarianism: Revisiting the dynamics between famine and celebrity'. *Third World Quarterly* 34 (3): 470–484. http://dx.doi.org/10.1080/01436597.2013.785342.

Mulligan, Martin, and Yaso Nadarajah. 2011. 'Rebuilding community in the wake of disaster: Lessons from the recovery from the 2004 tsunami in Sri Lanka and India'. *Community Development Journal* 47 (3): 353–368. https://doi.org/10.1093/cdj/bsr025.

Murdie, Amanda M, and David R. Davis. 2012. 'Shaming and blaming: Using events data to assess the impact of human rights INGOs'. *International Studies Quarterly* 56 (1): 1–16. http://dx.doi.org/10.1111/j.1468-2478.2011.00694.x.

Murphy, Sophia, and Kathy McAfee. 2005. *US food aid: Time to get it right*. Minneapolis, MN: Institute for Agriculture and Trade Policy.

Narang, Neil. 2016. 'Forgotten conflicts: Need versus political priority in the allocation of humanitarian aid across conflict areas'. *International Interactions* 42 (2): 189–216. https://doi.org/10.1080/03050629.2016.1080697.

NATO (North Atlantic Treaty Alliance). 2015. 'ISAF's mission in Afghanistan (2001–2014)'. Last modified 30 May 2022. Accessed 10 January 2023. https://www.nato.int/cps/en/natohq/topics_69366.htm.

NATO. 2020. 'NATO and Afghanistan'. Last modified 31 August 2022. Accessed 10 January 2023. https://www.nato.int/cps/en/natolive/topics_8189.htm.

New York Times. 1992. 'The well-fed dead in Bosnia'. *New York Times*, 15 July.

Newbury, Catharine. 1995. 'Background to genocide: Rwanda'. *Issue: A Journal of Opinion* 23 (2): 12–17. https://doi.org/10.2307/1166500.

Newbury, Catharine. 1998. 'Ethnicity and the politics of history in Rwanda'. *Africa Today* 45 (1): 7–24. Accessed 10 January 2023. http://www.jstor.org/stable/4187200.

NGOs (Non-governmental organizations). 2017. 'Non-governmental organisations collectively call for immediate cessation of armed conflict and protection of civilians in Kachin and Northern Shan States, Myanmar'. 25 January. Accessed 10 January 2023. https://www.rescue.org/press-release/non-governmental-organisations-collectively-call-immediate-cessation-armed-conflict.

Nicholls, Leah. 2006. 'The humanitarian monarchy legislates: The International Committee of the Red Cross and its 161 rules of customary international humanitarian law'. *Duke Journal of Comparative and International Law* 17: 223–252.

Niemann, Arne, and Natascha Zaun. 2018. 'EU refugee policies and politics in times of crisis: Theoretical and empirical perspectives'. *Journal of Common Market Studies* 56 (1): 3–22. https://doi.org/10.1111/jcms.12650.

Niland, Norah. 2014. 'Sri Lanka: Unrestricted warfare and limited humanitarian action'. *International Development Policy* 6 (1): 129–142.

Nunes, João. 2016. 'Ebola and the production of neglect in global health'. *Third World Quarterly* 37 (3): 542–556. https://doi.org/10.1080/01436597.2015.1124724.

O'Callaghan, Sorcha, and Leslie Leach. 2012. *Principles in action in Lebanon*. London: British Red Cross.

O'Callaghan, Sorcha, Adelina Kamal, Farouq Habib, Tetiana Stawnychy, and Fiona Terry. 2022. 'Beyond neutrality: Alternative forms of humanitarian action'. HPG/ODI. 1 December. Accessed 8 December 2022. https://odi.org/en/events/beyond-neutrality-alternative-forms-of-humanitarian-action/#Description.

O'Sullivan, Kevin. 2014. 'Humanitarian encounters: Biafra, NGOs and imaginings of the Third World in Britain and Ireland, 1967–70'. *Journal of Genocide Research* 16 (2–3): 299–315. http://dx.doi.org/10.1080/14623528.2014.936706.

Obrecht, Alice. 2017. *Using evidence to allocate humanitarian resources: Challenges and opportunities*. London: ALNAP/ODI.

OECD (Organisation for Economic Co-operation and Development). 1997. *DAC guidelines on conflict, peace and development co-operation*. Paris: OECD.

OECD. 2001. *The DAC guidelines: Helping prevent violent conflict*. Paris: OECD.

OECD. 2022. *DAC recommendation on the humanitarian–development–peace nexus, OECD/LEGAL/5019*. Paris: OECD.

Ogata, Sadako. 1992. 'Statement of Mrs. Sadako Ogata, United Nations High Commissioner for Refugees, to the International Meeting on Humanitarian Aid for Victims of the Conflict in the former Yugoslavia, Geneva, 29 July 1992'. Accessed 25 February 2020. https://www.unhcr.org/admin/ADMIN/3ae68fac1a.html.

Ogata, Sadako. 2005. *The turbulent decade: Confronting the refugee crises of the 1990s*. 1st edn. New York; London: W.W. Norton.

OHCHR (Office of the High Commissioner for Human Rights). 2010. 'Democratic Republic of the Congo, 1993–2003: Report of the Mapping Exercise documenting the most serious violations of human rights and international humanitarian law committed within the territory of the Democratic Republic of the Congo between March 1993 and June 2003 (unofficial translation from French original)'. August. Accessed 10 January 2023. https://www.ohchr.org/Documents/Countries/CD/DRC_MAPPING_REPORT_FINAL_EN.pdf.

OHCHR (Office of the High Commissioner for Human Rights). 2011. *International legal protection of human rights in armed conflict*. New York; Geneva: OHCHR.

Okoth-Obbo, George. 2001. 'Thirty years on: A legal review of the 1969 OAU Refugee Convention Governing the Specific Aspects of Refugee Problems in Africa'. *Refugee Survey Quarterly* 20 (1): 79–138. https://doi.org/10.1093/rsq/20.1.79.

Oliver-Smith, Anthony. 2010. 'Haiti and the historical construction of disasters'. *NACLA Report on the Americas* 43 (4): 32–36. https://doi.org/10.1080/10714839.2010.11725505.

Olsen, Gorm Rye, Nils Carstensen, and Kristian Høyen. 2003. 'Humanitarian crises: What determines the level of emergency assistance? Media coverage, donor interests and the aid business'. *Disasters* 27 (2): 109–126. https://doi.org/10.1111/1467-7717.00223.

Omaka, Arua Oko. 2016. *The Biafran humanitarian crisis, 1967–1970: International human rights and Joint Church Aid*. Madison, WI; Teaneck, NJ: Farleigh Dickinson University Press.

OSE (Office of the Special Envoy). 2012. *Can more aid stay in Haiti and other fragile settings? How local investment can strengthen governments and economies*. New York: United Nations Office of the Special Envoy for Haiti.

Oxfam. 2005. 'Making the case for cash: Humanitarian food aid under scrutiny Oxfam'. Oxford: Oxfam GB.

Oxfam. 2013. *Oxfam International Policy Compendium Note: Oxfam's role in humanitarian action*. Oxford: Oxfam GB.

Oxfam. 2021. *Oxfam biometric and foundational identity policy*. Oxford: Oxfam GB.

Oxfam. n.d. 'Our finances and accountability'. Accessed 12 March 2022. https://www.oxfam.org/en/what-we-do/about/our-finances-and-accountability.

Oztaskin, Murat. 2020. '"USA v Scott" and the fight to prove that humanitarian aid is not a crime'. The New Yorker. 8 July 2020.

Paddon, Emily. 2010. 'Beyond creed, greed and booty: Conflict in the Democratic Republic of Congo'. *Africa* 80 (2): 322–331. https://doi.org/10.3366/afr.2010.0207.

Paddon Rhoads, Emily, and Jennifer Welsh. 2019. 'Close cousins in protection: The evolution of two norms'. *International Affairs* 95 (3): 597–617. https://doi.org/10.1093/ia/iiz054.

Pain, Adam, and Simon Levine. 2012. *A conceptual analysis of livelihoods and resilience: Addressing the 'insecurity of agency'*. London: HPG/ODI.

Pallister-Wilkins, Polly. 2016. 'Personal protective equipment in the humanitarian governance of Ebola: Between individual patient care and global biosecurity'. *Third World Quarterly* 37 (3): 507–523. https://doi.org/10.1080/01436597.2015.1116935.

Pantuliano, Sara, and Sorcha O'Callaghan. 2006. *The 'protection crisis': A review of field-based strategies for humanitarian protection in Darfur*. London: ODI.

Pantuliano, Sara, Kate Mackintosh, Samir Elhawary, and Victoria Metcalfe. 2011. 'Counter-terrorism and humanitarian action: Tensions, impact and way forward'. In HPG Policy Brief 43. London: Overseas Development Institute.

Papataxiarchis, Evthymios. 2016. 'Being "there": At the front line of the "European refugee crisis"—part 1'. *Anthropology Today* 32 (2): 5–9. https://doi.org/https://doi.org/10.1111/1467-8322.12237.

Pape, Utz, and Ambika Sharma. 2020. 'Displacement camps: For better or for worse?' World Bank Blogs. 24 January. Accessed 19 November 2022. https://blogs.worldbank.org/dev4peace/displacement-camps-better-or-worse.

Parfit, Derek. 1997. 'Equality and priority'. *Ratio (new series)* X: 202–221.

Patrick, Jonathan. 2011. *Haiti earthquake response: Emerging evaluation lessons*. Paris: Network on Development Evaluation, OECD DAC.

Paulat, Lizabeth. 2014. 'Uganda sends 20 experts to aid in Ebola crisis'. Last modified 12 August 2014. Accessed 24 August 2021. https://reliefweb.int/report/liberia/uganda-sends-20-experts-aid-ebola-crisis.

Pécoud, Antoine. 2020. 'Death at the border: Revisiting the debate in light of the Euro-Mediterranean migration crisis'. *American Behavioral Scientist* 64 (4): 379–388. https://doi.org/10.1177/0002764219882987.

Pérouse de Montclos, Marc-Antoine. 2009. 'Humanitarian aid and the Biafra War: Lessons not learned'. *Africa Development* 34 (1): 69–82.

Peskin, Victor. 2005. 'Beyond victor's justice? The challenge of prosecuting the winners at the International Criminal Tribunals for the former Yugoslavia and Rwanda'. *Journal of Human Rights* 4 (2): 213–231. https://doi.org/10.1080/14754830590952152.

Petrova, Margarita H. 2018. 'Weapons prohibitions through immanent critique: NGOs as emancipatory and (de)securitising actors in security governance'. *Review of International Studies* 44 (4): 619–653. https://doi.org/10.1017/S026021051800013X.

Pew Research Center. 2010, 'Nearly half have donated or plan to give: Haiti dominates public's consciousness'. 20 January. Accessed 10 January 2023. https://www.pewresearch.org/wp-content/uploads/sites/4/legacy-pdf/580.pdf.

Phillips, Jason. 2019. *Working with detained populations in Greece and Libya: A comparative study of the ethical challenges facing the International Rescue Committee*. IRC.

Philo, Greg. 1993. 'From Buerk to Band Aid: The media and the 1984 Ethiopian famine'. In *Getting the message: News, truth and power*, ed. John Eldridge. London: Routledge, 104–125.

Pickering-Saqqa, Susannah. 2019. 'Why work "at home"? Oxfam's value-added and the UK Poverty Programme'. *Development in Practice* 29 (4): 477–488. https://doi.org/10.1080/09614524.2019.1574714.

Pictet, Jean. 1979. *The fundamental principles of the Red Cross: A commentary*. Geneva: IFRC.

Pilkington, Ed. 2020. 'UN response to Haiti cholera epidemic lambasted by its own rights monitors'. The Guardian, 4 May. Accessed 4 September 2021. https://www.theguardian.com/world/2020/may/04/united-nations-un-haiti-cholera-letter-rights-monitors.

Piot, Peter, Jean-Jacques Muyembe, and W John Edmunds. 2014. 'Ebola in west Africa: From disease outbreak to humanitarian crisis'. *The Lancet Infectious Diseases* 14 (11): 1034–1035. https://doi.org/10.1016/S1473-3099(14)70956-9.

Policinski, Ellen, and Jovana Kuzmanovic. 2019. 'Protracted conflicts: The enduring legacy of endless war'. *International Review of the Red Cross* 101 (912): 965–976. https://doi.org/10.1017/S1816383120000399.

Poole, Lydia. 2014. *Afghanistan beyond 2014: Aid and the transformation decade*. Bristol: Development Initiatives.

Poole, Lydia, and John Primrose. 2010. *Southern Sudan: Funding according to need*. Wells: Development Initiatives.

Poster, Alexander. 2012. 'The gentle war: Famine relief, politics, and privatization in Ethiopia, 1983–1986'. *Diplomatic History* 36 (2): 399–425. Accessed 10 January 2023. http://dh.oxfordjournals.org/content/36/2/399.abstract.

Powell, Colin L. 2001. 'Remarks to the National Foreign Policy Conference for Leaders of Nongovernmental Organizations, Washington DC, 26 October 2001'. Accessed 12 September 2020. https://avalon.law.yale.edu/sept11/powell_brief31.asp.

Power, Samantha. 2002. *A problem from hell: America and the age of genocide*. New York: Basic Books.

Powers, Matthew. 2018. *NGOs as newsmakers: The changing landscape of international news*. New York: Columbia University Press.

Price, Gareth, and Mihir Bhatt. 2009. *The role of the affected state in humanitarian action: A case study on India*. London: HPG/ODI.

Prunier, Gerard. 1998. *The Rwanda crisis: History of a genocide*. Rev. edn. London: Hurst.

Read, Róisín, Bertrand Taithe, and Roger Mac Ginty. 2016. 'Data hubris? Humanitarian information systems and the mirage of technology'. *Third World Quarterly* 37 (8): 1314–1331. https://doi.org/10.1080/01436597.2015.1136208.

Redfield, Peter, and Erica Bornstein. 2010. 'An introduction to the anthropology of humanitarianism'. In *Forces of compassion: Humanitarianism between ethics and politics*, ed. Erica Bornstein and Peter Redfield. Santa Fe, NM: School for Advanced Research Press, 3–30.

Reid-Henry, Simon, and Ole Jacob Sending. 2014. 'The "humanitarianization" of urban violence'. *Environment and Urbanization* 26 (2): 427–442. Accessed 10 January 2023. http://eau.sagepub.com/content/early/2014/09/09/0956247814544616.abstract.

Repo, Jemima, and Riina Yrjölä. 2011. 'The gender politics of celebrity humanitarianism in Africa'. *International Feminist Journal of Politics* 13 (1): 44–62. https://doi.org/10.1080/14616742.2011.534661.

Rever, Judi. 2018. *In praise of blood: Rhe crimes of the Rwandan Patriotic Front*. Toronto: Random House Canada.

Richards, Paul. 2016. *Ebola: How a people's science helped end an epidemic. African Arguments*. 1st edn. London: Zed Books.

Richey, Lisa Ann, and Alexandra Budabin. 2016. 'Celebrities in international affairs'. *Oxford Handbooks Online*. https://doi.org/10.1093/oxfordhb/9780199935307.013.3.

Roberts, Adam. 1995. 'From San Francisco to Sarajevo: The UN and the use of force'. *Survival* 37 (4): 7–28. https://doi.org/10.1080/00396339508442813.

Roberts, Adam. 1996. *Humanitarian action in war: Aid, protection and impartiality in a policy vacuum*. Oxford: Oxford University Press.

Robinson, Piers. 1999. 'The CNN effect: Can the news media drive foreign policy?' *Review of International Studies* 25 (2): 301–309. Accessed 10 January 2023. http://www.jstor.org/stable/20097596.

Robinson, Piers. 2000. 'The policy-media interaction model: Measuring media power during humanitarian crisis'. *Journal of Peace Research* 37 (5): 613–633. https://doi.org/10.1177/0022343300037005006.

Roepstorff, Kristina. 2020. 'A call for critical reflection on the localisation agenda in humanitarian action'. *Third World Quarterly* 41 (2): 284–301. https://doi.org/10.1080/01436597.2019.1644160.

Ross, Jacob, Simon Maxwell, and Margaret Buchanan-Smith. 1994. *Linking relief and development*. Brighton: Institute of Development Studies.

Rozakou, Katerina. 2017. 'Solidarity #humanitarianism: The blurred boundaries of humanitarianism in Greece'. *Etnofoor* 29 (2): 99–104. Accessed 10 January 2023. http://www.jstor.org/stable/26296172.

Rubenstein, Jennifer C. 2008. 'The distributive commitments of international NGOs'. In *Humanitarian in question: politics, power, ethics*, ed. Michael Barnett and Thomas G Weiss. Ithaca, NY: Cornell University Press, 215–234.

Rubenstein, Jennifer C. 2015. 'Emergency claims and democratic action'. *Social Philosophy and Policy* 32 (1): 101–126. https://doi.org/10.1017/S0265052515000096.

Sampaio, Antônio. 2019. *Illicit order: The militarized logic of organized crime and urban security in Rio de Janeiro*. Geneva: Global Initiative against Transnational Organized Crime.

Sanderson, David. 2019. *Urban humanitarian response*. London: HPG/ODI.

Sandoz, Yves. 1992. '"Droit" or "devoir d'ingérence" and the right to assistance: The issues involved'. *International Review of the Red Cross* 32 (288): 215–227. https://doi.org/10.1017/S0020860400070480.

Sandri, Elisa. 2018. '"Volunteer Humanitarianism": Volunteers and humanitarian aid in the Jungle refugee camp of Calais'. *Journal of Ethnic and Migration Studies* 44 (1): 65–80. https://doi.org/10.1080/1369183X.2017.1352467.

Save the Children. n.d. 'The history of Save the Children'. Accessed 20 September 2021. https://www.savethechildren.org/us/about-us/why-save-the-children/history.

Save the Children, and Oxfam. 2012. *A dangerous delay: The cost of late response to early warnings in the 2011 drought in the Horn of Africa*. Oxford: Oxfam GB.

Schaar, John. 2008. 'The birth of the Good Humanitarian Donorship initiative'. In *The humanitarian response index 2007: Measuring commitment to best practice*, ed. Silvia Hidalgo and Augusto Lopez-Claros. Basingstoke & New York: Palgrave Macmillan, 37–44.

Schenkenberg, Ed. 2016. *The challenges of localised humanitarian aid in armed conflict*. MSF.

Schenkenberg van Mierop, Ed. 2015. 'Coming clean on neutrality and independence: The need to assess the application of humanitarian principles'. *International Review of the Red Cross* 97 (897-898): 295–318. https://doi.org/10.1017/S181638311500065X.

Scheper, Elisabeth, Arjuna Parakrama, and Smruti Patel. 2006. *Impact of the tsunami response on local and national capacities*. London: Tsunami Evaluation Coalition.

Schuller, Mark. 2012. 'Haiti's bitter harvest: Humanitarian aid in the "republic of NGOs"'. In *The golden fleece: Manipulation and independence in humanitarian action*, ed. Antonio Donini. Boulder & London: Kumarian, 171–193.

Scott, Martin, Mel Bunce, and Kate Wright. 2022. 'The influence of news coverage on humanitarian aid: The bureaucrats' perspective'. *Journalism Studies* 23 (2): 167–186. https://doi.org/10.1080/1461670X.2021.2013129.

Seguridad Justicia y Paz. 2018. 'Ranking de ciudades 2017'. Accessed 3 November 2019. http://www.seguridadjusticiaypaz.org.mx/ranking-de-ciudades-2017.

Sen, Amartya. 1981. *Poverty and famines: An essay on entitlement and deprivation.* Oxford: Clarendon.

Sen, Amartya. 1999. *Development as freedom.* Oxford: Oxford University Press.

Sharpe, Marina. 2018. 'The Global Compact on Refugees and conflict prevention in Africa: "Root causes" and yet another divide'. *International Journal of Refugee Law* 30 (4): 707–710. https://doi.org/10.1093/ijrl/eey052.

Silva, Kalinga Tudor. 2009. '"Tsunami third wave" and the politics of disaster management in Sri Lanka'. *Norsk Geografisk Tidsskrift—Norwegian Journal of Geography* 63 (1): 61–72. https://doi.org/10.1080/00291950802712145.

Simm, Gabrielle. 2018. 'Disaster response in Southeast Asia: The ASEAN Agreement on Disaster Response and Emergency Management'. *Asian Journal of International Law* 8 (1): 116–142. https://doi.org/10.1017/S2044251316000205.

SIPRI (Stockholm International Peace Research Institute). 2021. *SIPRI factsheet: Trends in world military expenditure, 2020.* Stockholm: SIPRI.

Skran, Claudena. 2018. 'Gustave Ador, the ICRC, and leadership on refugee and migration policy'. Humanitarian Law & Policy (blog), ICRC. 30 January. Accessed 14 May 2020. https://blogs.icrc.org/law-and-policy/2018/01/30/gustave-ador-the-icrc-and-leadership-on-refugee-and-migration-policy.

Slaughter, Amy, and Jeff Crisp. 2009. *A surrogate state? The role of UNHCR in protracted refugee situations.* Vol. 168. New Issues in Refugee Research. Geneva: United Nations High Commissioner for Refugees (UNHCR).

Slim, Hugo. 1997. 'Relief agencies and moral standing in war: Principles of humanity, neutrality, impartiality and solidarity'. *Development in Practice* 7 (4): 342–352. Accessed 10 January 2023. http://www.jstor.org/stable/4029000.

Slim, Hugo. 2012. 'IASC real-time evaluation of the humanitarian response to the Horn of Africa drought crisis in Somalia, Ethiopia and Kenya: Synthesis report'. Inter-Agency Standing Committee.

Slim, Hugo. 2015. *Humanitarian ethics: A guide to the morality of aid in war and disaster.* London: Hurst.

Slim, Hugo, and Andrew Bonwick. 2005. *Protection: An ALNAP guide for humanitarian agencies.* London: ODI.

Slim, Hugo, and Miriam Bradley. 2013. *Principled humanitarian action and ethical tensions in multi-mandate organizations.* Geneva: World Vision International.

Slim, Hugo, and Emanuela-Chiara Gillard. 2013. 'Ethical and legal perspectives on cross-border humanitarian operations'. *Humanitarian Exchange* 59: 6–9.

Smillie, Ian. 2012. 'The emperor's old clothes: The self-created siege of humanitarian action'. In *The golden fleece: Manipulation and independence in humanitarian action*, ed. Antonio Donini. Boulder & London: Kumarian Press, 17–41.

Staunton, Enda. 1999. 'The case of Biafra: Ireland and the Nigerian civil war'. *Irish Historical Studies* 31 (124): 513–534. https://doi.org/10.1017/S0021121400014395.

Steets, Julia, Claudia Meier, Adele Harmer, Abby Stoddard, and Janika Spannagel. 2018. *Evaluation of WFP policies on humanitarian principles and access in humanitarian contexts.* Rome: WFP.

Steinhilper, Elias, and Rob Gruijters. 2017. 'Border deaths in the Mediterranean: What we can learn from the latest data'. Border Criminologies (blog), Faculty of Law, University of Oxford. 8 March. Accessed 10 January 2023. https://www.law.ox.ac.uk/research-subject-groups/centre-criminology/centreborder-criminologies/blog/2017/03/border-deaths.

Stewart, Megan A. 2018. 'Civil war as state-making: Strategic governance in civil war'. *International Organization* 72 (1): 205–226. https://doi.org/10.1017/S0020818317000418.

Stirrat, Jock. 2006. 'Competitive humanitarianism: Relief and the tsunami in Sri Lanka' *Anthropology Today* 22 (5): 11–16. http://dx.doi.org/10.1111/j.1467-8322.2006.00459.x.

Stockton, Nicholas. 1998. 'In defence of humanitarianism'. *Disasters* 22 (4): 352–360.

Stoddard, Abby, Adele Harmer, and Monica Czwarno. 2017. 'Aid worker security report 2017. Behind the attacks: A look at the perpetrators of violence against aid workers'. Humanitarian Outcomes.

Stoddard, Abby, Paul Harvey, Monica Czwarno, and Meriah-Jo Breckenridge. 2021. 'Aid worker security report 2021: Crime risks and responses in humanitarian operations'. Humanitarian Outcomes.

Stoianova, Velina. 2013. *Private funding for humanitarian assistance: Filling the gap?* Bristol: Development Initiatives.

Storey, Andy. 1997. 'Non-neutral humanitarianism: NGOs and the Rwanda crisis'. *Development in Practice* 7 (4): 384–394. http://dx.doi.org/10.1080/09614529754170.

Stremlau, John J. 1977. *The international politics of the Nigerian civil war, 1967–1970*. Princeton, NJ: Princeton University Press.

Strömberg, David. 2007. 'Natural disasters, economic development, and humanitarian aid'. *Journal of Economic Perspectives* 21 (3): 199–222. Accessed 10 January 2023. http://www.jstor.org/stable/30033741.

Suhrke, Astri. 2008. 'A contradictory mission? NATO from stabilization to combat in Afghanistan'. *International Peacekeeping* 15 (2): 214–236. https://doi.org/10.1080/13533310802041477.

Svoboda, Eva, Veronique Barbelet, and Irina Mosel. 2018. *Holding the keys: Humanitarian access and local organisations*. London: HPG/ODI.

Tanguy, Joelle, and Fiona Terry. 1999. 'Humanitarian responsibility and committed action: Response to "principles, politics, and humanitarian action"'. *Ethics & International Affairs* 13 (1): 29–34. http://dx.doi.org/10.1111/j.1747-7093.1999.tb00324.x.

Taylor, Adam. 2014. 'In the medical response to Ebola, Cuba is punching far above its weight'. Washington Post. 4 October. Accessed 10 January 2023. https://www.washingtonpost.com/news/worldviews/wp/2014/10/04/in-the-medical-response-to-ebola-cuba-is-punching-far-above-its-weight.

TDC (Transnational Development Clinic), GHJP (Global Health Justice Partnership), and AHDEN (Association Haitiénne de Droit de l'Environnment). 2013. 'Peacekeeping without accountability: The United Nations' responsibility for the Haitian cholera epidemic'. TDC, GHJP, and AHDEN.

Telford, John, and John Cosgrave. 2007. 'The international humanitarian system and the 2004 Indian Ocean earthquake and tsunamis'. *Disasters* 31 (1): 1–28. http://dx.doi.org/10.1111/j.1467-7717.2007.00337.x.

Terry, Fiona. 2000. *The limits and risks of regulation mechanisms for humanitarian action*. Paris: MSF Centre de Réflexion sur l'Action et les Savoirs Humanitaires (CRASH).

Terry, Fiona. 2002. *Condemned to repeat? The paradox of humanitarian action*. Ithaca, NY: Cornell University Press.

Terry, Fiona. 2011a. 'Myanmar: "Golfing with the Generals"'. In *Humanitarian negotiations revealed: The MSF experience*, ed. Claire Magone, Michaël Neuman, and Fabrice Weissman. London: Hurst, 109–126.

Terry, Fiona. 2011b. 'The International Committee of the Red Cross in Afghanistan: Reasserting the neutrality of humanitarian action'. *International Review of the Red Cross* 93 (881): 173–188.

Terry, Fiona. 2022. 'Beyond neutrality: Alternative forms of humanitarian action'. Accessed 10 January 2023. https://odi.org/en/events/beyond-neutrality-alternative-forms-of-humanitarian-action/#Description: HPG/ODI.

The Economist. 2016. 'The world's most violent cities'. *The Economist*, 3 February.

Thompson, Edwina. 2008. *Principled pragmatism: NGO engagement with armed actors*. Monrovia, CA: World Vision International.

Toom, Victor. 2020. 'Ontologically dirty knots: The production of numbers after the Srebrenica genocide'. *Security Dialogue* 51 (4): 358–376. https://doi.org/10.1177/0967010620902008.

Turner, Simon. 2016. 'What is a refugee camp? Explorations of the limits and effects of the camp'. *Journal of Refugee Studies* 29 (2): 139–148. https://doi.org/10.1093/jrs/fev024.

Turton, David. 2011. 'The politics of internal displacement and options for institutional reform'. *Deportate, esuli, profughe* 17: 1–24.

UCDP (Uppsala Conflict Data Program). 2012. One-sided Violence Dataset v 1.4-2012, 1989–2011.
Uche, Chibuike. 2008. 'Oil, British interests and the Nigerian civil war'. *Journal of African History* 49 (1): 111–135. https://doi.org/10.1017/S0021853708003393.
Ulysse, Gina Athena. 2010. 'Why representations of Haiti matter now more than ever'. *NACLA Report on the Americas* 43 (4): 37–47. https://doi.org/10.1080/10714839.2010.11725506.
UN CERF (Central Emergency Response Fund). 2016. 'About CERF'. UN CERF.
UN CERF. 2021. '15 years of saving lives together'. UN CERF.
UN General Assembly. 1991. 'Strengthening of the coordination of humanitarian emergency assistance of the United Nations'. 19 December, A/RES/46/182.
UN General Assembly. 2019. 'Report of the International Law Commission: Peremptory norms of general international law (jus cogens)'. Text of the draft conclusions and draft annex provisionally adopted by the Drafting Committee on first reading, 29 May, A/CN.4/L.936.
UN OCHA (Office for the Coordination of Humanitarian Affairs). 2010. 'Humanitarian funding analysis for Somalia'. September. OCHA.
UN OCHA. 2017. *New way of working*. OCHA.
UN OCHA. 2019. 'Building a culture of protection: 20 years of Security Council engagement on the protection of civilians'. OCHA.
UN OCHA. 2020a. 'About CBPFs'. OCHA.
UN OCHA. 2020b. 'OCHA on message: Humanitarian access'. OCHA.
UN OCHA. 2020c. 'What is the Cluster Approach?' Last modified 31 March 2020. Accessed 27 November 2022. https://www.humanitarianresponse.info/en/coordination/clusters/what-cluster-approach.
UN OCHA. n.d. 'Appeals and response plans 2021'. OCHA. Accessed 11 January 2022. https://fts.unocha.org/appeals/overview/2021.
UN Secretary-General. 1992a. 'Report of the Secretary-General pursuant to Security Council Resolution 757 (1992)'. UN Doc. S/24075, 6 June.
UN Secretary-General. 1992b. 'Report of the Secretary-General on the situation in Bosnia and Herzegovina'. UN Doc. S/24540, 10 September.
UN Secretary-General. 1994. 'Special report of the Secretary-General on the United Nations Assistance Mission for Rwanda'. UN Doc. S/1994/470, 20 April.
UN Secretary-General. 1999. 'Report of the Secretary-General pursuant to General Assembly resolution 53/35: The fall of Srebrenica'. UN Doc. A/54/549, 15 November.
UN Secretary-General. 2015. 'Statement attributable to the Secretary-General on recent refugee/migrant tragedies'. 28 August.
UN Secretary-General. 2016. 'Outcome of the World Humanitarian Summit'. UN Doc. A/71/353, 23 August.
UN Security Council. 1992a. Resolution 743. S/RES/743, 21 February.
UN Security Council. 1992b. Resolution 758. S/RES/758, 8 June.
UN Security Council. 1992c. Resolution 764. S/RES/764, 13 July.
UN Security Council. 1992d. Resolution 770. S/RES/770, 13 August.
UN Security Council. 1992e. Resolution 776. S/RES/776, 14 September.
UN Security Council. 1993a. Resolution 819. S/RES/819, 16 April.
UN Security Council. 1993b. Resolution 824. S/RES/824, 6 May.
UN Security Council. 1993c. Resolution 836. S/RES/836, 4 June.
UN Security Council. 1994a. Resolution 912. S/ RES /912, 21 April.
UN Security Council. 1994b. Resolution 918. S/RES/918, 17 May.
UN Security Council. 1994c. Resolution 925. S/RES/925, 8 June.
UN Security Council. 1994d. Resolution 929. S/RES/929, 22 June.
UN Security Council. 2001. Resolution 1386. S/RES/1386, 20 December.
UN Security Council. 2002. Resolution 1401. S/RES/1401, 28 March.
UN Security Council. 2009. 'Report of the Secretary-General on the protection of civilians in armed conflict'. S/2009/277, 29 May.

UNAMA (United Nations Assistance Mission in Afghanistan). 2015. 'Annual report 2014: Protection of civilians in armed conflict'. Kabul: United Nations Assistance Mission in Afghanistan.

UNDP (United Nations Development Programme). 1994. *Human development report 1994: New dimensions of human security*. New York: Oxford University Press.

UNDP. 2010. 'Human development report 2010'. In The real wealth of nations: Pathways to development, ed. UNDP. Basingstoke; New York: Palgrave Macmillan.

UNHCR (United Nations High Commissioner for Refugees), ed. 2010. *Convention and Protocol Relating to the Status of Refugees*. Geneva: UNHCR.

UNHCR. 2016. *Global trends: Forced displacement in 2015*. Geneva: UNHCR.

UNHCR. 2021a. *Global trends: Forced displacement in 2020*. Copenhagen: UNHCR.

UNHCR. 2021b. 'Refugee camps explained'. USA for UNHCR. 6 April. Accessed 19 November 2022. https://www.unrefugees.org/news/refugee-camps-explained/#What%20is%20a%20refugee%20camp?.

UNHCR. 2021c. 'Refugee data finder: Key indicators'. Last modified 18 June 2021. Accessed 6 October 2021. https://www.unhcr.org/refugee-statistics.

UNHCR. 2021d. *Report on use of flexible funding in 2020*. Geneva: UNHCR.

UNHCR. n.d.-a. 'Humanitarian principles'. 'UNHCR emergency handbook'. Accessed 10 January 2023. https://emergency.unhcr.org/entry/95307/humanitarian-principles.

UNHCR. n.d.-b. 'Nationality of arrivals to Greece, Italy and Spain: January–December 2015'. UNHCR.

UNHCR. n.d.-c. 'Shelter'. Accessed 19 November 2022. https://www.unhcr.org/shelter.html.

UNISDR (United Nations Office for Disaster Risk Reduction). 2005. *Hyogo framework for action 2005-2015: Building the resilience of nations and communities to disasters*. Geneva: UNISDR.

Uvin, Peter. 2003. 'Reading the Rwandan genocide'. *International Studies Review* 3 (3): 75–99. https://doi.org/10.1111/1521-9488.00245.

Vandevoordt, Robin. 2019. 'Subversive humanitarianism: Rethinking refugee solidarity through grass-roots initiatives'. *Refugee Survey Quarterly* 38 (3): 245–265. https://doi.org/10.1093/rsq/hdz008.

Vaux, Tony. 2001. *The selfish altruist: Relief work in famine and war*. London: Earthscan Publications.

Vaux, Tony. 2006. 'Proportion and distortion in humanitarian assistance'. In ALNAP Review of Humanitarian Action 2005. London: ALNAP/ODI, 35–88.

Verdirame, Guglielmo, and Jason Pobjoy. 2013. 'The end of refugee camps?' In *Research companion to migration law, theory and policy*, ed. Satvinder S. Juss. Farnham: Ashgate, 471–488.

Versluis, Anna. 2014. 'Formal and informal material aid following the 2010 Haiti earthquake as reported by camp dwellers'. *Disasters* 38 (S1): S94–S109. https://doi.org/https://doi.org/10.1111/disa.12050.

VSO (Voluntary Service Overseas). 2002. 'The Live Aid legacy: The developing world through British eyes—a research report'. VSO.

Waddell, Nicholas. 2006. 'Ties that bind: DfID and the emerging security and development agenda'. *Conflict, Security & Development* 6 (4): 531–555. https://doi.org/10.1080/14678800601066520.

Walker, Peter, Josh Glasser, and Shubhada Kambli. 2012. *Climate change as a driver of humanitarian crises and response*. Somerville, MA: Feinstein International Center, Tufts University.

Walsh, Sinead, and Oliver Johnson. 2018. *Getting to zero: A doctor and a diplomat on the ebola frontline*. London: Zed Books.

Waters, Ken. 2004. 'Influencing the message: The role of Catholic missionaries in media coverage of the Nigerian Civil War'. *Catholic Historical Review* 90 (4): 697–718.

Watson, Scott. 2015. 'The criminalization of human and humanitarian smuggling'. *Migration, Mobility & Displacement* 1 (1): 39–53. https://doi.org/10.18357/mmd11201513273.

Weiss, Thomas G. 1999. 'Principles, politics, and humanitarian action'. *Ethics and International Affairs* 13 (1): 1–22.

Weissman, Fabrice. 2011a. 'Silence heals . . . from the Cold War to the War on Terror, MSF speaks out: A brief history'. In *Humanitarian negotiations revealed: The MSF experience*, ed. Claire Magone, Michaël Neuman, and Fabrice Weissman. London: Hurst, 177–197.

Weissman, Fabrice. 2011b. 'Sri Lanka: Amid all-out war'. In *Humanitarian negotiations revealed: The MSF experience*, ed. Claire Magone, Michaël Neuman, and Fabrice Weissman. London: Hurst, 15–33.

Wenham, Clare. 2016. 'Ebola responsibility: Moving from shared to multiple responsibilities'. *Third World Quarterly* 37 (3): 436–451. https://doi.org/10.1080/01436597.2015.1116366.

WFP (World Food Programme). 2004. 'Humanitarian principles, Executive Board Annual Session, Rome, 24–26 May 2004', WFP/EB.A/2004/5-C. Rome: WFP.

WFP. n.d. 'Food assistance: Cash and in-kind'. Accessed 11 October 2022. https://www.wfp.org/food-assistance.

Wheeler, Nicholas J. 2000. *Saving strangers: Humanitarian intervention in international society*. Oxford: Oxford University Press.

WHO (World Health Organization). 2016a. 'Situation report: Ebola virus disease'. 10 June.

WHO. 2016b. *World Malaria Report 2016*. Geneva: WHO.

WHO. 2017. 'Measles vaccines: WHO position paper—April 2017'. *Weekly Epidemiological Record* 92 (17): 205–228.

Wilkinson, Annie, and Melissa Leach. 2015. 'Briefing: Ebola—myths, realities, and structural violence'. *African Affairs* 114 (454): 136–148. https://doi.org/10.1093/afraf/adu080.

World Bank. 2020. 'Urban development: Overview.' Last modified 20 April 2020. Accessed 2 January 2022. https://www.worldbank.org/en/topic/urbandevelopment/overview#1.

World Bank. n.d. 'The Humanitarian-Development-Peace Initiative.' Last modified 3 March 2017. Accessed 10 April 2022. https://www.worldbank.org/en/topic/fragilityconflictviolence/brief/the-humanitarian-development-peace-initiative.

World Vision Australia. 2017. 'Humanitarian principles: World Vision Australia's public policy position'. Accessed 10 January 2023. https://www.worldvision.com.au/docs/default-source/policies/wva_humanitarian-principles-policy---july-2017.pdf?sfvrsn=2.

World Vision International. n.d. 'Frequently asked questions.' Accessed 12 March 2022. https://www.wvi.org/about-us/faq.

Zeender, Greta. 2005. 'Engaging armed non-state actors on internally displaced persons protection;. *Refugee Survey Quarterly* 24 (3): 96–111. Accessed 10 January 2023. http://rsq.oxfordjournals.org/content/24/3/96.short.

Zetter, Roger. 2021. 'Theorizing the refugee humanitarian-development nexus: A political-economy analysis'. *Journal of Refugee Studies* 34 (2): 1766–1786. https://doi.org/10.1093/jrs/fez070.

Index